BIOLOGICAL TREATMENT OF SEWAGE BY THE ACTIVATED SLUDGE PROCESS

ELLIS HORWOOD BOOKS IN WATER AND WASTEWATER TECHNOLOGY

This collection of authoritative works reflects the awareness of the importance to the world of water, wastewater treatment, and closely related subjects. The titles are written and edited by experts from a wide range of countries closely concerned with research and development, monitoring and improving water quality and supplies, and treatment and disposal of wastewater.

Titles published in collaboration with the
WATER RESEARCH CENTRE, UK
AQUALINE THESAURUS 2
Editor: G. BOTHAMLEY, Water Research Centre
BIOLOGICAL FLUIDISED BED TREATMENT OF WATER AND WASTEWATER
Editors: P. F. COOPER, Water Research Centre, and B. ATKINSON, University of Manchester Institute of Science and Technology
ENVIRONMENTAL TOXICOLOGY: Organic Pollutants
J. K. FAWELL and S. HUNT, Water Research Centre
ENVIRONMENTAL PROTECTION: Standards, Compliance and Costs
Editor: T. J. LACK, Water Research Centre
WATER RESEARCH TOPICS
Editor: I. M. LAMONT, Water Research Centre
SMALL WATER POLLUTION CONTROL WORKS: Design and Practice
E. H. NICOLL, Scottish Development Department, Edinburgh
RIVER POLLUTION CONTROL
Editor: M. J. STIFF, Water Research Centre

Other books in Water and Wastewater
WATER POLLUTION BIOLOGY
P. ABEL, Sunderland Polytechnic
CHEMICAL PROCESSES IN WASTE WATER TREATMENT
W. J. EILBECK, University College of North Wales, and
G. MATTOCK, Waste Water Treatment and Resource Recovery Consultant
ALKALINITY–pH CHANGES WITH TEMPERATURE FOR WATERS IN INDUSTRIAL SYSTEMS
A. G. D. EMERSON, Independent Consulting Technologist
BIOLOGICAL TREATMENT OF SEWAGE BY THE ACTIVATED SLUDGE PROCESS
K. HÄNEL, Research Leader of Purification of Sewage Team, Leipzig, East Germany
ANALYSIS OF SURFACE WATERS
H. HELLMANN, Federal German Institute of Hydrology, Koblenz
WATER-ABSTRACTION, STORAGE, TREATMENT, DISTRIBUTION
J. JEFFERY, General Manager, North Surrey Water Company
POLLUTION CONTROL AND CONSERVATION
Editor: M. KOVACS, Hungarian Academy of Sciences
NATURAL WATER SAMPLING
J. KRAJCA, Research Institute of Geological Engineering, Brno, Czechoslovakia
GROUND RESOURCES AND YIELDS
H. KRIZ, Institute of Geography of the Czechoslovak Academy of Sciences, Czechoslovakia
PREVENTION OF CORROSION AND SCALING IN WATER SUPPLY SYSTEMS
L. LEGRAND, Former Head of the Water Dept. of City of Paris, and P. LEROY, Chemical Engineer of City of Paris, Chief of Corrosion Lab.
HANDBOOK OF WATER PURIFICATION Second Edition
W. LORCH, The Lorch Foundation
BIOLOGY OF SEWAGE TREATMENT AND WATER POLLUTION CONTROL
K. MUDRACK, University of Hanover, FRG, and S. KUNST, Institute for Sanitary Engineering, Hanover, FRG
IMPROVING WATER QUALITY
M. RAPINAT, Compagnie Générale des Eaux, France
DRINKING WATER MATERIALS: Field Observations and Methods of Investigation
D. SCHOENEN and H. F. SCHÖLER, Hygiene-Institut der Universität, Bonn, FRG
HANDBOOK OF LIMNOLOGY
J. SCHWOERBEL, University of Freiburg, West Germany
BIOTECHNOLOGY OF WASTE TREATMENT AND EXPLOITATION
Editors: J. M. SIDWICK and R. S. HOLDOM, Watson Hawksley, Buckinghamshire
INTERACTION OF MICRO-ORGANISMS AND CHEMICAL POLLUTION OF WATER
D. TOTH and D. TOMASOVICOVA, Institute of Experimental Biology & Ecology, Bratislava
BIOLOGICAL TREATMENT OF WASTEWATER, 2nd Edition
M. WINKLER, University of Surrey

ELLIS HORWOOD BOOKS IN AQUACULTURE AND FISHERIES SUPPORT

Series Editor: Dr L. M. LAIRD, University of Aberdeen
BACTERIAL FISH PATHOGENS: Disease in Farmed and Wild Fish
B. AUSTIN and D. AUSTIN, Heriot-Watt University
MICROBIAL BIOTECHNOLOGY: Freshwater and Marine Environments
B. AUSTIN and D. AUSTIN, Heriot-Watt University
MICROBIOLOGICAL METHODS FOR FISH AND SHELLFISH
B. AUSTIN and D. AUSTIN, Heriot-Watt University
AQUACULTURE, Volumes I & II
G. BARNABÉ, Université des Sciences et Techniques du Languedoc
AUTOMATED BIOMONITORING
D. GRUBER and J. DIAMOND, Biological Monitoring Inc., Virginia, USA
SALMON AND TROUT FARMING
L. M. LAIRD and T. NEEDHAM, University of Aberdeen
MARINE FISH FARMING: Warm Water Marine Culture
J. F. MUIR, University of Stirling

BIOLOGICAL TREATMENT OF SEWAGE BY THE ACTIVATED SLUDGE PROCESS

KLAUS HÄNEL
Leader of Research Team for Purification of Sludge
Dresden, West Germany

Translation Editor:
B. CHAMBERS
Water Research Centre

Translator:
B. D. HEMMINGS

ELLIS HORWOOD LIMITED
Publishers · Chichester

Halsted Press: a division of
JOHN WILEY & SONS
New York · Chichester · Brisbane · Toronto

This English edition first published in 1988 by
ELLIS HORWOOD LIMITED
Market Cross House, Cooper Street,
Chichester, West Sussex, PO19 1EB, England
The publisher's colophon is reproduced from James Gillison's drawing of the ancient Market Cross, Chichester.

Distributors:
Australia and New Zealand:
JACARANDA WILEY LIMITED
GPO Box 859, Brisbane, Queensland 4001, Australia

Canada:
JOHN WILEY & SONS CANADA LIMITED
22 Worcester Road, Rexdale, Ontario, Canada

Europe and Africa:
JOHN WILEY & SONS LIMITED
Baffins Lane, Chichester, West Sussex, England

North and South America and the rest of the world:
Halsted Press: a division of
JOHN WILEY & SONS
605 Third Avenue, New York, NY 10158, USA

This English edition is translated from the original German edition *Biologische abwasserreinigung mit belebtschlamm*, published in 1986 by VEB Gustav Fischer Verlag, Jena, © the copyright holders.

© 1988 English Edition, Ellis Horwood Limited

British Library Cataloguing in Publication Data
Hänel, Klaus, *1939–*
Biological treatment of sewage by the activated sludge process.
1. Sewage. Treatment. Activated sludge process.
I. Title
628.3'54

Library of Congress Card No. 88-8405

ISBN 0–7458–0295–8 (Ellis Horwood Limited)
ISBN 0–470–21252–7 (Halsted Press)

Phototypeset in Times by Ellis Horwood Limited
Printed in Great Britain by Hartnolls, Bodmin

COPYRIGHT NOTICE
All Rights Reserved. No part of this publication may be reproduced, stored in a retrieval system, or transmitted, in any form or by any means, electronic, mechanical, photocopying, recording or otherwise, without the permission of Ellis Horwood Limited, Market Cross House, Cooper Street, Chichester, West Sussex, England.

Table of contents

Preface to the original edition 9

1 Introduction 11
    1.1 Terminological details 13
    1.2 Essential features of the process 17
    1.3 Historical development of biological sewage treatment 20

2 Characterisation of the properties of sewage and activated sludge 24
    2.1 Chemical and physical methods 24
        2.1.1 Activated sludge 24
        2.1.2 Liquid composition 27
    2.2 Biochemical methods 33
        2.2.1 Biochemical oxygen demand (BOD) 33
        2.2.2 Respiration of activated sludge (OV) 35
        2.2.3 Toxicity 38
        2.2.4 Biomass 42
        2.2.5 Other biochemical methods 43
    2.3 Biological methods 44
        2.3.1 Problems and limits of application 44
        2.3.2 Sampling and transport of samples 45
        2.3.3 Conduct of the biological analysis 45
        2.3.4 The saprobic system 48
    2.4 Bacteriological methods 52
    2.5 Other biological methods 55

3 Organisms 56
    3.1 Viruses 56
    3.2 Schizomycetes (bacteria and blue algae) 57
    3.3 Mycophyta (fungi) 60
    3.4 Protozoa (primitive animalcules) 63
    3.5 Metazoa (multicellular animals) 70
    3.6 Algae 74
    3.7 Living communities 76
        3.7.1 Intake channel 76
        3.7.2 Aeration tanks 77
            3.7.2.1 Structure of the activated sludge flocs 77
            3.7.2.2 Colonisation of the activated sludge flocs .. 78
            3.7.2.3 Bulking sludge 87
        3.7.3 Outlet from the final settling tank 91

Table of contents

4 Biological, biochemical and biophysical processes 93
  4.1 Structure and function of important cell components 93
      4.1.1 The cell wall 94
           4.1.1.1 Structure 94
           4.1.1.2 Materials transport through the cell wall 96
           4.1.1.3 Floc formation 98
      4.1.2 Enzymes 99
  4.2 Fundamentals of metabolic processes 102
      4.2.1 Kinetics of enzyme-catalysed reactions 104
      4.2.2 Metabolic activity of the sludge flocs 113
      4.2.3 Growth and surplus sludge production 118
      4.2.4 Respiration and oxygen consumption 121
      4.2.5 Toxicity 127
      4.2.6 Effect of temperature 128
  4.3 Materials conversion processes 133
      4.3.1 Aerobic metabolism of carbon and hydrogen 136
      4.3.2 Anaerobic metabolism of carbon and hydrogen 137
      4.3.3 Metabolism of nitrogen compounds 140
           4.3.3.1 Ammonification 141
           4.3.3.2 Nitrification 142
           4.3.3.3 Denitrification 145
      4.3.4 Metabolism of phosphorus compounds 148
      4.3.5 Metabolism of sulphur compounds 151
      4.3.6 Waterborne toxic substances 152
      4.3.7 Roles of different groups of organisms 168
           4.3.7.1 Bacteria 168
           4.3.7.2 Fungi 169
           4.3.7.3 Protozoa and metazoa 170
  4.4 Pathogenic organisms 173
      4.4.1 Viruses 173
      4.4.2 Bacteria 175
      4.4.3 Protozoa 177
      4.4.4 Helminth ova 178
  4.5 Physicochemical consequences 179

5 Oxygen 185
  5.1 Oxygen input 185
  5.2 Measurement of oxygen input 187
  5.3 Aeration systems 189
  5.4 Regulation and automatic control of the oxygen input 192
  5.5 Use of technically pure oxygen 194
  5.6 Environmental pollution as a result of aeration 195

6 Reactor types and process variants 200
  6.1 Control methods 200
  6.2 Special methods 207

Table of contents 7

    6.3 Anaerobic activated sludge processes 210
    6.4 Arrangement of the activated sludge tanks 212

7 Final settling. 215
    7.1 Theoretical principles of activated sludge sedimentation 215
    7.2 Final settling tanks 218
    7.3 Recycled sludge. 221
    7.4 Biochemical processes in the final settling tank 223

8 Industrial effluents 224
    8.1 Continuity of incidence of the effluent 224
    8.2 Organic substrates 225
    8.3 Nutrient elements 225
    8.4 Oxygen demand. 226
    8.5 Bulking sludge. 226
    8.6 Surplus sludge 226
    8.7 Temperature. 226
    8.8 Toxic and abiotic constituents 226

9 Extended treatment 227
    9.1 Nutrient removal 228
        9.1.1 Importance of nutrient elements 228
        9.1.2 Origin of the nutrient elements 230
        9.1.3 Nitrogen removal. 230
        9.1.4 Phosphorus removal 233
    9.2 Disinfection. 236
    9.3 Other methods 237

10 Dimensional criteria 239

11 Treatment of surplus sludge 244
    11.1 Anaerobic digestion 244
    11.2 Aerobic stabilisation 245

12 Operation of activated sludge plants 247
    12.1 Running in phase 247
    12.2 Normal operation 249
        12.2.1 Treatment performance 250
        12.2.2 Effect of other technological processes. 252
    12.3 Operational disturbances. 254
    12.4 Automated operation 261

13 Future prospects 262

14 References 265

Plates 281

Index 290

Preface to the original edition

The rapid upsurge in manufacturing capacity during the last century has led to an historic advance in the interactions between man and nature. Striking pointers to the increasing utilisation of natural resources are to be seen in the growth and concentration of industrial and agricultural productivity, the development of synthetic organic, petrochemical and domestic chemical products, the development and use of agrochemicals and not least the improved standards of living for the population as a whole. Nature and effort are the basis of human existence and the availability of natural resources is limited. Based on the development laws for manufacturing capacity and production conditions, further economic growth that involves more intensive use of natural resources must leave them in a better condition. It lies within the range of capabilities of the human race to guide the direction and character of the interactions between man and nature towards a state of compatibility between ecology and the economy and to ensure that the natural environment not only acts as the foundation of national productivity, but also as the setting for human recreation and relaxation.

During the last hundred years, water as a natural resource has become qualitatively and quantitatively impaired to a considerable extent. As water is not only a raw material and an aid to manufacturing but also directly or indirectly forms an article of diet, its conservation within the natural cycle represents not only an essential part of environmental and human health care, but also an important political objective. The treatment of sewage thus occupies a central place in the national cultural legislation of the German Democractic Republic together with the associated regulatory provisions. In this way it embraces a complex web of social, economic and ecological criteria, of which we are concerned here only with the scientific factors governing the implementation of technological processes.

The activated sludge process has been widely adopted during the last 20–30 years as a method of treatment for organically polluted water and effluents. Despite the fact that the process does not completely purify the incoming liquid, as certain constituents can only be removed to a limited extent, it is nevertheless the method which has the greatest overall treatment performance. With the imposition of more stringent consent conditions and the growth of economically feasible methods, additional biological, physical and physicochemical processes must undoubtedly be linked with the activated sludge plant in future.

The principal aim of the present work is to present the biochemical processes involved in the activated sludge process in conjunction with other relevant branches of science and technology and to describe the methods available for operational control of the process. It is written from the standpoint of water quality management and as a contribution to biotechnology, and should serve to demonstrate the immense capabilities and versatility possessed by microorganisms in a biotope of anthropogenic origin and design.

The choice of subject matter is of course subjective; it has been heavily influenced by discussion with operators and managers of activated sludge systems, scientists in one's own and other disciplines, and colleagues in the official Inspectorate of Water Quality and Hygiene. This is also a good indication of the expected circle of readers. The detailed information should also be of use to students of hydrobiology and to others with technical qualifications and scientific interests. Chapter 12 is particularly intended for the managers and operators of activated sludge plants. They are the group with a decisive influence on the successful performance of such installations, and have hitherto been inadequately provided with literature on the subject of sufficient quality and depth.

This very diverse circle of readers and the nature of the vocabulary drawn from sewage technology, hydrobiology, ecology, physiology, medicine and other biological sciences has stometimes given rise to a few semantic problems. The foreign words however, apart from the species names for specific organisms, are rapidly gaining in worldwide recognition (in particular the ecological terms) on account of the increasing importance of water quality and environmental protection everywhere.

It is my special concern to thank all those who have assisted me by means of expert advice, literature searching and preparation of illustrations. Among these are Dr G. Seltmann, from the Institute for Experimental Microbiology, Wernigerode, who helped in the preparation of section 4.1.1 on the cell wall. Special thanks are also due to the staff of the Leipzig Sewage Laboratory forming part of the Water Research Centre under the direction of Frau Ch. Anton, for making available many of the analytical results quoted in the book. For a critical review of the manuscript I am indebted to Professor R. Birr (Merseburg), Dr G. Breilig (Berlin), Professor L. Hussel (Leipzig), Dipl. Ing. K. H. Lang (Dresden), and Dipl.-Chem. B. Mau (Leipzig). For typing the manuscript I am grateful to Frau Th. Muller and I am also indebted to Herrn B. Schreiber for preparation of the drawings. The help of VEB Gustav Fischer Verlag Jena, in particular Frau Schluter in the preparation of the manuscript, and their constructive collaboration is also gratefully acknowledged.

1
Introduction

As a result of the vigorous development of productive capacity since the middle of the nineteenth century and the introduction of piped water supplies to cities and municipal districts, the volume of sewage and other effluents has continued to increase. The water closet, the kitchen sink and the associated sewer network which transports the domestic sewage with all its constitents in a hygienically acceptable manner, thus eliminating a centuries-old source of infection, led to a very substantial improvement in the sanitary and hygienic conditions of towns and cities. However, during the past century serious adverse effects on natural waters were already becoming apparent, such as unpleasant odours, fish kills and limitations of use, due to the introduction of sewage effluents. The inherent 'self purification capacity' of natural waters was in many places so badly overloaded, that some form of artificial treatment of sewage became a pressing necessity for the maintenance of acceptable human living conditions. In paragraph 15 of the GDR Law of 7 October 1974 it is stated that 'The purity of natural waters and of the atmosphere and the protection of plant and animal life as well as the scenic beauty of the environment must be ensured by the relevant official bodies and are moreover the concern of every responsible citizen'.

Sewage consists of 'water which has been adversely affected by its use for domestic, trade or industrial purposes as well as by the presence of surface runoff orginating from rainfall on residential property and industrial premises', according to the definition given in TGL 55032/04†. The diverse origins of such wastewaters mean that they differ in their composition both qualitatively and quantitatively: in addition, diurnal, day-to-day and seasonal fluctuations occur. The curves shown in Fig. 1.1 indicating the variation in time of the volume of sewage and the BOD_5 loading (of particular significance for biological treatment) for three municipal treatment plants of different sizes, reflect the peculiar nature of each of the relevant installations. Industrial installations, by reason of the shift system of operations, the week-end shut-down period and discontinuous production cycles usually exhibit even more pronounced fluctuations in volumetric flowrate and composition of their effluent.

† A list of the East German Standards cited in the text is given on p. 280.

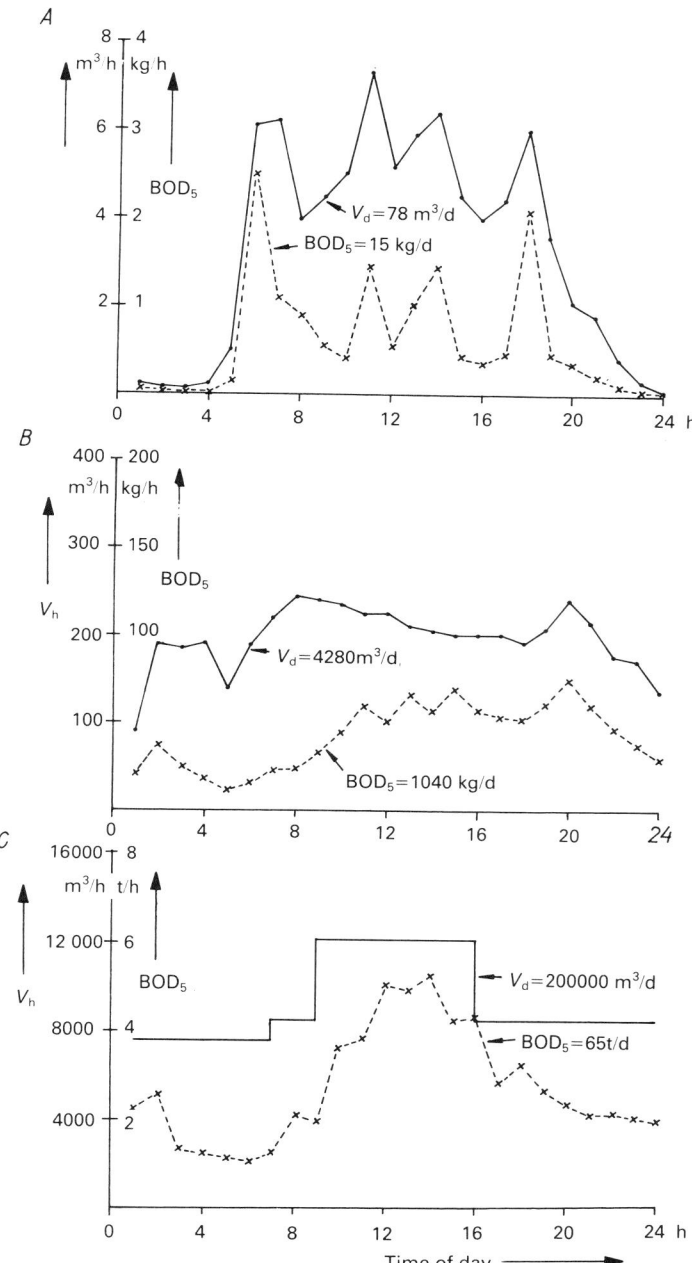

Fig. 1.1 — Diurnal profiles for raw sewage flowrate and BOD at three sewage treatment plants of different capacity. A: Pflegeheim for 300 inhabitants. B: Municipal treatment plant for 17 000 inhabitants. C: Public treatment plant for 500 000 PE with a pumping station upstream. For B and C local industrial and trade effluents of various kinds are discharged to sewer in quantities varying from place to place.

For municipal sewage, the following ranges for the most important parameters may be adopted:

Biochemical oxygen demand (BOD_5)	100–600 mg/l
Chemical oxygen demand (COD–Mn) using $KMnO_4$ as oxidising agent	60–200 mg/l
Chemical oxygen demand using $K_2Cr_2O_7$ as oxidising agent (COD–Cr)	150–900 mg/l
Organic carbon (C)	50–200 mg/l
Nitrogen (N)	25– 80 mg/l
Phosphorus (P)	5– 20 mg/l
Specific sludge volume (SV)	1– 8 ml/l
Sludge solids content (TS)	100–1000 mg/l

A comparison of these figures with those of Table 2.3 shows that the discharge of untreated sewage to natural waters will cause a serious impairment of water quality, to an extent depending on the natural flow. Since the total renewable water supply per capita in the GDR is only 880 m$_3$/ annum, and also due to the high level of industrialisation and modern agricultural production techniques, a 44% utilisation of the available resource occurs in years of average rainfall, rising to 90% utilisation in particularly dry years. In river basins with heavily exploited water resources the water is used seven times over (Reichelt 1981). According to [1] the quantity of sewage requiring treatment will increase between the years 1980 and 2000 from 2.32 to 3.35×10^9 m^3/annum. Around 80% of this total originates from industry and agriculture. The biological treatment of the effluent by means of activated sludge is the most effective process, which brings about a far reaching improvement in the quality of natural waters and hence allows for the practice of multiple-reuse.

1.1 TERMINOLOGICAL DETAILS

The activated sludge process is defined in TGL 55032/04 as 'biological sewage treatment in artificially aerated vessels, in which the organic constituents of the sewage are metabolised by the activated sludge, being partially or completely degraded and/or mineralised.

The microorganisms growing in or on the flocs are termed activated sludge or sludge biomass, and the treatment process is accordingly known as the activated sludge process. The term 'activation process' which was formerly employed in Germany, is not adopted in this definition as it does not provide a unique or sufficiently comprehensive description, thus the numbers of bacteria in primary settled municipal sewage may amount to 10^8–10^9 organisms/ml, while well-treated sewage effluents will only contain from 10^5 to 10^8 organisms/ml — which is not activation in the true sense of the word. Owing to the misleading name, differing opinions used to be held (Imhoff and Imhoff 1972) and it was debatable whether the term 'activation' related to the sewage or the biomass contained in it.

The terms and abbreviations employed in this work are mainly drawn from sewage technology and not from the biotechnology field. Table 1.1 contains the most

Table 1.1 — Terms and symbols used in sewage treatment (based in part on TGL 55032/04)

Term	Symbol	Units	Description
General terms			
Discharge	a	—	Biological treated sewage effluent
Effluent concentration	C	mg/l	Measure of a constituent of sewage per unit vol.
Effluent load	L	kg/d, kg/h	Measure of a constituent of sewage in the sewage flow per unit of time, $L = C \times V_d$ or $C \times V_h$
Aeration tank	BB	—	Compartment or tank in which the sewage/sludge biomass suspension is aerated
Removal performance	Δ	various	Elimination of certain sewage constituents during treatment (e.g. BOD_5)
Reaeration tank	RB	—	Tank for separate aeration of recycled sludge
Final settling tank	NKB	—	Settling tank for gravity settling of sludge from the sewage sludge mixture (mixed liquor)
Removal efficiency	η	%	Percentage reduction in the concentration of a particular constituent of sewage $= \dfrac{C_z - C_a}{C_a} \times 100$
Influent	z	—	Raw sewage prior to biological treatment
Loading-input conditions			
Retention time	t	h	Theoretical retention time of sewage in a particular treatment stage or plant component
Flowrate	V	m³/h, m³/d	Rate of liquid flow to any plant item or treatment step
Areal loading rate	V_A	m³/m² h	Feed rate per unit of surface area and time to, for example, final settling tank
Sludge area loading	V_{SV}	m³/m² h	Volume of sludge entering unit area of the final settling tank per unit of time $V_{SV} = (SV_{BB} + V_h)/A_{NKB}$
Areal loading	B_A	kg/m² h	Mass of sludge solids entering unit area of the final settling tank per hour $B_A = (TS_{BB} V_h)/A_{NKB}$
Recycled sludge flowrate	V_{RS}	m³/h, m³/d	Rate of return of recycled sludge from final settling tank to aeration tank
Volumetric loading rate	B_R	kg/m³ d	Mass of pollutant (as BOD_5) which is supplied to unit volume of the aeration tank per day $B_R = L_z/V_{BB}$
Sludge loading rate	B_{TS}	kg/kg d	Mass of pollutant (as BOD_5) supplied to unit wt. of sludge solids per day $B_{TS} = B_R/TS_{BB}$

Sec. 1.1] Terminological details

Sludge degradative performance	ΔB_{TS}	kg/kg d	Mass of pollutant (as BOD_5) removed by unit wt. of sludge solids per day ($B_{TS} = \eta B_{TS}$)
Activated sludge biomass			
Sludge age	SA	d	mean residence time of an activated sludge floc in the aeration tank $SA = (TS_{BB} V_{BB}) US$ or $(SV_{BB} V_{BB}) Us$
Sludge solids (filtrable solids)	TS	mg/l, g/l, kg/m^3	Mass of solids per unit volume resulting from filtration or centrifugation of sludge suspension followed by drying at 105°C
Organic sludge solids	oTS	mg/l, g/l, kg/m^3	Difference between TS and the residue on ignition at 550°C
Sludge volume	SV	ml/l	Volume of sediment per litre of sludge suspension after settling for 30 min
Corrected sludge vol.	kSV	ml/l	Volume of sediment per litre of sludge suspension after dilution to such an extent that the SV after settling for 30 min is 150–250 mg/l $kSV = SV(1+m)$
Sludge volume index	I_{SV}	ml/g	Sludge volume given by 1 g of sludge solids after settling for 30 min $I_{SV} = SV/TS$
Surplus sludge wastage rate	US	m^3/d, kg/d	Quantity of activated sludge wasted from the system per day
Surplus sludge production	USP	kg/kg	Quantity of biomass generated form the elimination of 1 kg BOD $US = L_z - L_a$
Aeration			
Oxygen input	OC_{10}	g/m^3 h	Mass of oxygen taken up per h by 1 m^3 of initially deoxygenated clean water at 10°C and 1 atm
	OC'_{10}	g/m^3 h	Value of OC_{10} for the sewage/sludge suspension $OC'_{10} = OC_{10}$
	OC	kg/h	Mass of oxygen introduced per h by an aeration device into clean water under the same conditions as for OC_{10}
	OC'	kg/h	Value of OC for the sewage/sludge suspension (mixed liquor) in the aeration tank
Oxygen uptake efficiency	OC/P	kg/kWH	Ratio of oxygenation performance to the gross energy consumption
Energetic efficiency	EO	W/m^3 or kWH/m^3 d	Hourly or daily input of energy per unit volume either as net energy or gross energy for evaluation of turbulence or economic performance
Nitrogen metabolism			
Nitrification rate	—	g/kg h	Mass of ammonium ions oxidised per unit wt of organic solids per h
Denitrification rate	—	g/kg h	Mass of nitrate or nitrate reduced per unit wt. or organic solids per h

Table 1.2 — Terminology of biological processes as applicable to sewage engineering

Term	Symbol	Description
Acclimatisation (adaptation)	—	Functional readjustment of microorganisms to changed external conditions
Aerobic	—	Biochemical processes taking place in the presence of physically dissolved oxygen
Anaerobic	—	Biochemical processes taking place in the absence of physically dissolved oxygen
Anoxic	—	Biochemical processes taking place in the absence of physically dissolved oxygen but in association with nitrite or nitrate as sources of oxygen
Respiration	OV (mg/mg h)	Process of obtaining energy for aerobic organisms by the enzyme-controlled oxidation of hydrogen to water with the aid of oxygen in solution or nitrite/nitrate ions
Denitrification	—	Enzymatically controlled reduction of nitrite and/or nitrate to gaseous nitrogen
Diauxy	—	Sequential (i.e. non-simultaneous) metabolism of two different substrates
Enzyme	—	Biological catalyst consisting of protein of high-molecular weight
Fermentation	—	Process for production of upgrading of a product by microbiological action
Anaerobic fermentation	—	Process by which anaerobic organisms obtain energy
Carbohydrate	—	Organic compound having the empirical formula $C_n(H_2O)_n$, and compounds derived from such
Lipid	—	Organic substance of widely varying structure, insoluble in water but soluble in lipophilic solvents, for example chloroform
Nitrification	—	Enzymatically controlled oxidation of ammonia to nitrite or nitrate
Peptide	—	Organic compound composed of amino-acids
Polyauxy	—	Sequential metabolism of several different substrates
Protein	—	Organic compound of high molecular weight, composed principally of amino acids
Metabolism	—	Rearrangement of the constituent substances of living organisms as a means of supporting life
Substrate	S(mg/l)	Organic and/or inorganic constituents of sewage capable of utilisation for biosynthesis or biodegradation by enzyme-controlled reactions

Sec. 1.2] **Essential features of the process** 17

important sewage engineering technical expression, while Table 1.2 lists some of the frequently used biochemical terms.

1.2 ESSENTIAL FEATURES OF THE PROCESS

The activated sludge process constitutes one phase of the sewage treatment process. It exerts a pronounced purification effect, for which some form of mechanical pretreatment is an essential prerequisite, and a subsequent sludge treatment stage is also required in addition. A complete sewage treatment plant of conventional design comprises the various technical stages shown in Fig. 1.2.

With the activated sludge process we are concerned with a form of *applied microbiology*, which however, differs from all other biochemical processes in its objectives. Other processes are based on the principle of employing a special nutrient medium and a pure culture of microorganisms to produce an end product with certain desirable properties, such as alcohol, cheese, organic acids, etc. In some cases the organisms themselves may constitute the end product, for example, in yeast production. These processes operate with only *one* species of organism and are purely biological in nature, the metabolic performance of the specialised strains of organisms often being very high. The non-utilised dilute substrate, which because of its change in properties is no longer effective as a source of the desired end product, is usually regarded as a waste product and is mostly discarded.

In the activated sludge process, however, the dilute nutrient medium known as 'sewage', is by means of many groups and strains or species of different organisms, used to produce 'waste activated sludge' which represents the product in the applied microbiolgial sense, but which is usually contaminated with various toxic residues or pathogenic organisms and in practice is reckoned to be a troublesome waste product, for which costly disposal measures are required. It must also be borne in mind that the activated sludge process is not a process of a purely biological nature, but that numerous physical and chemical reactions proceed at the surface of the activated sludge flocs. The terms applicable in the context of biological process engineering, such as substrate utilisation, biomass synthesis, growth rate and so on are thus only partly appropriate for most types of sewage. The true end product of the activated sludge process in the form of treated sewage effluent has no exact counterpart in the field of industrial applied microbiology.

Sewage is usually a multicomponent substrate, that is, it is composed of many different and constantly changing substrates which are mainly outside the operator's control and may even occasionally be of a toxic nature. The smaller the size of the community, or the more unbalanced the industrial component (e.g. due to shift operation) the more the nutrient properties of the sewage are of a non-uniform and problematic nature. Only mixed populations composed of numerous species are capable of coping with this native variability on the substrate.

Following the initial start-up, a gradual multiplication of organisms present in the sewage (or derived from the atmosphere or inoculated sludge biomass) takes place in the activated sludge tank, for which the organic substrates present in the sewage act as the nutrient source. The oxygen necessary for the respiration of the organisms is supplied by aeration equipment which also induces an internal circulation of sewage and suspended microorganisms. As the aim of biological sewage treatment is

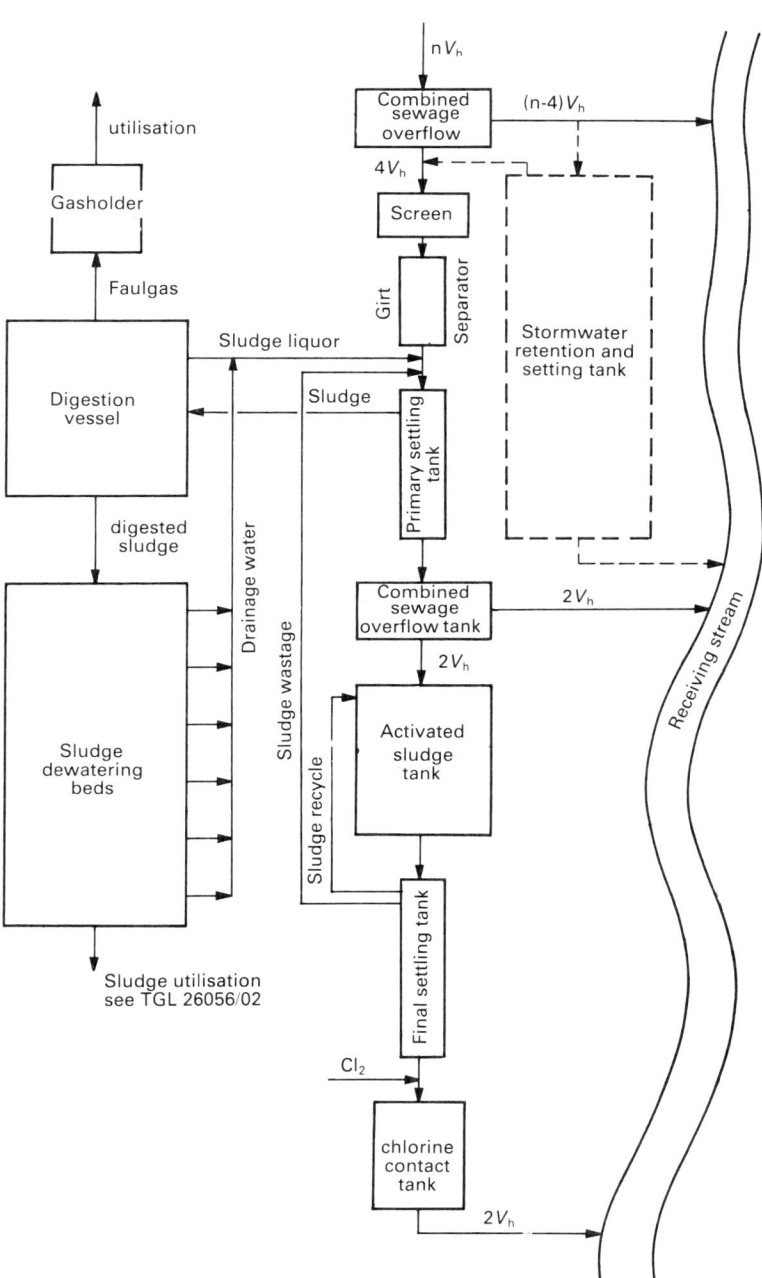

Fig. 1.2 — Diagram showing a typical layout for a conventional sewage treatment plant.

to produce a treated effluent which is almost substrate-free, the activated sludge organisms must be induced to accept all the substrates capable of utilisation as nutrients; in practice this is possible only in the case of low nutrient availability or very prolonged aeration times.

Following preliminary pretreatment municipal sewage generally has a BOD_5 of about 240 mg/l, from which as a result of biosynthesis and the accumulation of inert substances, an activated sludge biomass of 150–200 mg/l is produced. The content of bacteria in this activated sludge, is however, quite small, so that unacceptably long aeration times would be necessary for treatment at this level of biomass (Fig. 1.3).

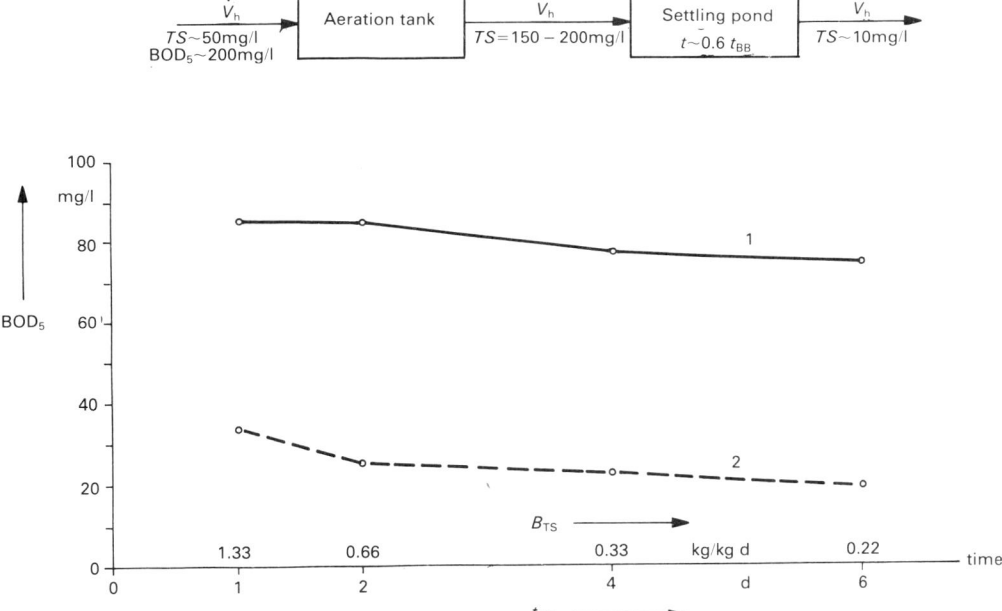

Fig. 1.3 — Process arrangement and treatment performance for an activated sludge plant without sludge recycle. Curve 1: BOD_5 in the aeration tank, with the BOD_5 resulting chiefly from endogeneous respiration of the organisms present. Curve 2: BOD_5 in the effluent from the final settling pond. Adapted from [21].

Even concentrated industrial effluents require residence times of a day or so for complete biological treatment, as the growth of the organisms very often proceeds more slowly than in the case of municipal sewage.

In order to curtail the aeration time it is necessary to increase the sludge biomass content in the aeration tank by a factor of 10 or 20. In order to achieve this in practice, the activated sludge biomass is separated from the liquid phase by gravity settling or by flotation in the final settling tank, and recycled in a more concentrated form to the aeration tank. The concentration of the sludge biomass in the aeration tank is usually of the order of 2–4 g/l; excess growth over and above this level is bled off in the form of so-called waste activated sludge, and is usually introduced into the primary settling tank and anaerobically digested along with the primary sludge. With an activated sludge biomass concentration of 2–4 g/l the treatment of municipal sewage is virtually complete at the end of an aeration period lasting six to eight hours. An illustration of the principle of the activated sludge process is provided by the diagram of the small-scale activated sludge plant according to TGL 22767 shown in Fig. 1.4.

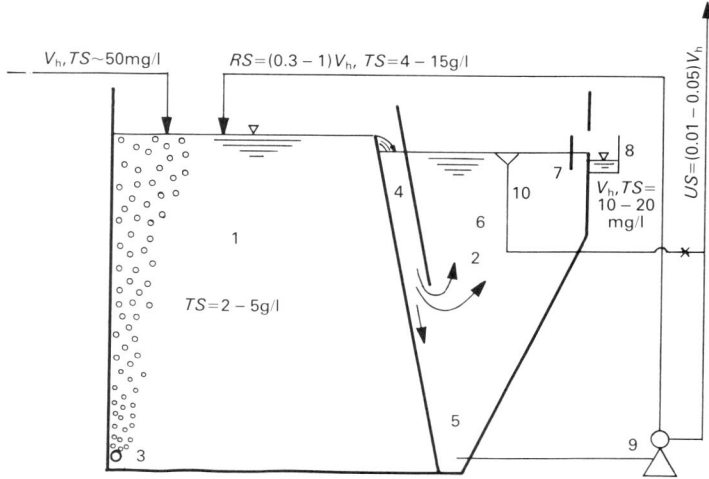

Fig. 1.4 — Sketch showing principal features of an activated sludge system according to the specification for compact systems contained in TGL 22767. Key: 1: aeration compartment; 2: settling compartment; 3: air inlet; 4: degassing and flocculation zone; 5: thickening zone for recycled and waste activated sludge; 6: sedimentation zone; 7: submerged baffle for retention of floating solids; 8: outlet duct; 9: recycle and waste sludge pump; 10: scum take-off.

The activated sludge process can be represented as a materials conversion process expressed in the following simplified overall equation:

$$\text{Substrate} + O_2 \xrightarrow[\text{biomass}]{\text{activated sludge}} \underset{C_{106}H_{180}O_{45}N_{16}P}{\text{surplus sludge}} + H_2O + CO_2 + NH_4^+ + NO_2^- + NO_3^- + PO_4^{3-} + SO_4^{2-} + \text{energy}.$$

1.3 HISTORICAL DEVELOPMENT OF BIOLOGICAL SEWAGE TREATMENT

With the evolution of life in water and the subsequent colonisation of the Earth's surface by plant and animal organisms, organic substances were produced which were subject to the general laws of nature concerning their decomposition and ultimate mineralisation. The microbiological processes which occur in today's activated sludge tanks evolved on Earth thousands of millions of years ago in their basic form.

As a consequence of the concentration of people and animals within narrowly restricted areas the disposal of the excreta and waste products formed a vital question. The discharge of these to natural waters gave rise to devastating epidemics, which from time to time decimated the population. The catastrophic hygienic conditions in the water systems during the Middle Ages and in the cities of early capitalist times have been impressively described by Gilsenbach (1971) as 'pestilential schoolmasters' for sanitary hygiene. According the the World Health Organization (WHO), three-quarters of the population of the Third World are still without sanitary facilities and 80% of outbreaks of infectious disease are attributable to contaminated drinking water.

The first technical sewage treatment plants were simple mechanical sedimentation plants, where the settling tanks were analogous to the tailings ponds of the mining industry. The subsequent infiltration of sewage into the soil, as in the Berlin irrigation fields, can be regarded as the very first biological treatment process and in principle very much older than even the mechanical treatment.

Even before the turn of the century, biological methods of sewage treatment had become an urgent necessity. Research was concentrated on stimulating the natural purification processes in flowing waters. As pebbles occurring in natural waters presented ideal supports for the growth of attached sewage-treatment organisms, attempts were made to reproduce these naturally-occurring conditions in compact sewage treatment installations. The coarse packing material was at first flooded with sewage for a certain period of time and then drained off for the purpose of admitting oxygen to the so-called 'biological film' Later it was found to be more effective to sprinkle previously well-settled sewage onto the packing material from above. These biological systems were known as trickling filters, and are still widely found today, for example, in conjunction with an activated sludge plant connected in series (Plate 15). In comparison to activated sludge plants they have the advantage of lower energy consumption, but the disadvantages of a seasonally variable treatment performance, higher capital costs and more severe operating problems due to clogging and attrition of the packing material, insect plagues and so on. They are now used only in the smaller treatment installations for populations of up to 20 000.

Activated sludge process
Attempts to purify sewage by the introduction of air were made around the turn of the century by, among others, Mairich in the sewage treatment plants at Aschersleben and Neustadt (Brix *et al.* 1934). The development of the activated sludge process is, however, of Anglo-American origin. Clark of the Massachusetts Department of Public Health performed discontinuous jar tests on the aeration of sewage in 1912 in which the resulting activated sludge flocs were separated from the aerated liquid and used to innoculate fresh sewage. The jars were operated in principle like the present-day single-compartment plants or holding tanks, such as oxidation ditches. The experiments were observed by an Englishman named Fowler in the course of a study tour, and subsequently developed by his co-workers Lockett, Jones and Ardern into

a continuous flow-through system at Manchester in 1914. Since that time engineers, biologists, chemists, process engineers, machinery manufacturers and civil engineers have striven continuously for improvements in the tank configuration, the aeration systems and process design, as well as for the extension of the process to industrial effluents.

Aeration systems (Plates 7–12)
The first practical activated sludge plants operated with long aeration times of 6–15 hours. These long times were necessary on account of the low level of efficiency of the aeration equipment with a low rate of specific oxygen uptake, as a result of which simultaneous nitrification was often attained. Aeration devices consisted of paddle wheels in channels as at Erfurt, special pumps as at Rabenstein/Karl Marx Stadt, and even compressed air which was employed at the first German activated sludge installation, at Essen-Rellinghausen. From the original paddle wheels variants such as the bar and cage rolls and also the Mammoth rotors were developed. Special attention was given to the development of compressed air systems in Germany, which were adopted for the larger installations, such as those at Stahnsdorf (1932). The centrifugal aeration system also originated in Manchester. The original type of rotary device named the Bolton wheel, after its inventor, gave rise to a series of widely publicised versions after the Second World War which resulted in a variety of centrifugal rotor designs. The use of heavily loaded activated sludge systems also required combinations of several different forms of aeration equipment, or the use of intensive aeration devices such as suction rotors, jet entrainment and submerged jet systems.

Process configuration (Fig. 6.2)
In the classic type of activated sludge plant the sewage and the recycled sludge were both introduced at the head of an elongated aeration tank, since it was assumed that a lengthy contact period between the sludge and the sewage to be treated was called for. It very soon became apparent that a massive oxygen demand occurred in the first part of the tank which could not be supplied from the air injection points distributed uniformly along the length of the tank. This deficiency was overcome by enhanced air input to the forward section of the tank, the so-called stepwise aeration system. In the USA the first activated sludge plants were operated without any form of pretreatment, but this practice was soon abandoned for purely mechanical reasons. From studies of the progress of treatment along the length of the tank it was established that relatively little treatment effect occurred in the final sections of the tank. Attempts to shorten the aeration time were often unsuccessful owing to bulking sludge formation. However, in the USSR and in the USA in particular, the practice of sludge regeneration was widely adopted, according to which the contact time between the sludge and the sewage being treated was reduced to one to two hours, while the recycled sludge was reaerated for a period of several hours in a separate tank.

During the 1920s the addition of asbestos to the aeration tank, in the Zigerli process, was proposed in order to uprate the treatment performance by acting as an inert support and as a means of increasing the contact surface area. Also the addition of iron salts, which resulted in the formation of flocs of hydrated iron oxides, was tried for the same reason, and is still frequently used today for different reasons, for example, in the simultaneous precipitation of phosphate.

When treating industrial effluents of low nutrient content, the important role of *nitrogen* and *phosphorus* was soon recognised. The Magdeburg-P process (Nolte 1934) was developed around this time and is still used in many industrial effluent treatment plants.

The aforementioned front-end feed system was largely abandoned later on, being replaced, for example, by the step-feed system of Bazjakina, 1936, in Moscow, and Gould, 1939, in New York (Imhoff 1949), or by reactor systems with a constant sludge loading (Beuthe 1967), and also by the nowadays widely-used completely mixed reactor. The heavy-duty activated sludge process with aeration times of two hours was first employed by Gould at the Jamaica, New York, sewage treatment works. Today this method is used with residence times of only 20 to 30 minutes and BOD_5 volumetric loadings of $20-30 \, kg/m^3$. In addition the two-stage process was devised just prior to the Second World War.

After the war research was concentrated initially on the use of high sludge loadings, and later on the extemely lightly loaded systems with simultaneous aerobic sludge stabilisation. During the last decade many plants employing oxygen in place of air have been constructed, and much activity devoted to the study of nutrient elimination, with erection of the first plants of this kind.

Tank configuration (Plates 7–12)
The aeration system determines the geometry of the basin. Paddle wheels, rolls, Mammoth rotors, etc., are installed in trench-like channels such as Haworth channels, oxidation ditches and carrousel plants; compressed air systems require elongated rectangular shaped tanks of greater depth, while centrifugal surface aerators are used in tanks of square-shaped horizontal section. In the last 10 to 20 years, intensive aeration systems of various special types have been devised, for example, tall column or deep-shaft systems, closed tanks, etc.

From numerous patents and scientific publications of recent years it appears that many strains of bacteria and fungi, such as *Acinetobacter*, *Thiobacillus*, *Pseudomonas*, *Vibrio*, *Nocardia*, etc., have demonstrated the ability for performing special metabolic tasks at incxeased conversion rates, for example, phosphate and nitrogen removal, chromate and chlorate reduction, liberation of heavy metals from sludges, decomposition of persistent substrates such as mineral oils, chlorinated hydrocarbons, aromatics, nitrites and so on. In addition the immobilisation techniques, which foster and retain specialised organisms for certain specific tasks, as a basis for later inoculation into activated sludge biomass for dealing with specific substrates, are in the early stages of development. Such investigations and their results prove that the scope for development of the activated sludge process has not yet been exhausted.

2

Characterisation of the properties of sewage and activated sludge

Determinations of the constituents of sewage and activated sludge are performed according to the 'Selected methods of water examination' [2, 3]: see TGL 28400. In some cases modifications tailored to the needs of the activated sludge plant are given here.

2.1 CHEMICAL AND PHYSICAL METHODS

2.1.1 Activated sludge

Samples for assessing the properties of the activated sludge are obtained by withdrawal from the aeration tank or the recycled sludge pipeline.

Sludge volume (SV)

The sludge volume provides a rough indication of the quantity of organisms. As this plays an important role in determining the treatment performance and also the oxygen and energy requirements, and moreover influences the sedimentation behaviour in the final settling tank, the measurement of the sludge volume is one of the most important tasks for the study and optimisation of an activated sludge plant. When starting up the plant, or dealing with upsets in plant performance or controlling the rate of sludge recycle, a knowledge of the sludge volume in the recycled sludge is required. In the case of dimensional design of the final settling tank according to the Standard WAPRO 2.26 the sludge volume-areal loading rate is used as the governing factor.

The sludge volume specifies the quantity of sludge which settles out within a period of 30 minutes in a 1-litre measuring cylinder of 6-cm diameter. The settling time, the depth of fill and the diameter of the vessel all exert an influence on the result and hence these stipulations must be adhered to for comparability of results; alternative containers should be avoided except for sludge volumes of less than 100 ml/l when a conical funnel may be used.

The sludge volume should be determined immediately after sampling; delays or long transport times lead to incorrect readings. During the settling period the measuring cylinder should be kept at the same temperature as the aeration tank, in order to prevent the formation of convection currents; direct sunlight should be avoided for the same reason. For liquid temperatures greater than 18°C, nitrate-containing and very active sludges, flotation of sludge at the surface often occurs after 15 to 20 minutes as a result of denitrification.

For sludge volumes of over 250 ml/l, the diameter of the measuring cylinder may be such as to give rise to hindered settling, a physical process which leads to artificially high readings. In such cases *serial dilutions* (1:1 to 1:4) must be performed and the true sludge volume obtained by means of back-calculation. The dilution water should consist of biologically treated effluent from the final settling tank of the same sewage treatment plant. The sludge volumes so obtained are designated as correct, dilution or relative sludge volumes. With bulking sludges the bacterial threads may become even more distended as a result of dilution, so that the corrected sludge volume may have values of over 1000 ml/l.

From an observation of the sludge during the settling process it is often possible to draw conclusions regarding the nature of the organisms present, the properties of the activated sludge, and the nature of the biological transformations taking place, for example denitrification.

As measurements of the sludge volume are needed several times per day in a plant with a high rate of waste activated sludge production, and the measurements are time-consuming, automatic measuring equipment has been developed, but has so far not been widely adopted. Where the waste activated sludge is bled off automatically, the determination of the sludge volume ceases to be of importance.

Sludge solids content (TS)

The sludge solids content is a measure of the mass of the organisms present. It is used in determining the size of the aeration tank on the basis of the BOD_5-sludge loading and hence the treatment performance, and also for determining the areal loading of the final settling tank.

The sludge solids content is the amount of biomass obtained following separation from the liquid phase by filtration through medium-grade filter paper or by centrifuging at 5000 rev/min for five min, and drying at 105°C. It contains all the coarse solids and hence includes biologically inert material such as hairs, textile fibres and inorganic hydroxides. As the determination of the dry solids is a relatively lengthy process there are efforts towards the development of correlation between turbidity and sludge solids.

In order to determine the content of organic matter in the sludge solids, they are heated in a furnace to 550°C. The difference between the residual ash, the so-called residue on ignition, and the sludge solids content is termed the organic solids content. Depending on the type of sewage, nature of coagulants used and the sludge age, the organic solids content may constitute from 60% to 90% of the sludge total solids. At 550°C, however, not only is the organic fraction of the biomass volatilised but also all other organic matter and some inorganic compounds like hydroxides and even some heavy metals. Hence the organic solids content does not provide an exact measure of the biomass.

The residual mass following drying of a sample at 105°C is termed the residue on evaporation. As this residue contains the dissolved salts present in varying amounts in the sewage it cannot be used as a measure of the activated sludge concentration.

Sludge volume index I_{SV} or SVI)
For characterising the settling properties of the activated sludge the sludge volume index is employed, also called the sludge index or the Mohlmann index after its inventor. It represents the sludge volume obtained following a settling period of 30 minutes from 1 gram of activated sludge. The calculation is based on the equation:

$$I_{SV} = \frac{SV}{TS} \text{ (ml/g).}$$

It is to be strenuously noted that the sludge volume used in the calculation is derived from a settled sample which contained less than 250 ml/l of sludge; otherwise spuriously high values and possibly also widely scattered results from determinations performed at closely spaced intervals will be obtained. Where the sludge volume index is below 100 ml/g the sludge exhibits very good settling behaviour; for values in the range 100–150 ml/g settling is still good, but sludges with values of over 150 ml/g are denoted by the term bulking sludge. They exhibit very poor settling behaviour and hence frequently give rise to sludge carryover in the effluent from the final settling tank. Sludges with indices of 40–60 ml/g possess poor flocculation characteristics, so that these also lead to high levels of filterable solids in the final effluent.

Sludge age (SA)
The term sludge age is employed to denote the interval between the formation of an activated sludge floc and its removal in the form of waste activated sludge. It is calculated using the following equation (symbols as listed in Table 1.1):

$$SA = \frac{TS_{BB} \times V_{BB}}{WS} \text{ (d) or } \frac{SV_{BB} \times V_{BB}}{WS} \text{ (d)}$$

This, however, takes no account of the fact that from 5–20 mg/l of sludge biomass is carried over in the final effluent; for very lightly loaded activated sludges and for hydraulically overloaded final settling tanks and poorly flocculating sludges this wash-out effect may sometimes even result in the total absence of any waste activated sludge, and hence an infinitely large value of the sludge age would be obtained from the calculation. This error may be eliminated by the use of the following equation (symbols as listed in Table 1.1):

$$SA = \frac{TS_{BB} \times V_{BB}}{WS + V_d TS_a} \text{ (d)}$$

Both equations ignore the activated sludge contained in the final settling tank. The true sludge age is usually 5–25% greater than that given by the above equations under normal loading conditions (Table 10.1). The sludge age varies from 0.5 days for heavily-loaded plants, to 100 days for lightly-loaded activated sludges. It is correlated with the BOD_5-sludge loading and hence also with the treatment effect.

Sludge level
Disturbances in the activated sludge balance are often brought about by the formation of sludge deposits in the aeration tank, pump intakes and final settling tanks. For detecting such deposits, and also for determination of the sludge level in the final settling tank as a basis for automatic wasted sludge removal, or for indication of a certain operational state, submerged sensors are usually employed which operate on the principal of turbidity measurement. The Russian turbidity meters M 101 and Suf 42 are used for this purpose. In addition optic fibres are suited for many purposes. Sludge deposits in the aeration tank may also be detected with the aid of oxygen electrodes, as such deposits are invariably anoxic.

2.1.2 Liquid composition
Oxygen (O_2)
The measurement of the dissolved oxygen content in the aeration tank is a prerequisite both for the control of the biochemical processes and for the economic operation of the plant. Frequently the presence of a certain level of dissolved oxygen in the biologically treated effluent is a requirement of the Water Quality Inspectorate in the GDR.

The widely used Winkler method of water analysis often leads to erroneous values for sewage owing to the presence of certain constituents, for example nitrite, iodine-binding organic matter and so on. The modified Winkler method developed for use on activated sludge suspensions [2], including the inactivation of the organisms to prevent oxygen uptake, and precipitation of the activated sludge with metal hydroxides, also contains numerous possible sources of error, which are liable to occur in particular at low levels of dissolved oxygen. For these reasons, electrochemical methods have been in use for more than 20 years, which are certainly less accurate for use in pure water, but are nevertheless more reliable for use in sewage liquids than the Winkler method. The measuring process utilises the principle of polarography, for which the electrode systems $Pt/KOH/Ag/Ag_2O$ without temperature compensation and $Pt/KCl/Ag/AgCl$ with temperature compensation are commonly employed in East Germany. With these electrodes and the relevant instruments, both discontinuous and continuous measurement of the dissolved oxygen content can be performed and hence the conditions needed for an oxygen-related aeration control system are fulfilled. In addition they are also used in connection with respiration rate measurements and for the determination of BOD_5. Both electrode systems are, however, severely affected by the presence of hydrogen sulphide, both as regards their accuracy and service life (Hale 1983). It is therefore advisable, especially for long-term usage, to select measuring points at which there is no risk of hydrogen sulphide occurring. Other electrode systems such as $Au/Na_2S/Ag/Ag_2S$ exist which are not adversely affected by hydrogen sulphide and are hence more suited for use in activated sludge installations.

Hydrogen sulphide (H₂S)

Hydrogen sulphide and its compounds may be formed in sewage as a result of the anaerobic decomposition of sulphates and proteins or occur as a waste product of industry. Anoxic organic effluents with long transit times in the sewer network before reaching the treatment plant invariably contain H_2S. It is recognisable by its smell and by the greyish-black colour of the liquid, due to the presence of iron sulphide. In addition sludge liquors from the sludge treatment plant frequently exhibit high concentrations of H_2S. It is poisonous to humans (maximum acceptable concentration (MAC value) 15 mg/m³) and also to many microorganisms on account of its cytotoxic behaviour. In the aeration tank it is oxidised to sulphur and sulphuric acid. If H_2S is detectable in the aeration tank or in the effluent from the final settling tank, this is a clear indication of insufficient oxygen input or the presence of localised septic pockets in the aeration or final settling tanks.

Chemical oxygen demand (COD)

The COD is the most widely used and relatively rapidly measurable parameter for estimating the organic constituents in the sewage liquid. It is the weight of oxidising agent, expressed in terms of molecular oxygen, needed for the oxidation of the organic substrates. It thus represents a bulk analytical method of determination. Depending on the nature of the oxidising agent and the type of organic substrate, only certain fractions of the total are included (Lienig 1979).

Either $KMnO_4$ or $K_2Cr_2O_7$ may be used as oxidising agents; in exceptional cases persulphate may also be used. The values for COD-Mn and COD-Cr for municipal sewage are in a ratio of from 1:2 to 1:5, the actual ratio being naturally dependent on the type of sewage concerned. Any pretreatment of the samples such as homogenisation, filtration, sedimentation or centrifugation may affect the results, a problem discussed more fully in connection with the BOD_5. The COD is one of the most important criteria of water quality. It is, however, less suited to an assessment of the biological treatment effect of an activated sludge plant since it embraces organic constituents which are not susceptible to elimination by biological methods. For this reason biological treatment of municipal sewage results in a shift of the BOD:COD ratio which for BOD:COD-Mn may be from 2–3:1 to 1:2–6, and for BOD:COD-Cr from about 1:2 to 1:5–9. For industrial activated sludge plants the extent of elimination with respect to BOD_5 is often as great as 95% while that with respect to COD-Mn is only about 50%, providing the BOD_5 determination is not affected by the presence of toxins. The presence of a high COD in the effluent from an activated sludge plant coupled with low BOD_5 values is indicative of the limit attainable with biological treatment.

Organic carbon (TOC, total organic carbon)

The combined organic carbon is a measure of the level of organic contamination of sewage and of the organic fraction of the sludge. In a normal situation 1 g carbon corresponds to 1.724 g biomass. The TOC value is determined by means of wet chemical or high-temperature oxidation, whereby the product of oxidation, namely carbon dioxide, is measured either conductometrically or by means of infra-red spectrometry. In the principle of measurement and in its predictive value, the parameter resembles the COD; the analytical procedure, however, entails more stringent reaction conditions, so that the more chemically stable compounds are taken into account in the result.

High-temperature oxidation is used in seveal modern analysing instruments, which are placed in the sewage. However, there are still problems, especially because of large particles of sewage material.

From the TOC the dissolved organic carbon (DOC) can be calculated; following filtration, the sewage probe is used in the same manner as when measuring the TOC.

Settleable solids (SV)

The settleable solids are a measure of the sludge content of the sewage; they are therefore of special significance when monitoring the performance of settling tanks. As the sludge content at this stage is normally very small, settling funnels or so-called Imhoff-funnels are used, with a total volume of 1 litre, but which permit the reading of volumes of less than 1 ml. Settling times of 15 or 30 minutes or two hours may be required. Prior to reading the volume, a slight twisting motion is given to the funnel so that the particles clinging to the sides are induced to settle to the bottom. With properly functioning final settling tanks a nil result is obtained after settling periods of such duration, even though the effluent may be quite cloudy. For separation of the microflocs intervals of at least 12–24 hours are necessary; in order to induce these flocs to settle, careful stirring several times with a glass rod is advisable.

Filterable solids and total solids (TS)

Besides BOD_5, the filterable solids are one of the most important properties of the treated sewage effluent. The requirements of the Water Quality Inspectorate generally call for a maximum of 20 mg/l for biologically treated sewage. The determination is performed in the same way as for sludge solids in the sludge biomass. In order to minimise analytical errors, one or more litres of fully treated sewage effluent must be filtered until a mass of at least 10 mg after drying is obtained. In general 1 mg/l TS_a represents from 1.5 to 3.5 mg/l of COD-Cr and 0.4 to 0.6 mg/l of BOD_5.

Nitrogen

Nitrogen occurs in sewage principally as combined organic nitrogen (org N) and as ammonia nitrogen (NH_4^+-N). In the course of biological treatment the organic nitrogen is converted into ammonia nitrogen, and given the appropriate conditions NH_4^+-N may be further metabolised to nitrite (NO_2^-) and ultimately to nitrate (NO_3^-). As these two oxidation reactions are the work of very delicate microorganisms, the attainment of the final oxidation stage is a positive indication of the absence of toxic substances.

The analytical detemination of the nitrogenous components is very time-consuming. A qualitative indication of the presence of oxidised nitrogen in the activated sludge can be obtained with the aid of the settled sample. If this is left for several hours (sometimes even for only as long as 15 min) at room temperature the onset of denitrification causes the sludge to rise to the surface as evidence of the formation of nitrogen gas bubbles. The presence of NO_2^- and NO_3^- in the sludge is, however, no guarantee that nitrification has occurred. Surprisingly enough many raw effluents contain these compounds, occasionally in substantial concentrations.

In recent times, measurement methods for electrochemical estimation of NH_4^+, NO_2^- and NO_3^- have been devised, but have not yet become general practice for the routine control of sewage treatment plants. For monitoring and control of denitrification in larger installations the use of nitrate-selective electrodes has already been found satisfactory.

The concentrations of the nitrogenous components are normally expressed in terms of the mass of nitrogen. Conversion factors are as follows:

>1 mg NH_4^+-N   1.29 mg NH_4^+
>1 mg NO_2^--N   3.29 mg NO_2^-
>1 mg NO_3^--N   4.43 mg NO_3^-

Turbidity

The oldest, most widely used criterion of water quality, is the measurement of turbidity, often in the hands of a non-skilled operator. In the context of water analysis, however, its importance is debatable, as there is no exact correlation with the other constitutents of the sample. Manual tests are performed using a visibility disc (a 20×20 cm square with a centrally placed black cross, the interesecting lines measuring 5 cm×1 cm). The limit of visibility is expressed as the depth (in cm) at which the cross just becomes invisible. The relationship between the limit of visibility and the BOD_5 of biologically treated sewage effluent is shown in Fig. 2.1(A).

For the continuous measurement of turbidity a number of devices have been developed during recent years based on a variety of different principles. The apparently quite close correlation illustrated in Fig. 2.1(B) between turbidity and both BOD_5 and filterable solids suggests that the turbidity is primarily due to solitary microorganisms and microflocs of bacteria which resemble each other in their optical properties. An important advantage of this method is the immediately available result, in contrast to all the other water quality criteria.

pH value, alkalinity, acidity

As most organisms will only tolerate pH fluctuations in the range pH 6–9 without their metabolic performance being impaired, a knowledge of the pH value and also a determination of the buffering capacity towards acids and alkalis is necessary in many cases. For the measurement of alkalinity, that is the acid-combining capacity, the quantity of 0.1 N HCl needed to adjust a 100 ml sample to a pH of 8.35 (p-value, phenolpthalein end point) or to a pH of 4.5 (m-value, methyl orange end point) is determined. The alkalinity arises chiefly from the presence of carbonates, bicarbonates and hydroxides; in sewage there may be interference from ammonium, phosphate and sulphide ions, as well as from organic substances. From the alkalinity of a pure water sample one can calculate:

>combined CO_2: m-value×22 (mg/l)
>bicarbonate: m-value×44 (mg/l)
>temporary hardness: m-value×2.8 (deg H).

The acidity is measured similarly to the alkalinity, but 0.1 N NaOH is used for the

Sec. 2.1] Chemical and physical methods 31

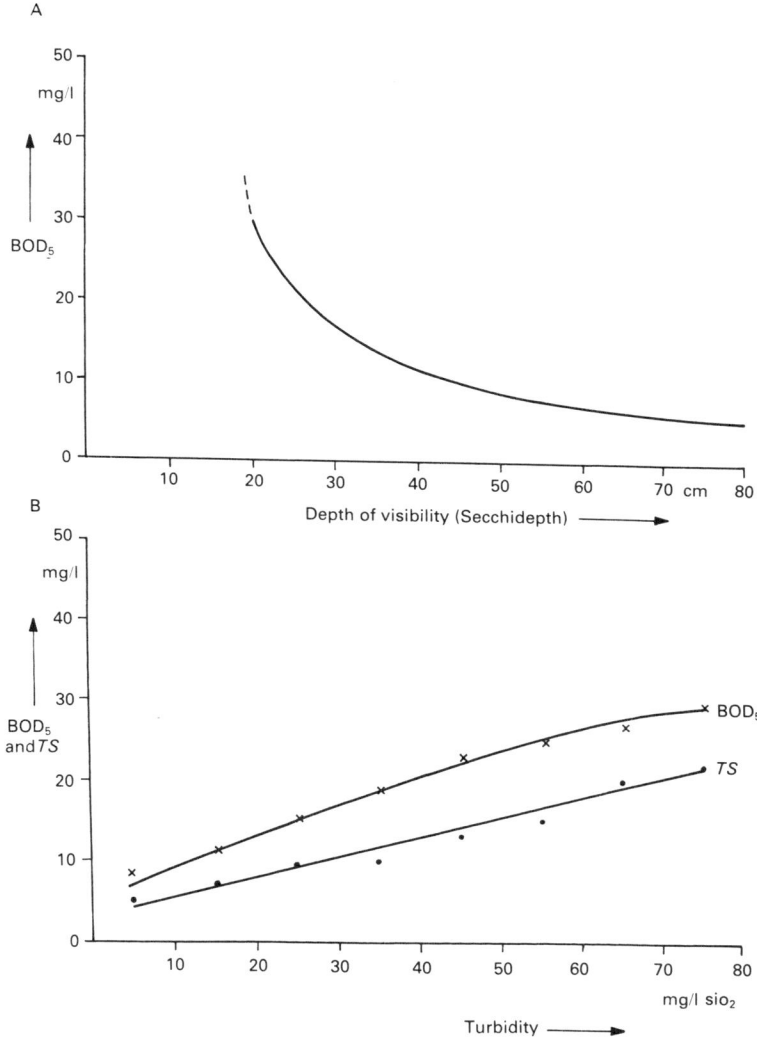

Fig. 2.1 — Dependence of BOD$_5$ and filtrable solids on turbidity in the biological treated effluent of municipal origin. A: Dependence of Secchi depth on BOD$_5$; the correlation is only valid as a guide. B: Dependence of turbidity obtained by nephelometry on the BOD$_5$ and filtrable solids. From Hänel *et al.* (1981) with modifications.

titration. The p-acidity value indicates the content of free carbon dioxide, while the m-acidity value indicates the amount of free mineral acids; 1 ml of 0.1 N NaOH corresponds to:

    1.008 mg/l of H^+
    36.5   mg/l of HCl
    49   mg/l of H_2SO_4

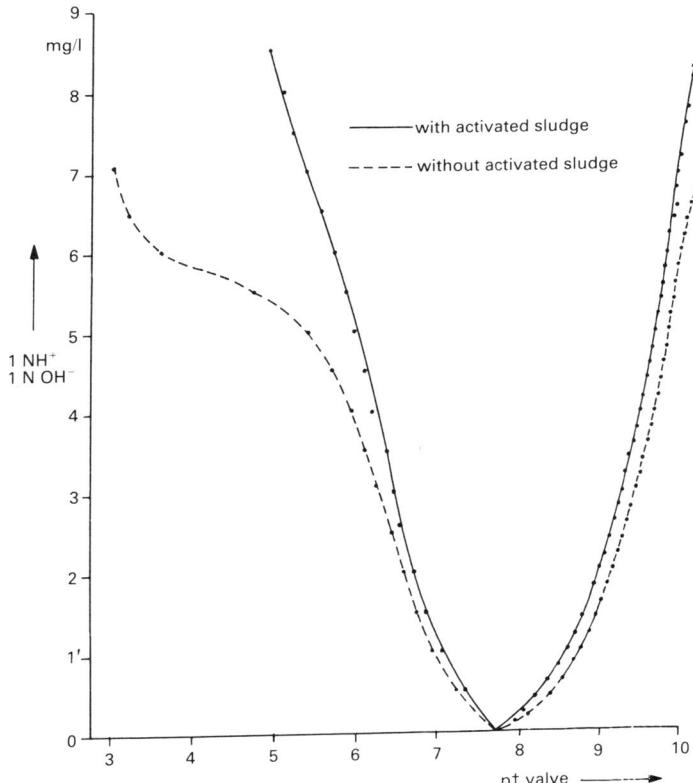

Fig. 2.2 — Change in the pH of sewage and sewage/sludge biomass suspension in response to the rapid addition of acids and alkalis, in order to assess the relevant buffering capacity.

    63   mg/l of HNO_3
    47   mg/l of HNO_2.

The values for alkalinity and acidity in the filtered liquid from the aeration tank can be used to calculate the actual acid and alkali buffering capacity, or alternatively this may be obtained from the titration curve showing the change in pH as a function of the amount of added acid or alkali. If activated sludge samples are similarly titrated, the results provide an approximation of the potential buffering capacity available in the aeration tank (Figs 2.2 and 4.31).

Extractable matter
For the extraction of fats and oils and other fatty substances (e.g. lipids or lipophilic substances) the samples are shaken with ether, petroleum ether or chloroform. The organic solvent to which the fatty material is thus transferred is then distilled off and the residue weighed. The actual procedure, and the solvent to be employed, are specified in the current publications of the Water Quality Inspectorate.

Extraction with organic solvents is also used to separate other organic constituents, such as herbicides, using hexane, chloroform or methanol according to TGL 27796.

Miscellaneous constituents
Recently a number of special elements, compounds or groups of materials have achieved prominence, and which are not covered by the conventional methods of water analysis. For their determination some very expensive equipment and demanding operations are required such as UV-absorption, gas-chromatography, thin-layer chromatography, atomic absorption spectroscopy, separation of mixtures into polar and non-polar fractions (the latter being characteristic of certain toxic constituents), and the determination of total organocholorine and adsorbable organochlorine components.

2.2 BIOCHEMICAL METHODS

2.2.1 Biochemical oxygen demand (BOD)
The biochemical oxygen demand (BOD) is the mass of dissolved molecular oxygen which is required for the decomposition or conversion of the biochemically oxidisable constituents in a litre of the sample under the action of acclimatised microorganisms, given specified conditions and within a certain period of time (denoted by subscript n in days or hours). The processes occurring in the course of the BOD estimation are biochemically analogous to those taking place in the activated sludge process, although in contrast to the latter, the biomass cannot be bled off in the form of waste activated sludge, but must be retained as an oxygen-consuming ballast during the incubation period. Since as a rule a different synthesis:oxidation balance obtains (Fig. 4.2) from that in the activated sludge the BOD_5 value does not provide a direct measure of the oxygen requirement of an activated sludge plant.

The BOD serves as a bulk parameter indicating the level of organic pollution of the sewage although it does not represent the entire organic pollution load, but rather the quantity of oxygen needed for the respiration of all the organisms contained in the BOD bottle during the prescribed incubation period, assuming that the consumption of oxygen for nitrification is inhibited by the addition of allyl thiourea [2]. Its value as a design parameter for biological sewage treatment plants must be treated with caution. However, a clear indication of the functional behaviour of the plant is obtained from the reduction in the BOD, in contrast to changes in the COD and organic carbon.

The BOD is made up largely from three reaction stages.

Stage 1: Oxygen demand for decomposition and metabolism of *carbon and hydrogen compounds*

In the case of the sewage samples used in connection with Fig. 2.6 this phase lasted 4.5 days for Curve 2, 3.5 days for Curve 3, and over 10 days for Curve 4; following this endogenous respiration commenced. Curves 2 and 4 indicate sequential metabolising of different substrates. For many years one has been accustomed to express the BOD as the BOD_5 value, implying a five-day test period, corresponding to the average transport times of English rivers, measured at 20°C. The result is thus not obtained until a considerable time has elapsed since the sewage effluent left the treatment plant — an indisputable drawback to the use of the method. As a result there have been many attempts to determine the BOD by means of rapid BOD methods, giving the results as BOD_1 or BOD_2. Even longer incubation times are sometimes used, as in Scandinavian laboratories where the BOD_7 is adopted with considerable advantages for the weekly rhythm of working, and the BOD_{20} in Russia.

Stage 2: Oxygen demand for *endogenous respiration*
Organisms experiencing a substrate deficiency carry out what is termed endogenous respiration, which involves the consumption of about 100 mg O_2/day per gram of activated sludge, that is about 0.5 mg O_2 for each milligram of biomass during a five-day period. A biologically treated water containing 20 mg/l of filterable solids will therefore exhibit an apparent BOD_5 of 10 mg/l over and above that required for the oxidation of C-, H- and N-compounds in the treated effluent. This endogenous respiration continues until the physiological death or sporulation of the organisms concerned.

Stage 3: Oxygen demand for *oxidation of ammonia and nitrite*
Following the example of Fig. 2.3, for raw sewage the conversion of ammonia to nitrite occurs primarily between the 12th and 20th days, while oxidation to nitrate takes place between the 38th and 50th days, an effect which also occurs when starting up activated sludge plants. As municipal sewage ordinarily contains around 50 mg

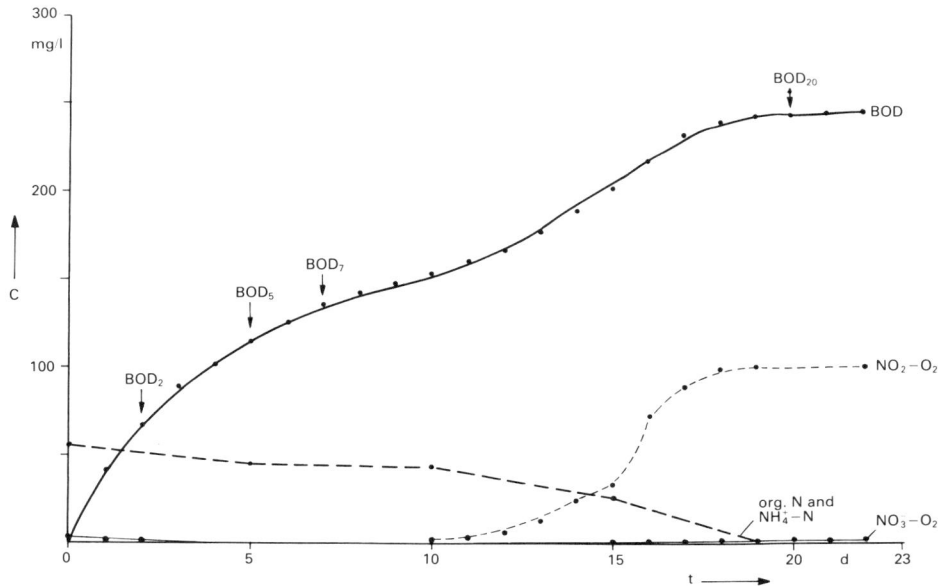

Fig. 2.3 — Oxygen demand of a primary treated sewage by reason of respiration of organic substrates and nitrite oxidation.

N/l, oxidation theoretically requires 222 mg/l of O_2. As the nitrogen-oxygen balance of Fig. 2.3 shows, however, considerable denitrification takes place during the incubation of the BOD sample, which helps to reduce the actual oxygen demand, an effect which occurs in the aeration tank of the treatment plant.

During the BOD determination, nitrification usually occurs only in already partially nitrified sewage samples. According to [2], nitrification is regarded as an interference in the BOD determination and is thus suppressed by addition of n-allyl thiourea which acts as a competitive inhibitor of nitrification. With regard to the oxygen balance of natural waters, however, it is immaterial whether the oxygen deficit is occasioned by the oxidation of organic substrates or the oxidation of nitrogen compounds. It is essential to determine the nitrogenous components of biologically treated effluents separately, in order to be able to assess their possible influence on the oxygen balance in the receiving stream.

The pretreatment of the samples can exert an important influence on the level of BOD. The instructions mention homogenisation, sedimentation, filtration and centrifugation. The last three treatments result in a partial elimination of micro-organisms and coarse solids which ultimately leads to a reduction in the BOD. For biologically treated effluents filtration reduces the BOD by 50–80%, and for primary treated sewage by up to 30%. In the present work all BOD values refer to homogenised or well-agitated samples, where no other form of pretreatment is expressly stated. The methods for BOD determination also encompass numerous variants, for example the dilution method, oxygen method, manometric method, which likewise will give various differing values.

As the BOD_5 represents the combined total of substrate and endogenous respiration, and owing to its long measurement time is unsuited as a means of process control, proposals have often been made to use the oxygen demand obtained from measurements of respiration rate of activated sludge during the endogenous and substrate utilisation phases and to determine the oxygen demand for substrate decomposition from the difference between the two respiration rates. However, the rapid-BOD values obtained by this method are often only about 10–30% of the theoretical oxygen demand (Fig. 2.4). The nutrient status of the sludge, the level of acclimatisation, and the BOD_5 sludge loading all have an important influence on the ratio between rapid BOD and BOD_5. While for the BOD_5 a high rate of synthesis is ensured by the choice of test conditions, in the respiration-BOD the ratio of respiration to synthesis (Fig. 4.2) varies, added to which the level of endogenous respiration determined after the substrate is exhausted is usually greater than that prior to substrate addition and continues so for a long time. The reason for this lies in the utilisation of cell reserve material, the previously activated metabolic processes, and also the conversion of the intermediate products of metabolism. Where the oxygen demand is substantially less than that expected from the COD-Cr values, then prolonged respiration rate measurements should be performed supported by measurements of biomass growth rate, COD-elimination, or the decrease in concentration of selected substrates.

2.2.2 Respiration of activated sludge (OV)
The respiratory activity of the activated sludge per unit of time represents the oxygen demand which is required for the oxidation of hydrogen (Fig. 4.18). The respiration

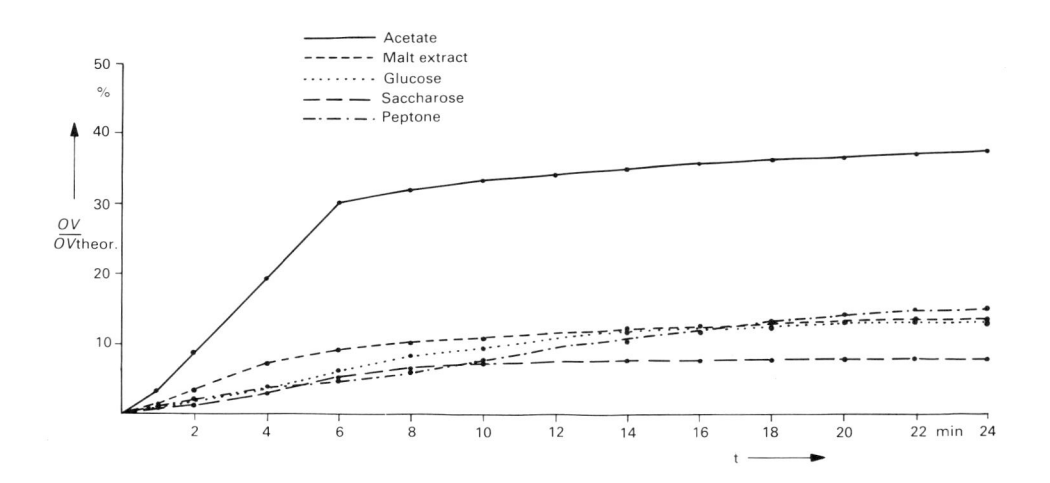

Fig. 2.4 — Amount of the substrate respired during substrate-induced respiration as a percentage of the theoretical total oxygen demand of the substrate. Substrate dose 12 mg/l; total solids 3 g/l; temperature 16°C.

rate determination, unlike any other method, gives an insight into the enzymatically controlled metabolic processes. It reflects in a quantifiable manner not only the biochemical oxygen demand but also any toxic effects, the sequential decomposition of different substrates, biochemical sub-processes and the metabolism of multicomponent substrates. It provides information on the type of substrate utilisation and hence on the physiological condition of the organisms, and even a deficiency of nutrients or trace substances can be demonstrated in a very short period of time (Fig. 2.5, Curve 2). For practical purposes the respiration rate measurement may be used to ascertain the necessary oxygen input to the aeration tank, and in the case of samples of effluent from the tank, to assess the need or otherwise for an additional aeration time. Where special discharge consents for particular effluent types are concerned, the method can give a clue as to the presence of suitable enzymes or the adaptation periods required for their development.

The determination of respiration rate is carried out on a continuously mixed activated sludge sample, hermetically sealed and isolated from the external environment, at a constant temperature and ideally with an initial dissolved oxygen content

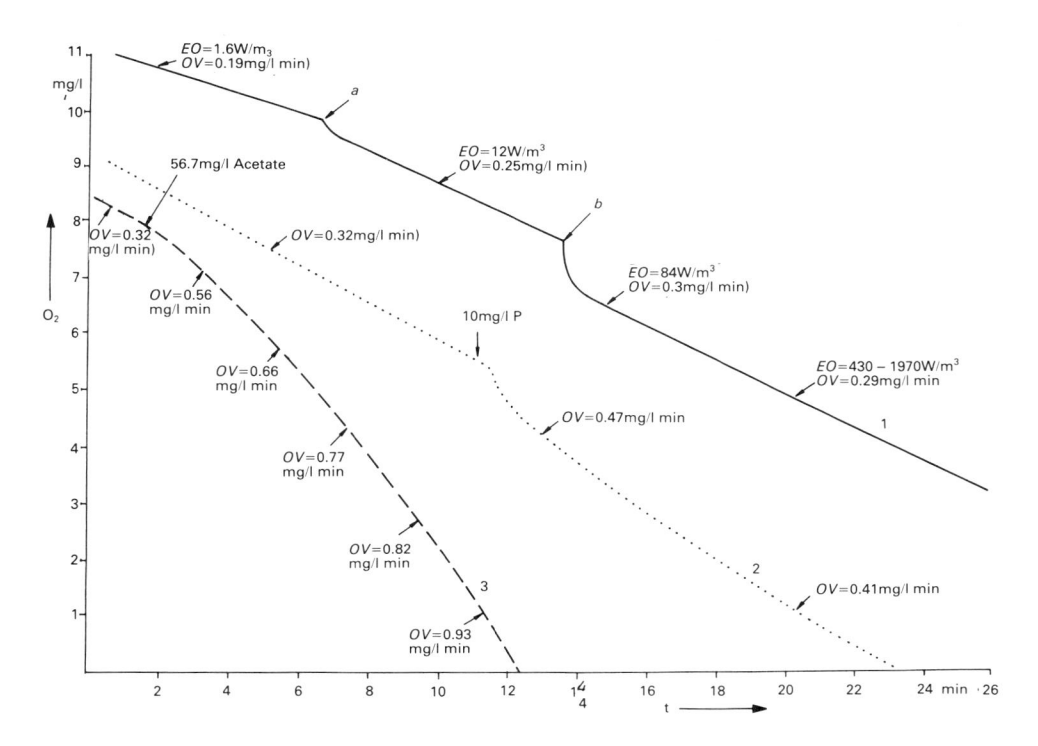

Fig. 2.5 — Respirometer-derived oxygen profiles. Curve 1: showing the effect of turbulence on the respiration rate of a sludge biomass as a consequence of particle size differences: prior to b: particle size >1 mm, after b: microflocs only. Points of inflexion occurred at a and b due to liberation of the oxygen-free water occluded within the flocs; from calculations allowing for the turbulence-dependent response of the oxygen electrode, the corresponding amounts; were 3.8% and 13% of the reactor volume at a and b respectively; slightly enhanced respiratory activity after floc breakdown arises from the respiration of products of anaerobic metabolism within the interior of the flocs. $T=10°C$, $TS=4.35$ g/l. Absolute values are reactor-specific and hence cannot be transferred to other tanks. Curve 2: showing the effect of phosphorus deficiency for a sludge biomass from the brown coal processing industry. Curve 3: showing the possibility of adaptation to high acetate concentration for a municipal sludge biomass.

close to the saturation value. An apparatus for the measurement of respiration rate under laboratory conditions is illustrated in Plate 5.

The respiratory activity may be subdivided as follows:

— Maximum respiration rate (OV_{max}). The oxygen demand under conditions of substrate excess, that is when the substrate availability exceeds about 100 Km. Either a non-toxic sewage or a synthetic sewage as in [3] may be used. As Figs 2.4 and 2.7 show, sodium acetate gives the most useful results in a period of only eight to ten minutes. Acetate has the advantage that in the presence of coenzyme A it is immediately incorporated into the citric acid cycle and respired at up to 50% of the theoretical maximum oxygen demand during the OV_{max} phase.
— Substrate respiration — oxygen demand in the presence of substrates.
— Basal respiration (endogenous respiration (OV_{end})) — oxygen demand in a

substrate-free activated sludge. Freedom from substrate may be achieved by a one to two hour period of aeration of a sample taken from the aeration tank. The endogenous respiration, often considered as the respiration of cell substance after all reserves have been exhausted, is often not clearly distinguishable from substrate respiration owing to the multiplicity of organisms and species with a corresponding variety of food chains and utilisation pathways.

For both the OV_{max} and the OV_{end} the oxygen demand per unit of time and sludge mass is relatively constant and with substrate respiration there is a general transition to OV_{end}. For long-term measurements, however, owing to a gradual induced acclimatisation the OV_{max} may show a slight increase, while OV_{end} exhibits a steady decrease on account of the progressive depletion of cell reserves. Where curves showing the oxygen demand over a period of several days are desired, the *respirometer* of Peukert and Thomsch (1981), the Sapromat, the Pollumat, the Warburg apparatus and so on, are recommended. Should the carbon dioxide produced in the latter be removed in its entirety, however, a disrupting increase in pH may occur.

As in large-scale aeration tanks, the floc size and hence the oxygenated parts of the flocs will vary with the degree of turbulence within the respirometer (Fig. 2.5, Curve 1). A particular oxygen demand obtained from respirometer measurements must therefore be confined strictly to installations with equal turbulence on the large scale.

2.2.3 Toxicity
Toxicological investigations are concerned with:

— No-effect concentration (NEC): maximum concentration of a toxic substance at which no adverse effect occurs.
— Threshold concentration (TC): lowest concentration of a toxic substance at which the first definite adverse effect is observed.
— Median lethal concentration (LC_{50}): the concentration of a toxic substance at which 50% of the test organisms are killed; the terms LC_{10} and LC_{90} are also in common use.
— Lethal concentration (LC): the concentration of a toxic substance at which all the organisms are killed.

For aquatic toxicological investigations, as a rule, certain particular species [3] are specified from which a practically uniform response to toxic substances may be expected. In the case of a mixed biocoenosis such as occurs in activated sludge, selective actions of toxic substances on certain species are to be expected, so that the result obtained always represents an average in respect of the complete biocoenosis; for example the value of LC_{10} may in fact be equal to the LC value for the nitrifying bacteria.

The toxic concentrations observed in tests with activated sludges are thus not transferable to other organisms and biocoensoses; in comparison with algae for example there are often differences amounting to several powers of ten.

In the case of toxicity tests on activated sludges it is usually a question of the toxicity of a particular effluent or of a certain constituent or group of constituents. The basis of toxicity test determinations lies in an estimation of the degradability of

Sec. 2.2]	**Biochemical methods**	39

the toxic substance, the adaptive capability of the biocoenosis towards the substance and its chronic effects.

The inhibition of oxygen uptake as a consequence of the presence of substances of a toxic nature is used as a measure of the toxic activity. Toxic effects on the flucculation properties of the bacteria can only be demonstrated with the aid of continuous-feed test systems.

In [3] the 'BOD-Warburg Method' for determination of toxicity towards bacteria is advocated, such that the duration of the test is at least 48 hours. The following measurements should be performed (see Fig. 2.6):

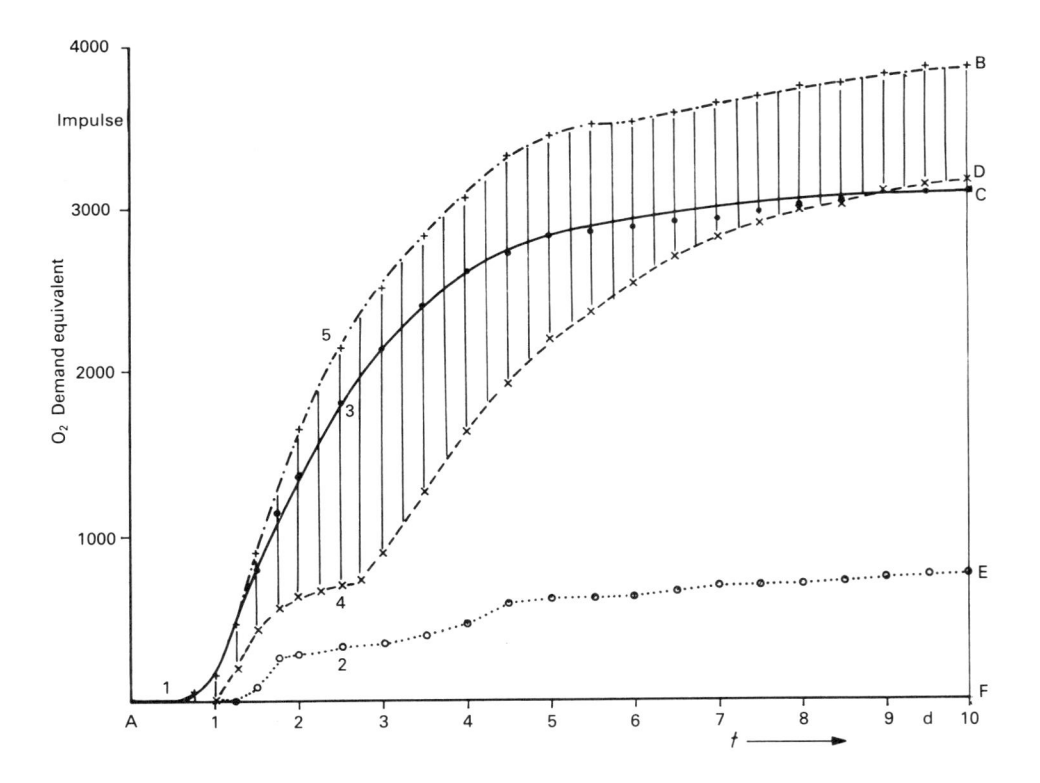

Fig. 2.6—Study of the inhibitory effect of an industrial effluent using a Sapromat after Offhaus (1965). Curve : Inherent demand of organisms from BOD_5 dilution water (after [21]), in this case negligibly small; the inoculating bacteria were derived from an industrial sludge biomass which had not been acclimatised to peptone and the sewage under test, hence the pronounced lag phase. Curve 2: BOD of the effluent under test, derived from the brown coal industry. Curve 3: BOD of the reference substrate, namely peptone. Curve 4: BOD of the effluent under test and the reference substrate. Curve 5: Theoretical oxygen demand; the hatched area represents the extent of inhibition which amounted to 100% after one day, 62.8% at two days, 36.4% at five days and 20.1% at ten days. This decrease in inhibition with time implies that the sludge biomass is becoming acclimatised to the toxic effluent or that the toxin is being metabolised.

— endogeneous respiration (Curve 1);
— substrate-induced respiration for the effluent under test (Curve 2);
— substrate-induced respiration for a reference substance (Curve 3);
— combined substrate respiration for the effluent under test and the reference substance (Curve 4);
— from Curves 2 and 3 a theoretical additive curve (Curve 5) is obtained.

The degree of toxic inhibition (H) at time F is calculated from the equation

$$H = 100 - \frac{DF \times 100}{BF} \ (\%).$$

The more meaningful inhibition for the entire period is obtained by means of planimetry for the surface areas ADF and ABF, according to the equation

$$H = 100 - \frac{ADF \times 100}{ABF} \ (\%).$$

A toxicity measurement performed in this manner cannot normally be used as an estimate of the toxic effects on activated sludge because of the possible adaptation to, and metabolism of, the toxins.

For activated sludge plants the danger always exists that toxic substances will be introduced abruptly. A determination of the values for NEL, TC, LC_{50} and LC for short-term activity is therefore often more relevant than a long-term toxicity test: for this purpose the *respirometer* used for respiration studies is suitable. The substance or effluent to be tested is added at different concentrations, and the respiration rate compared with that of an untreated sample (Figs 2.7 and 2.8). In order not to subject the activated sludge sample to an additional temperature shock, such respirometric toxicity measurements should be carried out at the temperature to which the sludge biomass has been adapted, especially as in technical activated sludge installations there is no standard temperature for the occurrence of toxic effects. Toxicity measurements for inorganic toxins may also be performed, as distinct from the BOD Warburg Method, without the addition of substrate, as the endogenous respiration of the sludge biomass also reflects the toxic effect of the substance, often even better than the substrate-induced respiration, as the organisms are in a physiologically sensitive starvation state.

Toxicity measurement in respirometers, however, contains one major failing with respect to the conditions obtaining in activated sludge plants. Any large-scale aeration tank is to a certain extent an approximation to a stirred tank reactor, while the respirometer toxicity test simulates the behaviour of a plug-flow reactor with its unfavourable buffering characteristics with regard to sudden inputs of toxic substances. The test result may indicate the need for a special pretreatment in the case of degradable toxic substances, or even in extreme cases the use of non-biological treatment methods, even though the toxin may become completely innocuous as a result of sufficient dilution and metabolism. Where a toxic effect is detected as a

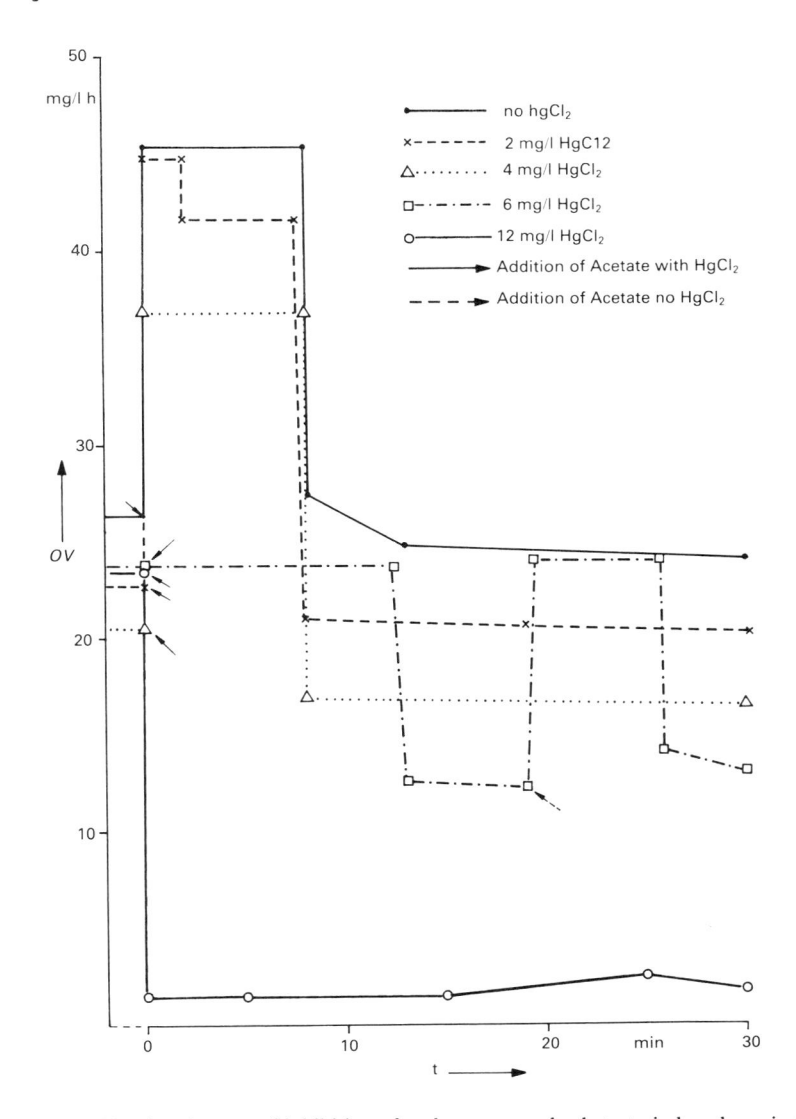

Fig. 2.7 — Varying degrees of inhibition of endogenous and substrate-induced respiration in response to increasing HgCl$_2$ concentrations, in schematic form. At ↓ addition of 11.3 mg/l Na acetate; $TS=2.4$–2.6 g/l, $T=17$–17.5°C. The curve for 6 mg/l HgCl$_2$ demonstrates the possibility of overcoming toxic shock by addition of substrate.

consequence of respirometer tests, then further confirmatory studies under the conditions obtaining in a stirred reactor should be undertaken.

These conditions may be approximated in a respirometer provided the toxic-substance-containing effluent is added in small amounts extending over a period of time roughly corresponding to the theoretical residence time in a full-scale aeration tank, so that the limiting concentration is not reached. It is safer to set up one or more laboratory aeration tanks, fill them with an activated sludge suspension, aerate them

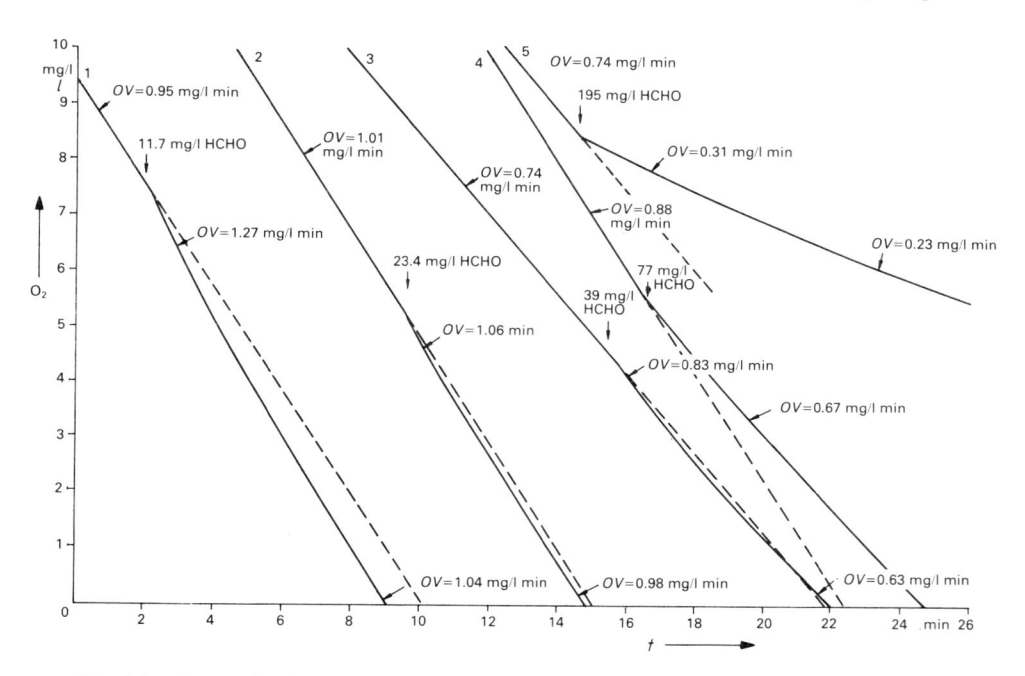

Fig. 2.8 — Determination of the short-term toxicity of formaldehyde toward a non-acclimatised nutrient-deprived municipal sludge biomass, using a respirometer; OV in mg/l min. $TS=4.25$ g/l, $T=13.2°C$. Curve 1: utilisation of HCHO as substrate. Curves 2 and 3: utilisation of HCHO as substrate is inhibited although the inhibition is initially obscured by the endogenous respiration. Curves 4 and 5: severe toxicity; toxic effect increasing with time.

and operate them as test and control systems. The test substrate is introduced into the test vessel, over a period of time corresponding to the residence time in the full-scale aeration tank, by means of a dropping funnel. The respiration rate is measured continuously (or at least every 15 minutes) in the two test vessels. Where the toxic substance is biologically degradable the test vessel exhibits substrate-induced respiration and the control vessel exhibits endogenous respiration.

For a determination of the adaptive capability of the microorganisms towards a toxic material, the activated sludge is gradually acclimatised to the toxin in a flow-through system. The toxic substrate is added in increasing concentration over a period of time corresponding to the time-related adaptive capacity of the biomass. Adaptation periods of up to three months are not exceptional.

2.2.4 Biomass

As the sludge volume and sludge solids are not accurate measures of the active biomass, more exact indices are continually being sought. Besides the actual counting of the organisms, in many laboratories either the protein content or the nitrogen content of the sludge biomass is used as this stays relatively constant at around 6–7% of the biomass. In recent years several methods have been proposed which are based on the metabolic regulators of the organisms, namely adenosine triphosphate, deoxy- and ribonucleic acids, proteins, dehydrogenase and reductase activities as well as other specific enzymes. All these methods are relatively costly and are only recommended for special investigations. Moreover, as fairly rigorous extraction methods are often used, one also determines the inactive and dead organisms present in the interior of the flocs. These methods accordingly give the potential amount of biomass, but not its active fraction. The active fraction can be determined with sufficient accuracy by means of respiration and growth measurements.

Biomass growth
The increase in weight of the activated sludge biomass includes the effect of absorption and adsorption processes as well as the attachment of coarse particulates. The actual growth can be inferred from the nitrogen uptake: 62.5 mg N are required for the synthesis of 1 gram biomass.

With this incorporation of nitrogen, it may be remarked that the term originating abroad of the 'adsorption-activated sludge process' is scientifically untenable. The treatment process according to this heavily loaded activated sludge system cannot remain at the stage of adsorption or absorption behaviour, but leads to conversion of the substrate into biomass of normal composition.

Dehydrogenase activity
Many authors apply the TTC-test proposed by Bucksteeg and Thiele (1959) for the determination of the activity of the activated sludge. In this method the compound 2,3,5-triphenyltetrazolium chloride (TTC) is reduced under anaerobic conditions to the red 2,3,5-triphenylformazan, and its concentration determined colorimetrically. It is therefore assumed that the presence of reductases is a side-effect of the metabolism of aerobic organisms as well, and that the activity of the microorganisms under aerobic conditions is proportional to that under anaerobic conditions.

According to the studies of Gottschaldt (1968), Riedel (1971), and others, the TTC method, owing to uncontrollable extraneous effects, does not give meaningful results.

2.2.5 Other biochemical methods
Putrefactive activity
The property of a treated sewage effluent used to be measured by reference to the putrefactive tendency. The principle of this depends on the fact that in the absence of oxygen, methylene blue is reduced by oxygen-consuming biological processes, with the formation of leuco-compounds, so that samples incubated at 20°C are decolorised. The time for decolorisation, expressed in hours, was used as a measure of the oxygen-depleting constituents; where the colour was retained for 120 h the liquid was classed as non-putrefactive.

Redox potential (rH or E_h)
The redox potential is a possible means of assessing the quality of a particular sample of effluent or activated sludge. According to Mortimer (1942/3) the different potential stages are characterised by the following redox potentials:

NH_4^+ stage $\qquad \leqslant 100$ mV

NH_4^+/O_2 stage	200–350 mV
NO_2^-/O_2 stage	350–400 mV
NO_3^-/O_2 stage	≤450 mV.

Measurements performed by Zimmermann *et al.* (1977) at the Berlin activated sludge plant confirmed these values, for example:

mechanically settled sewage	<0 mV
biologically treated sewage	~500 mV
activated sludge with adequate oxygen supply	250–400 mV
activated sludge with oxygen deficiency	<250 mV.

Correlations between redox potential and other quality criteria have so far not been observed.

2.3 BIOLOGICAL METHODS

2.3.1 Problems and limits of application

The physicochemical properties of the activated sludge flocs and their colonising organisms are determined by the ecological conditions prevailing in the aeration tank, the final settling tank and their connecting pipework, and also by the nature of the sewage to be treated. By means of a microscope for examination of the activated sludge the following factors, which are ultimately of overriding importance for the performance of the activated sludge plant, may be evaluated.

— *Saprobic state:* the overall ecological situation relating to the sludge biomass and the fixed biocoenosis at the sampling point.
— *Oxygen supply:* affects the course and rate of sewage treatment.
— *Floc geometry:* affects the settling behaviour and transfer surface of sludge flocs.
— *Bulking sludge.*
— Adverse effects on microorganisms due to *toxic substances*.

For assessing the saprobic state of the treated sewage and hence the biological effectiveness of the treatment plant, a biological investigation of the surface growth in the outlet from the final settling tank or the outfall from the treatment works is a useful guide. The suspended organisms flushed out of the settling tank in the final effluent are not suitable for this purpose, as they are not directly correlated with the biocoenosis of the activated sludge biomass.

Where toxic sewage constituents are suspected as the cause of upsets in the operation of an activated sludge plant, then a biological investigation of the surface growth in the sewer system, or the intake duct prior to the admixture with the waste activated sludge, may provide clues as to the time and place of entry of the interfering substances.

Compared with other methods of investigation, a biological examination has the following advantages:

— as the communities of living organisms reflect all the relevant ecological factors of

the biotope, they provide the most comprehensive indication for an assessment of the activated sludge and sewage quality;
— as the relevant communities become established over a relatively long period they may supply information concerning ecological conditions that have previously existed, even though at the time of sampling the actual conditions may be quite different;
— as the relevant communities represent an average over a period of time of all ecological factors, a single examination is sufficient. It can therefore be performed more expeditiously and much more cheaply than any other kind of analysis.

Disadvantages of biological evaluation are:

— among the qualitative factors which govern the sewage treatment process, only the hydrogen sulphide/oxygen regime, the toxic substances as a lumped factor, and in special cases, the salt content, can be distinguished;
— precise quantitative indications concerning the magnitude of the factors affecting treatment performance cannot be given.

2.3.2 Sampling and transport of samples

Samples are obtained from the aeration tank by dipping in a bucket or other container. Because of the small qualitative and quantitative differences in the community of microorganisms from one part of the tank to another, the sampling point can be chosen at will.

Attached growths from the intake and outlet channels and from the tank walls are removed by scraping. For very thick growths, only the superficial layer which is in direct and continuous contact with the water body should be sampled. Either sanitised brushes or scrapers may be used; in addition the areal extent of any predominant organisms should be estimated, for example 10% of the surface covered with a brown deposit of diatoms, 40% with green thread-like algae, etc.

The samples must be dealt with immediately after collection. Owing to an oxygen demand of the sludge of >10 mg/lh., transport or lengthy storage may cause anaerobic conditions to set in, as a result of which organisms may become unrecognisable or even die out. Where a lengthy interval between sampling and examination is unavoidable, about 10 ml of the sample is poured into a 250 ml flask and after proper labelling is kept in a cool place, preferably in a refrigerator. With this method the large headspace in the flask is sufficient to ensure the maintenance of aerobic conditions for around two to three days. Any *fixation,* that is preservation of organisms by killing with toxic chemicals, is ruled out owing to possible morphological changes in the protozoa inhabiting the sludge biomass. Only for the later determination and counting of metazoa, the associated rhizopods and the microscopically detectable bacteria should fixing with $HgCl_2$ (1 ml of 2% solution to 0.1 litre) or formaldehyde (1 ml 35% solution to 0.1 litre) be attempted.

2.3.3 Conduct of the biological analysis
Microscope
For routine analysis transmission optical microscopes, for example Ergaval or

Amplival instruments from VEB Carl Zeiss, Jena, with objectives of 6×, 16× or 20× and 40× magnification and eyepieces of 6 to 20× magnification are suitable. For size determinations a micrometer eyepiece is indispensable, whichis calibrated using a micrometer objective. A phase-contrast system of illumination facilitates the recognition of special organelles such as cilia of flagellates. For observation of the large metazoa and the floc morphology a stereoscopic microscope is useful.

Determination of numbers and species of organisms
Macrofauna (organisms >500 μm)
Sixty millilitres of activated sludge or surface growth is placed on a white plastic or enamel dish with a surface area of 400 cm². Organisms greater than 3 mm are determined by direct counting with the naked eye, those between 0.5 and 3 mm using a stereo-microscope.

Macroscopically visible organisms in the activated sludge are also profitably observed during and after determination of the sludge volume in a measuring cylinder. They escape from the settled sludge after the onset of anaerobic conditions and swim like plankton in the supernatant liquid.

Microflora and fauna (organisms <500 μm)
Two to four millilitres of a well-mixed sample are placed in a Petri dish of 7–9 cm diameter with a flat transparent base. The sample is examined under a magnification of 60 to 100×. In order to obtain an overall impression of the nature and size of the sludge flocs as well as of the presence of the larger protozoa (>15 μm) a total of 100 fields must be inspected. If necessary a sample that is too dense may be diluted in a certain ratio using flat (that is chlorine-free) tap water.

For determination of the smaller organisms (<15 μm) a drop of activated sludge is applied to a microscope slide and a cover glass applied, the excess water being mopped up with filter paper. Again 100 fields must be inspected under a magnification of 200×.

For calculation of the saprobic index it is sufficient to determine the numbers of individuals for each species by reference to a six-stage frequency table, having regard to the size of the organisms (Table 2.1).

Should the individual counts have to be obtained precisely, then the use of counting chambers (eg. Thema, Cyrus, etc.) with thicknesses of 0.1–0.5 mm becomes necessary.

The characteristics of the activated sludge flocs, with reference to size, shape and compactness, are described verbally.

Determination of unknown organisms
For a quality assessment of liquid and sludge samples with respect to the organisms present, accurate species identifications are required. As most of the protozoa become unrecognisable as a result of the action of fixing agents, live observations are necessary. These are ordinarily performed on samples entrained between the slide and the cover glass. For very agile organisms it is advisable to minimise their activity by the addition of quiescence agents such as 0.02% solution of $NiSO_4$.

For long-term investigations of rarely-encountered organisms liable to exhibit rapid motion, these are captured from the sample contained in the Petri dish using

Table 2.1 — Determination of frequency of occurrence

	h	Individual nos./100 fields			Individuals/scan	
Magnification		200	(60–)100	20		0
Size category		I: <15 μm	II: 15–500 μm	III: 500–3000 μm		IV: >3000 μm
Very scarce	1	≤2	≤2	1		1
Scarce	2	3–10	3–10	2		2
Common	3	11–40	11–40	3–10		3–5
Numerous	5	41–100	41–100	11–25		5–10
Very numerous	7	101–200	101–200	26–50		10–20
Abundant	9	200	>200	>50		>20

hair or capillary pipettes directed under microscopic observation. Several small drops of about 2-mm diameter are placed on a cover glass and the presence of the desired species is checked with the aid of a microscope. When the particular species can be seen in one of the drops then the cover glass is lowered into one of the indentations in a hollow-ground microscope slide. By means of one or two drops of paraffin or cooking oil applied to the edge of the cover glass, the sample is hermetically sealed from the ambient environment. Samples in hanging drops can be kept for a week, and are suitable for many types of special investigation.

In order to render taxonomically relevant cell structures visible, they may be stained. The *staining methods* for bacteria have been summarised by Seeliger (1978). For a study of bulking sludge organisms, dyes for detection of the sheath and also both Gram and Neisser stains are required. For detection of the ability of organisms for oxidising sulphur, 20 ml activated sludge is added to 10 ml Na_2S solution (100 mg Na_2S in 1 litre distilled water), shaken, and the presence of free sulphur in the cells confirmed after 10 min with the aid of a microscope.

2.3.4 The saprobic system

In the course of the historical development of aquatic organisms on the Earth's surface some species have become adapted to differing water properties and constitutents in such a way that they can be used as indicator organisms for these properties or constituents. Such organisms have been identified as indicators for temperature and for the presence of iron, manganese, salts, oxygen, hydrogen sulphide and for putrescible organic matter. For the very recent (relative to the time-scale for organism development) artificial sewage constituents derived from industry, agriculture and urban sewage treatment plants, and hence for activated sludge, there are as yet no specific indicator organisms. For their assessment it is necessary to employ the indicator organisms for organic matter, the so-called saprobic organisms (derived from Greek, sapros=foul, bios=life). Those other indicators occasionally present in the activated sludge in general offer no indication of the level of the treatment performance.

'Saprobic organisms comprise plants and animals which are involved in the natural self-purification process for water, which through their interactions with the environment at every stage of decomposition of organic matter reach that optimal level of deployment, which offers the conditions most favourable to their survival, and hence provides a characteristic community of organisms for any particular ecological set of conditions, that serves as a pointer for the degree of self-purification apparent during biological water quality investigations' [3] (1982).

The saprobic system was worked out by Kolkwitz and Marsson (1908, 1909) and later much refined. According to their original system four *saprobic states* were differentiated:

— polysaprobic: the stage involving primary cleavage processes
        water quality category: 4;
— α-mesosaprobic: the stage involving secondary cleavage processes
        water quality category: 3;
— β-mesosaprobic: the stage involving advanced mineralisation
        water quality category: 2;

— oligosaprobic: the stage of complete mineralisation
water quality category: 1.

Currently there are trends toward further subdivision of these stages, especially the oligosaprobic and polysaprobic categories (Table 2.2).

Table 2.3 shows the classification of flowing waters according to the saprobic system, together with significant correlations between the saprobic indices and other water quality criteria. There are of course some very marked differences between flowing waters, activated sludge and secondary sewage effluent with respect to certain parameters, notably NO_3^-, NH_4^+ and bacterial counts.

In the activated sludge biomass and in surface growths from the inlet and outlet channels, indicator organisms from several saprobic states are normally present. For a mathematical expression of the quality status, the 'weighted mean' of all types of indicator is employed. For this purpose the frequency values h (Table 2.1) obtained for each of the saprobic states are summed and multiplied by the value for the saprobic state s. The sum of the products is then divided by the sum of the frequencies, namely:

$$S = \frac{\Sigma\,(s\,h)}{\Sigma\,h}.$$

The saprobic index S is a dimensionless quantity, which fluctuates between 6 (xenosaprobic) and 4, or in activated sludges between 2.3 and 4.

The value of a particular type of organism as an indicator of water quality is much greater, if it is highly dependent on a particular condition of the water. Species with a high-degree of adaptive ability to various water properties and constituents are poorly suited for use as indicator organisms. In order to be able to express these variations on ecological significance, Marvan and Zelinka (1961) introduced the term 'indication weight'. They assigned to every species 10 points according to the statistical frequency of the findings in one or more saprobic states. Where a species was always found only in one of the saprobic states it received a score of 5 for the indication weight. Species with a greater or lesser degree of scatter among the saprobic states were assigned values from 4 to 1. Organisms with scores of 3 are only serviceable as indicator organisms to a limited extent, those with scores of 2 or 1 are unsuitable.

For a calculation of the saprobic index according to [3] it is necessary to have at least 12 indicator species in the biocoenosis, with an overall frequency total $\Sigma h = 30$; as a rule such a number of indicators is not reached in activated sludge. Further sources of error are the drift of organisms from other biotopes, such as the sewer network or digestion vessels, severe depopulation by reasons of sewage properties, and massive proliferations of a single species. As similar interferences arise in the estimation of the saprobic index for other biotopes, other types of water quality rating procedures based on the use of species abundance or deficiencies in the species inventory are in course of development.

For an accurate application of the saprobic system to specialised systems such as

Table 2.2 — Classical saprobic grades according to Kolkwitz and Marsson (1908, 1909) and the Sládeček (1973) proposals of higher saprobic categories for sewage

Saprobic grade	Symbol	Description
Oligosaprobic	0 or 1	Mineralisation of organic constituents complete; not attainable in an activated sludge system
β-mesosaprobic	β or 2	Extensive mineralisation, no H_2S, plenty of O_2, water clear with no extraneous smell. Numerous species of metazoa and protozoa, but few bacteria ($<10^5$ per ml)
α-mesosaprobic	α or 3	Oxidising reactions, H_2S only occasionally, O_2 in widely fluctuating amounts, many organic substances present, water cloudy, musty smell and with high O_2-demand. Bacteria numerous ($>10^5$/ml). Protozoa abundant, also metazoa
Polysaprobic	p or 4	Reducing processes (septicity), H_2S often present at high concentration. O_2 usually scarce. Water very turbid with decaying or faecal odour for example domestic sewage. Large numbers of facultative anaerobic bacteria ($>10^6$/ml) and bacteria-feeding protozoa
Isosaprobic	1 or 5	BOD_5: 200–300 mg/l; O_2 absent, traces of H_2S, coliform bacteria present up to 3×10^6/ml. Organisms: ciliates
Metasaprobic	m or 6	BOD_5:200–700 mg/l; O_2 absent, H_2S in amounts of >10 mg/l organisms: colourless flagellates
Hypersaprobic	h or 7	BOD_5:500–1500 mg/l; O_2 absent, H_2S in amounts of a few mg/l, for example industrial effluents in the degradative stage. Organisms: Bacteria up to 10^3 coliforms/ml, some fungi
Ultrasaprobic	u or 8	$BOD_5>60\,000$ mg/l; O_2 and H_2S both absent, for example industrial effluents prior to decomposition. No organisms
Antisaprobic	a	Effluents containing toxic ingredients. No organisms
Cryptosaprobic	c	Effluents containing non-toxic ingredients, for example coal slurry, mineral suspensions, mineral oils, etc.; no organisms
Radiosaprobic	r	Effluents containing radioactive substances

Table 2.3 — Selected criteria for classification of typical flowing water (From TGL 22764/01). Use categories according to Table 5 of the same standard

Water quality parameter		Class saprobicity	1 oligo	2 β-meso	3 α-meso	4 poly	5 hyper	.6 abiotic
Saprobic index			≤1.75	≤2.5	≤3.25	≤4	—	
Oxygen	mg/l		≤7	≤6	≤4	≤2	<2	
BOD$_5$	mg/l		≤4	≤10	≤20	≤40	>40	
COD-Mn	mg/l O$_2$		≤5	≤10	≤30	≤50	>50	
COD-Cr	mg/l O$_2$		≤8	≤25	≤80	≤120	>120	
NH$_4^+$	mg/l		≤0.5	≤2	≤4	≤10	>10	
NO$_3^-$	mg/l		≤10	≤20	≤40	>40	—	
Filtrable solids	mg/l		≤10	≤20	≤50	≤100	≤200	
Dissolved sulphide H$_2$S	mg/l		n.d†	n.d	n.d	≤0.1	>0.1	
Total bacterial count	ml^{-1}		≤10^6	≤3.10^6	≤10^7	≤5.10^7	>5.10^7	
Colony count	ml^{-1}		≤10^3	≤10^4	≤10^5	≤10^6	>10^6	
Coliforms	ml^{-1}		≤10^2	≤5.10^2	≤5.10^3	≤5.10^4	>5.10^4	

†n.d = not determined.

activated sludge the 'Selected Methods' [3] advocates a calibration. This had not been proposed previously. Buck (1968) assigns the most prominent activated sludge ciliates, on the grounds of statistically-based population studies, the following saprobic indices:

Saprobic index	Ciliate indicators
4	Colpidium campylum
3.6	Vorticella putrina
3	Vorticella convallaria
	Aspidisca costata
2.6	Vorticella campanula

Curds and Cockburn (1970) from subsequent applications of the fact that the occurrence of the organisms is a function of the organic contamination, derived a method for estimating the BOD_5 of the treated sewage. Corresponding to the percentage occurrence of ciliates in the BOD_5 ranges of 0–10, 11–20, 21–30 and >30 mg/l, a 10-point distribution for the individual species was proposed. Table 2.4 shows the most frequently occurring forms and their ranking according to this proposal. Using this method the effluent BOD_5 can be predicted with about 85% certainty.

2.4 BACTERIOLOGICAL METHODS

For bacteriological investigations, rigorous cleanliness of the working area, together with cleaning and sterilisation of the apparatus, are essential provisos. In addition, collection and transport of samples can only be carried out using sterile equipment and containers, while the task of sampling should only be performed by specially trained persons.

Sewage presents a potential health risk both to animals and humans. Bacteria provide both qualitative and quantitative indications of the quality of the liquid. But while in the case of water investigations the bacterial numbers are largely correlated with the content of organic matter, substantial deviations occur in the case of raw sewage, as a result of its low level of colonisation by bacteria, and for treated sewage effluent because it originates froma biocoenosis extremely rich in bacteria. Effluents from activated sludge plants often have a higher bacterial count that the influent even though the plant may be performing satisfactorily. Correlations between specific bacteria in secondary effluent and the BOD_5 sludge loading can be gathered from Table 3.3.

Bacteriological methods are employed for determination of bacterial counts, for assessing the degree of colonisation of the activated sludge flocs, for identification of thread forming bacteria, and also in special cases for toxicity testing. From the epidemiological standpoint, the pathogenic organisms are of particular interest. However, as it is practically impossible to check for the presence of individual pathogens because of their enormous variety, the probability of pathogenic bacteria being present must be inferred from the occurrence of certain indicator organisms.

Table 2.4 — Distribution table for the most commonly encountered ciliates in activated sludge systems as a means of estimating the BOD_5 in the treated effluent. From Curds and Cockburn (1970) who list 66 species

Species	0–10	11–20	21–30	>30
Acineta cuspidata	10	0	0	0
A. foetida	0	0	10	0
Aspidisca costata	3	3	2	2
A. lynceus	5	5	0	0
A. turrita	10	0	0	0
Carchesium polypinum	3	5	2	0
Chilodonella cucullula	4	4	1	1
Ch. uncinata	3	6	1	0
Cinetochilum margaritaceum	7	3	0	0
Coleps hirtus	10	0	0	0
Colpidium campylum	2	2	2	4
C. colpoda	0	0	4	6
Discophrya elongata	0	10	0	0
Drepanomonas revoluta	1	4	5	0
Epistylis plicatilis	0	4	4	2
E. rotans	10	0	0	0
Euplotes affinis	6	4	0	0
Eu. carinatus	2	4	4	0
Eu. eurystomus	2	4	4	0
Eu. moebiusi	3	3	3	1
Eu. patella	4	3	3	0
Glaucoma scintillans	2	2	3	3
Hemiophrys fusidens	3	4	3	0
H. pleurosigma	10	0	0	0
Linotus carinatus	10	0	0	0
L. fasciola	0	10	0	0
Opercularia coarctata	2	2	4	2
Paramecium caudatum	2	5	3	0
P. trichium	4	3	2	1
Podophyrya fixa	0	2	7	1
P. maupasi	0	10	0	0
Tokophrya quadripartita	4	3	3	0
Trachelophyllum pusillum	3	3	3	1
Trochilia minuta	0	10	0	0
Vorticella aequiliata	2	2	3	3
V. alba	3	3	3	1
V. campanula	8	2	0	0
V. convallaria	3	4	2	1
V. fromenteli	5	4	1	0
V. hamata	7	2	1	0
V. microstoma	2	4	2	2
V. octava	3	3	2	2
No ciliates	0	0	0	10

Total bacterial count (numbers/ml)
The total bacterial count gives the total number of all the bacteria present in the liquid; it is obtained by direct microscopic counting in counting chambers, and by the use of a phase-contrast microscope. The enumeration of the bacteria may, however, be complicated by the presence of large amounts of debris and bacterial flocs.

Colony count (colonies/ml)

The colony gives the number of aerobic, saprophytic bacteria which are detectable on nutrient agar after incubation for 44 h±4 h at 20±2°C, with sixfold lens magnification. As sewage samples exhibit high colony counts, serial dilutions of 1:10, 1:100, etc., are desirable. The colony count normally lies around 1 to 2 powers of ten below the total bacterial count.

Anaerobes (bacteria/ml)

The anerobic colony count permits deductions concerning the presence of anaerobic and facultative organisms. The ratio of the aerobic to anaerobic colony count as a rule reflects the oxygen regime of the biotope. With activated sludge plants, however, it must be noted that about 50% of the bacteria in the influent and an even higher percentage of the bacteria attached to the flocs can undergo anaerobic metabolism if subjected to oxygen deficiency. A high anaerobic colony count is thus not a reliable indication of a poor oxygen balance in activated sludge plants.

In order to ensure anoxic conditions during the incubation period of 48 hours±2 hours at 37°C±1°C, cysteine-casein nutrients of low redox potential are employed.

Faecal bacteria (bacteria/ml)

Testing a water sample for bacteria of faecal origin is a regular part of the task of hygienic water quality monitoring.

Faecal bacteria comprise:

— *Coliforms:* aerobic or facultative anaerobic rod-shaped bacteria which decompose lactose at 37°C with production of acid and gas. Coliforms can be of faecal origin.
— *Faecal coliforms:* as for coliforms, however, lactose decomposition also takes place at 44°C. They are of faecal origin in 95% of cases.
— *Escherichia coli:* as for faecal coliforms, but additionally produce indole from tryptophan. *E. coli* are definitive indicators of faecal contamination.

Evidence of possible faecal pollution is also obtained from the determination of *Enterococci* (inhabitants of the intestines of warm-blooded animals, which also occur on plants), of *Pseudomonas aeruginosa* (bowel inhabitants), and *Clostridium perfringens* (obligate anaerobic spore-former, causative agent of gas gangrene.

Besides colony counts the so-called titre is also measured for coliforms, faecal coliforms, and enterococci. It represents the smallest volume of water in which these groups of bacteria can be detected. The titre therefore has the dimensions of millilitres, for example coliform titre 0.4 ml means that one organism is present in 0.4 ml. The titres are poorly verifiable statistically. More reliable predictions can be obtained from four sample volumes graduated in powers of 10, giving a 'probable titre' [3].

Pathogenic bacteria

Investigations with pathogenic organisms may only be performed in special laboratories under the terms of legal regulations (in East Germany G Bl II, 1966, No. 16, DB, of the Law for Prevention and Counteraction of Transmissible Diseases in

Humans 25 Jan 1966). The reference laboratories in E. Germany for handling specific pathogenic organisms are listed in [5].

For sewage-related investigations the presence or elimination of the following bacteria are checked:

— Salmonella, div. spec.: causative agents of diarrhoeal diseases like typhus, etc.;
— Mycobacterium, div. spec.: causative agents of tuberculosis;
— Clostridium, div. spec.: causative agents of gastroenteritis;
— Staphylococcus, div. spec.: toxin-producing bacteria.

The elimination of pathogenic bacteria inclusive of faecal bacteria is a principal task of biological sewage treatment.

2.5 OTHER BIOLOGICAL METHODS

Worm eggs

Worm eggs (parasitic ova) are separated from the remainder of the sediment by means of sedimentation, centrifugation and flotation (density of flotation liquid 1.3 kg/l) and then enumerated on a counting plate. The morphology of the eggs (Table 3.2, Figs 3.9 and 3.11–3.17) frequently permits their identification.

Effluent samples from the final settling tank are settled for 8 h and the sediment enumerated in a counting chamber without further treatment.

Viruses

Investigation with viruses can only be performed in special laboratories (Reference laboratories for pathogenic viruses in the GDR are given in [5], studies on water-borne viruses in the Regional Hygiene Institutes).

In the course of water investigations at present only the epidemiologically important enteroviruses, reoviruses and adenoviruses (Table 4.21) are detected.

Fungi

For determination of fungi, culturing on special agar media is necessary. The rapid growth and early manifestation of multiplication states favours the isolation of moulds from activated sludge. In order to prevent interfering growths of bacteria a pH of 4.5 is selected, or alternatively pH 7.0 with addition of antibiotics.

3
Organisms

Taxonomic identification of the organisms in activated sludge installations is often difficult, firstly because of an incomplete state of knowledge concerning the variability of different forms, and secondly because the existing knowledge of specialists in the particular groups of organisms has not been applied to any great extent to activated sludge, or has not been widely published; reference often has to be made to out-of-date literature. For some groups of organisms, such as the *Vorticella*, certain rhizopods and the thread-forming bacteria, further taxonomic studies are required. A summary of the literature on identification can be found in [3]. The most frequently occurring and abundant organisms are represented in Figs 3.1 to 3.10; other forms which occur sporadically, or in individual plants or lightly loaded systems are not covered.

The saprobic states and the indication weights quoted are take from [3]. The stated size refers, where not otherwise indicated, to the cell length, without projections, outer casings, etc.

3.1 VIRUSES

Viruses are composed of ribonucleic or deoxyribonucleic acids (RNA and DNA viruses) protected by means of a protein envelope.

The size of viruses varies from 10 to 400 nm; their structure and morphology are specific to the individual forms; rod-shaped, elongated, spherical and polyhedral forms are known. The bacterial viruses, the so called bacteriophages, consist generally of head and body parts. The latter is formed of a hollow stem with a sheath and end plate, from which host-specific tentacles and spikes are extended, with the aid of which the phages attach themselves to the bacterial cells. Both cells and viruses usually have receptive parts and means of attachment which engage with each other. Following attachment either the nucleic acid alone, or the entire virus, is forced through cell membrane into the interior of the bacterial cell. The multiplication which ensues within the cell lasts for periods ranging from a few minutes to several hours; the building material necessary for this viral multiplication is made available as a result of enhanced bacterial metabolism within the cell. After a ripening stage the daughter viruses are eventually released, when the bacterial cell either dies immediately, or remains in a virus-producing state for a still longer period. Many bacterial species have their own host-specific phages, of which there are over 100 different phage types already known (Lindner 1978).

Upsets in the activated sludge process due to phage activity might be suspected, but have not yet been demonstrated. Reasons for this are the difficulty of identifying phages in the activated sludge, the maintenance of BOD_5-elimination rates following the lapse of a particular species by the mixed biocoenosis comprising numerous bacterial species, and possibly sometimes a mistaken identification of the causes of plant upsets, for example due to toxic substances. The inactivation of pathogenic bacteria by the action of phages is however a desirable effect.

3.2 SCHIZOMYCETES (BACTERIA AND BLUE ALGAE)

Under the term 'bacteria' in the context, the non-photoautotrophic blue algae are also included, in agreement with Häusler (1982) and [16].

Bacteria constitute the most primitive group of plants, with about 6000 different species. Their cell characteristics may be assigned to three basic categories.

— Cocci (Fig. 3.1(1))

Fig. 3.1 (p. 58) — *Bacteria*. After accounts by Lindner (1978), Liebmann (1962), Eikelboom (1975), Häusler (1982), Cyrus and Sládeček (1973). 1. *Micrococcus*, various spp. 2–7. Rods; 2. coccoids, 3. short plump, 4. long, 5. slender, 6. spindle shaped, 7. distended thigh-shaped. 8. *Vibrio*, various spp. 9. *Pseudomonas* various spp, 1.5–4 μm. 10. *Spirillum volutans*, Ehrenberg: 14–70 μm. 11. *Spirillum imdulla* (Müller) Ehrenberg: 5–20 μm. 12. *Bacterium cyrusii* Cyrus and Sládeček, 7–14 μm. 13. *Archromatium volutans* (Hinze) van Niel: H_2S-indicator 10–15 (3–87.5) μm. 14. *Macromonas mobilis* (Lauterborn) Utermohl: H_2S-indicator, 15–30 μm. 15. *Sarcina paludosa* Schracter p, v, h 2–3 μm. 16. *Pseudoromeria cyrusiana* Cyrus and Sládeček 2.5–4.5 μm. 17. *Lampropedia hyalina* (Eurenberg) Schrocter 2.5–3.8 μm. 18. *Zooglea ramigera* Itzigsohn 1–3 μm. 19. *Zooglea filipendula* Beger 1–2 μm. 20. *Zooglea uva* Kutzing 1–3 μm. 21. *Nocardia* various spp 3–5 μm. 22. *Beggiatoa alba* Trevisan H_2S indicator, width 2.5–5 μm. 23. *Beggiatoa leptomitiformis* Trevisan H_2S indicator width 1–2.5 μm. 24. *Haliscomenobacter hydrosis* van Veen *et al.* bulking sludge former width 0.2–0.45 μm. 25. *Leucothrix mucor* Oerstedt bulking sludge former, width 1.5–3 μm. 26. *Achronoma spiroideum* Skuja width 0.3–0.5 μm. 27. *Microscilla agilis* Pringsheim width 0.5–0.8 μm. 28. *Spirochaeta plicatilus* Ehrenberg width 0.5–0.75 μm. 29. *Microthrix parvicella* Eikelboom bulking sludge former width 0.4–0.6 μm. 30. *Nostocoidea limicola* Eikenboom bulking sludge former, width 0.7–1.5 μm. 31. *Peloploca fibrata* Skuja bulking sludge former, width 0.4–0.6 μm.

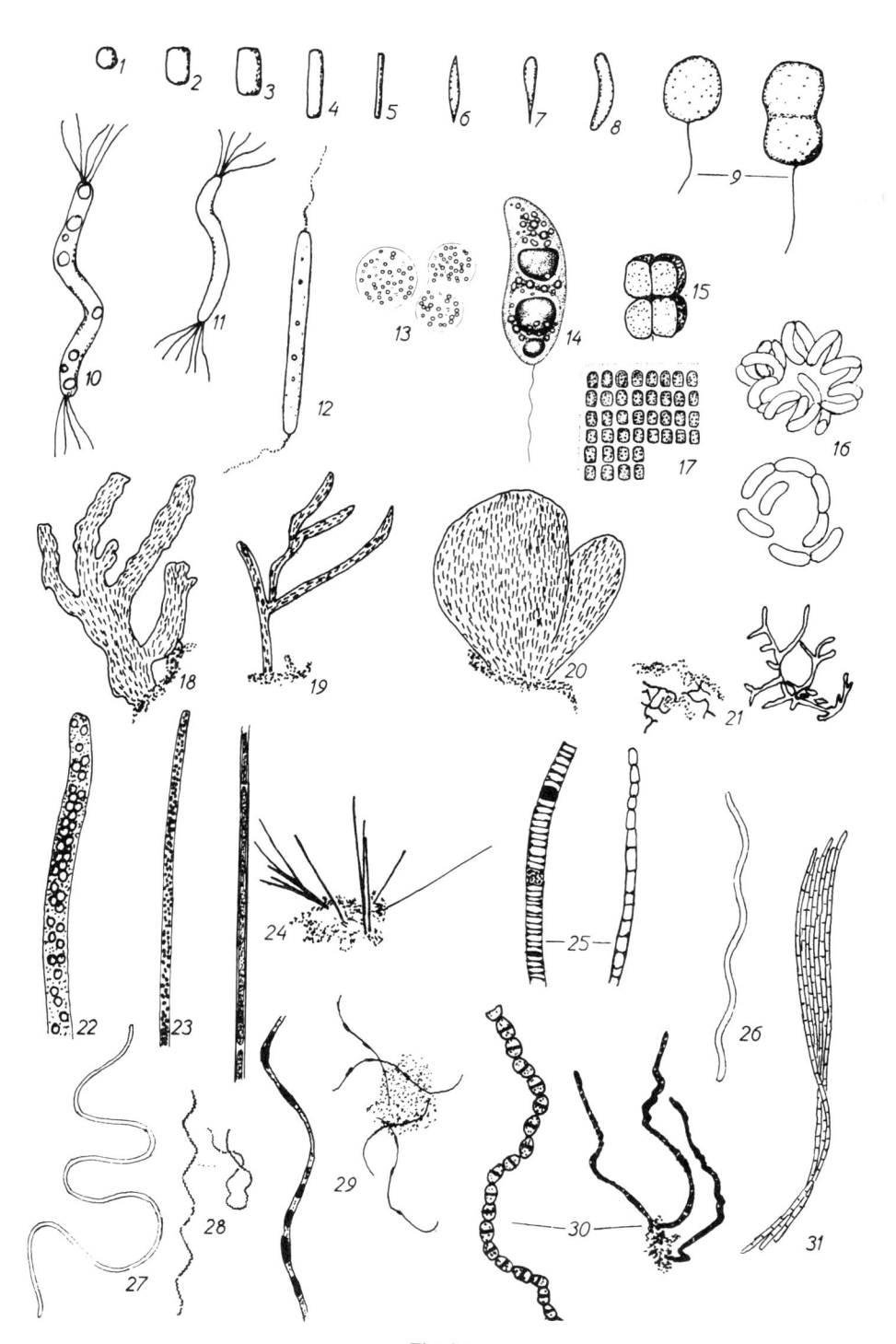

Fig. 3.1

Diameter 0.5–1.5 μm, spherical, occasionally elliptic or otherwise bun- or kidney-shaped. By means of cell-division at different levels multiple forms may be produced, such as chains (e.g. *Streptococcus* (Fig. 3.2(5)), doublets (e.g. *Diplococcus*) bundles (e.g. *Sarcina* (Fig. 3.1(15)) and clusters (e.g. *Staphylococcus*)

— Rods (Fig. 3.1(2–7))
  0.6–200 μm long, 0.4–0.5 μm thick; structure coccoid, short and fat or long and slender, spindly or even swollen to a club-shape. Often thread-forming, sometimes with true branches.

— Spirillae (Fig. 3.1(10–11))
  0.5–70 μm long, spiral-shaped but sometimes only slightly curved, such as *Vibrio*. Spirochaetes are extremely flexible, up to 500 μm long but may be only 0.1 μm thick.

Some bacteria possess cilia for forward propulsion, or can develop these when required. These cilia are up to 20 μm long, but owing to their very small diameter of only 50 nm (max.) are invisible under an optical microscope. Numbers and arrangement of cilia are species-specific. Multiplication takes place by means of cell division, rarely by sprouting or budding.

Many bacterial species exhibit differently-shaped individuals (pleomorphy); this may be brought about by a variety of different environmental conditions such as effect of chemicals, bacteriophages, temperature, etc. Stunted forms occur in nutrient-starved cultures, for example during aerobic sludge stabilisation; L-forms, involuted forms and gonidia represent further morphological variants of a species which may be only partially capable of reverting to the typical original version on the restoration of normal environmental conditions. Some species form very resistant and protracted stages in the form of spores and cysts.

Owing to the relatively few morphological distinguishing features and their variability, species identification is only rarely possible. Where a species diagnosis is called for, then pure cultures and biochemical test methods must be applied.

Depending on their growth habit, the bacteria occurring in activated sludge may be subdivided as follows:

— solitary-growing bacteria Fig. 3.1(1–14)
— colony-forming bacteria Fig. 3.1(15–20)

By colonies we mean associations of single organisms of the same species, not the mixture of species present in many activated sludge flocs.

— *Mycelium-forming bacteria*
  (Actinomycetales) (Fig. 3.1(21)) Rod-shaped or coccoid bacteria which can form a multi-branched segmented or non-segmented pseudomycelium. The mycelium resembles that of true fungi; however, owing to their small cell diameter of 0.5–2 μm and the absence of chitin and cellulose in the cell wall, and also the cell nucleus, they are readily distinguishable.

— *Thread-forming bacteria*
(Fig. 3.1(22–31)), Fig. 3.2(1–12))
Although thread-forming bacteria are often of major importance for activated sludge plants, and despite certain progress during recent years, the taxonomy and hence the biology of these organisms is still largely unfamilar. The most frequently-occurring thread-forming organisms recorded by Eikelboom (1975) are indicated in Table 3.1.

3.3 MYCOPHYTA (FUNGI) Fig. 3.2(13–17)

Fungi are heterotrophic lower plant forms with a true cell nucleus. The vegetative bodies are either unicellular or multicellular. Multicellular fungi exhibit thread-like growth; they may be either branched or unbranched, segmented or unsegmented. The individual fungal threads are termed *hyphae*; their collective total as mycelium. The diameter of the hyphae may lie between 2 μm and over 100 μm. Their propagation takes place by means of naked flagellated swarm cells or spores.

Out of the 35 000 different species of fungi which have been described, more than 50 occur in sewage. In the activated sludge biomass the dominant forms are, according to studies by Cooke *et al.* (1970), *Endomyces lactis*, *Trichosporon* spp, *Penicillium* spp, *Cephalosporium* spp, and *Cladosporium cladosporioides*, and according to Ottová (1962). *Fusarium roseum*, *Endomyces lactis*, *Mucor spinosus*, and members of the genus *Candida*. The fungal hyphae resemble each other very closely, so that microscopic identification is possible only in exceptional cases; in addition the hyphae often assume an atypical form in the liquid. For a precise classification, culturing on synthetic media is a necessity. As several species are cultivated for industrial purposes (*Saccharomyces*, *Torula*, *Rhodotorula*) as well as in domestic situations, and hence may be discharged to sewers, their occurrence in the activated sludge is not a firm indication that they are native to this particular biocoenosis. In screened sewage from Prague, up to 96 000 yeast cells/ml were recorded (Sladká and Hänel 1964) and at Karl-Marx Stadt, 125 000 cells/ml.

In general, fungi occur only rarely and in small numbers. Some thread-forming species give rise to bulking sludge development (Batek 1979, Jones 1964, Cooke and Ludzack 1958).

Fig. 3.2 (p. 61) — Bacteria (1–12), fungi (13–17), viruses (18). From articles by Cyrus and Sládeček (1973), Eikelboom (1975), Haussler (1982), Niethammer (1947), Sládeček (1963), Wollenweber and Reinking (1935), Cooke and Ludzack (1958), Lindner (1978). 1. *Peloploca pulchra* Skuja, treadwidth 0.4–0.6 μm with sheath up to 1.8 μm. 2. *Lineloa longa* Pringsheim Bulking sludge former threadwidth 1.2 to 1.4 μm. 3. *Saprospira albida* (Kolkwitz) Lewin, threadwidth 0.7–1 μm. 4. *Streptobacterium zavisii* Cyrus and Sládeček, threadwidth 2 μm. 5. *Streptococcus margaritococcus* Schroeter, threadwidth 1.5–3 μm. 6. *Sphaerotilus natans* Schroeter Bulking sludge former, threadwidth 0.7–3 μm. 7. *Thiothrix nivea* (Rabenhorst) Vinogradskij H_2S-indicator, bulking sludge former, threadwidth 1.4–2.5 μm. 8. *Thiothrix tenuis* Venogradskij H_2S-indicator, bulking sludge former, threadwidth 1 μm. 9. *Thiothrix tenuissima* Vinogradskij H_2S-indicator bulking sludge former, threadwidth 0.3–0.5 μm. 10. Type 0961 Eikelboom Bulking sludge former, threadwidth 1.1–1.5 μm. 11. Type 0092 Eikelboom Bulking sludge former, threadwidth 0.4–0.6 μm. 12. Type 0803 Eikelboom Bulking sludge former threadwidth 0.6–0.8 μm. 13. *Fusarium aquaeductum* Lagh threads 2–3.5 μm. 14. *Endomyces lactis* (Fres) Wind. Bulking sludge former, threadwidth 2.2–3.6 μm. 15. *Leptomitus lacteus* Agardh Bulking sludge former, threadwidth 6–20 μm. 16. *Saccharomyces cerevisiae* Hansen threadwidth 5.5–9.8 μm. 17. *Zoophagus insidans* Sommerst Bulking sludge former, threadwidth 4–6 μm. 18. Bacteriophages T_2-phage, T_3-phage, tailed phage pictured in 3-D format, up to 400 nm.

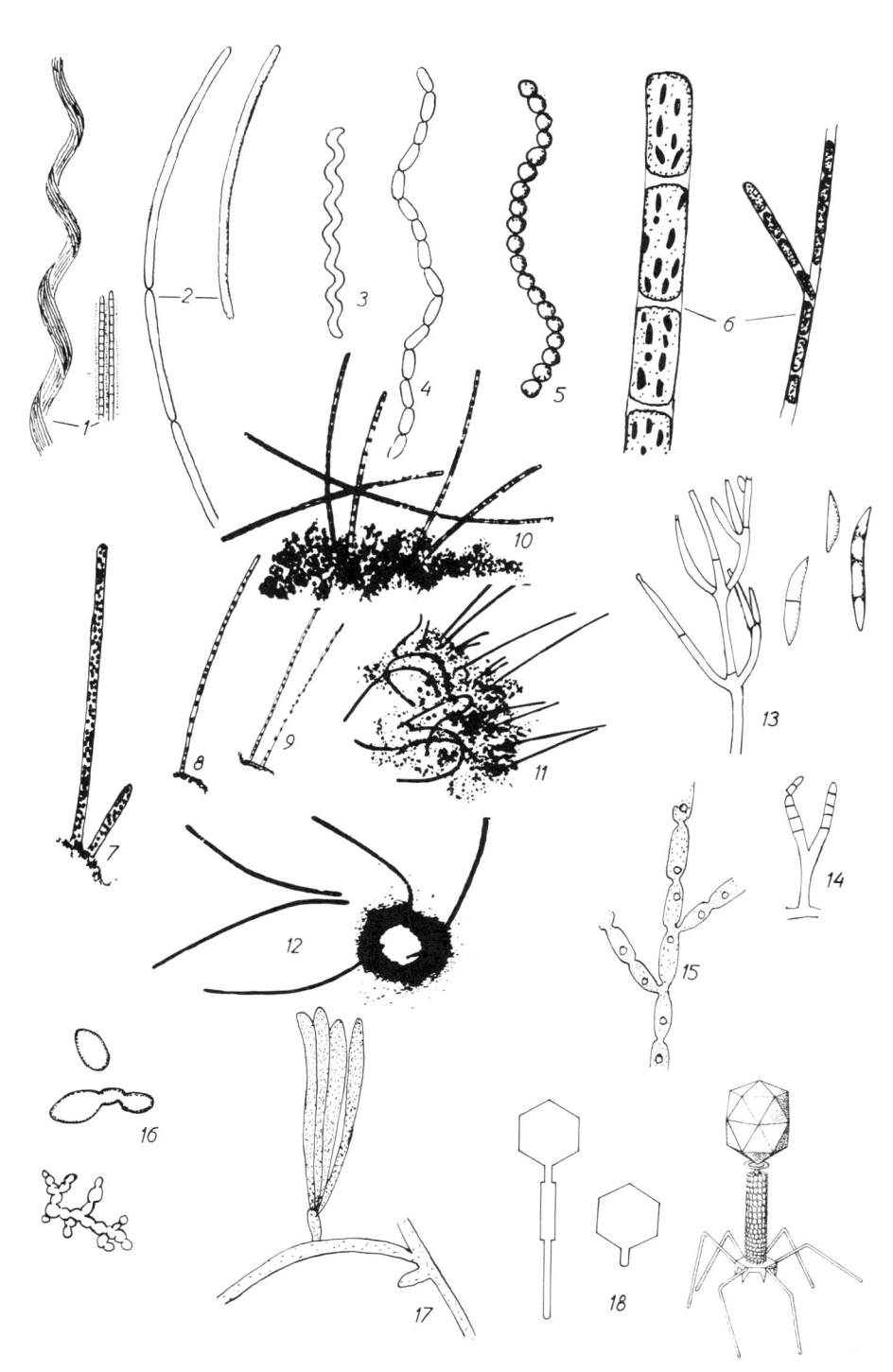

Fig. 3.2

Table 3.1 — Decision table for identification of thread-forming organisms encountered in activated sludge — tabulated according to [6] with modifications

1	Branching threads	see 2
	Non-branching threads	see 4
2	Branching spurious, with sheath	*Sphaerotilus*
	True branching	see 3
3	Thread width $<1\ \mu m$	*Actinomycetes* spp. *Nocardia*, etc.
	Thread width $>1.5\ \mu m$	Fungi
4	Motile threads	see 5
	Non-motile threads	see 6
5	With sulphur inclusions	*Beggiatoa*
	without S-inclusions (after S-test)	*Flexibacter Microscilla*
6	With sulphur inclusions (possibly after S-test)	see 7
	without sulphur inclusions	see 8
7	Gram-positive	Type 0014
	Gram-negative, sessile	*Thiothrix*
8	Cell septa barely visible	see 9
	Septa clearly distinguishable	see 13
9	Threads straight or slightly curved	see 10
	Threads strongly curved	see 12
10	Neisser stain positive	Type 0092
	Neisser stain negative	see 11
11	Gram-negative	*Haliscomenobacter hydrosis*
	Gram-positive	Type 1851
12	Gram-negative	Type 0581
	Gram-positive	*Microthrix parvicella*
13	Neisser stain — positive	*Nostocoidea limicola*
	Neisser stain — negative	see 14
14	Cell width 1–2 μm	see 15
	Cell width $<1\ \mu m$	see 17
	Cell width $>2.5\ \mu m$	Blue algae
15	Gram-positive, usually sessile	Type 0041
	Gram-negative, free-swimming	see 16
16	Square to rectangular cells	Type 0961
	Rounded cells, spherical, ellipsoidal or cylindrical	Type 021 N $\simeq$ *Leucothrix mucor*
17	With sheath, usually sessile	Type 1701
	Without sheath, usually free-swimming	see 18
18	Cells square or rectangular	Type 0803
	Cells rounded, spherical, ellipsoidal or cylindrical	Type 1863

3.4 PROTOZOA (PRIMITIVE ANIMALCULES)

Under the microscope protozoa appear as small unicellular and chlorophyl-free organisms, of heterotrophic nature with one or more cell nuclei. Propagation may occur through cell fusion, but mostly takes place asexually by cell-division.

Flagellata (ciliated forms) Figs 3.3 and 3.4

A non-uniform group from a systematic viewpoint. Cells have from one to very many cilia; the cilia may occasionally regress. Some species or genera are able to form pseudopodia. Multiplication occurs by the process of binary fission. Colony formation may occur in some species. A common characteristic is that of cyst-formation in order to survive unfavourable conditions. The flagellates which occur regularly in activated sludges are colourless (that is, they do not possess chromatophores).

Rhizopoda

Cells with naked protoplasm; lacking a cell mouth and any rigid structure and without enduring projections; often having an outer sheath. Movement and nutrient intake occur via false feet (pseudopodia). Multiplication occurs by division or budding but may also take place by sexual means.

Class — Amoebae: pseudopodia are lobe or thread like.

Order — Gymnamoebae (naked amoebae, no outer casing) Fig. 3.5(1–12).

Poorly-distinguishable organisms, so far no special preparative methods. The free-

---

Fig. 3.3 (p. 64) — Colourless flagellates: Distomatinae (1–6), Bodonidae (7–17) Cercobodonidae (18). From papers by Hänel (1979) and Skuja (1956). 1. *Tropomonas agilis* Dujardin 7–30 μm. 2. *Trigonomonas compressa* Klebs 24–33 μm. 3. *Trepomonas rotans* Klebs 10–13 μm. 4. *Hexamitus pusillus* Dujardin 10–13 μm. 5. *Hexamitus rostratus* (Klebs) Limmermann 16–25 μm. 7. *Ancyromonas sigmoides* Kent 2.5–6 μm. 8. *Bodo augustus* (Duj) Butschli 4–15 μm. 9. *Bodo caudatus* (Duj) Stein 8–16 μm. 10. *Bodo edax* Klebs 7–15 μm. 11. *Bodo erectus* (Ruhle) Hänel 4–10 μm. 12. *Bodo globosus* Stein 6–14 μm. 13. *Bodo minimus* Klebs 3–8 μm. 14. *Bodo repens* Klebs 8–15 μm. 15. *Bodo saltans* Ehrenberg 5–9 μm. 16. *Pseudobodo minimus* Hollande 4–12 μm. 17. *Rhyncomonas nasuta* (Stokes) Klebs 5–10 μm. 18. *Helkesimastrix faecicola* Woodcock and Lapage 5–10 μm.

Fig. 3.4 (p. 65) — Colourless flagellates. From papers by Skuja (1956), Hänel (1979), Liebmann (1962), Cyrus and Sládeček (1973). 1. *Monas ocellata* (Scherffel) pringsheim 5–10 μm. 2. *Monas termo* (Müller) Hänel 3–12 μm. 3. *Monas guttula* Ehrenberg 8–20 μm. 4. *Anthophysa steini* Senn (Synonymous with *Monas sociabilis*) 7–12 μm. 5. *Anthophysa vegetans* (Ehrenb) Stein 7–12 μm. 6. *Paraphysomonas veslita* (Stokes) De Saedeler 8–20 μm. 7. *Oicomonas mutabilis* Kent 5–10 μm. 8. *Paramastix conifera* Skuja 10–18 μm. 9. *Astasia klebsii* Lemmermann 40–60 μm. 10. *Peranema cuneatum* Playfair 25–70 μm. 11. *Petalomonas pusilla* Skuja 5–12 μm. 12. *Petalomonas mediocannilata* Stein 22–25 μm. 13. *Notosolenus apocamptus* Stein 6–11 μm. 14. *Bicoeca lacustris* Clark 4–7 μm. 15. *Bioeca petiolata* (Stein) Pringsheim 21–35 μm. 16. *Codonosiga botrytis* (Ehrenb) Kent 8–30 μm. 17. *Pachysocca obliqua* Fott casing 6–7 μm. 18. *Polytoma uvella* Ehrenberg 15–20 μm. 19. *Chilomonas paramecium* Ehrenb 20–40 μm.

Fig. 3.5 (p. 66) — Rhizopods. From papers by Page (1976), Harnisch (1963), Decloitre (1962). 1. *Vahlkampfia aberdonica* Page 18–17 μm. 2. *Vahlkampfia avara* Page 15–30 μm. 3. *Vahlkampfia limax* Dujardin, up to 100 μm. 4. *Vahlkampfia* spp. 5–10 μm. 5. *Hartmaniella cantabrigensis* Page 16–33 μm. 6. *Hartmaniella vermiformis* Page 12–33 μm. 7. *Thecamoeba striata* Penard 30–80 μm. 8. *Thecamoeba verrucosa* Ehrenberg 80–350 μm. 9. *Vanella mira* Schaeffer 20–55 μm. 10. *Vanella platypodia* Glaser 10–20 μm. 11. *Amoeba proteus* Leidy 220–3000 μm. 12. *Trichamoeba cloaca* Bovee 80–150 μm. 13. *Arcella dentata* Ehrenberg shell width 132–184 μm. 14. *Arcella vulgaris* Ehrenberg shell width 48–152 μm. 15. *Euglypha acanthophora* (Ehrenb) Perty shell length 30–150 μm. 16. *Euglypha ciliata* Ehrenberg shell length 56–100 μm. 17. *Chlamydophrys minor* Belaf shell length 15–17 μm. 18. *Chlamydophrys stercocea* Cienkowski shell length 30–40 μm. 19. *Diplophrys archeri* Bark shell diameter 8–20 μm. 20. *Cochliopodium bilimbosum* Averbach 24–56 μm. 21. *Actinophrys sol* Ehrenberg 20–120 μm.

64 **Organisms** [Ch. 3

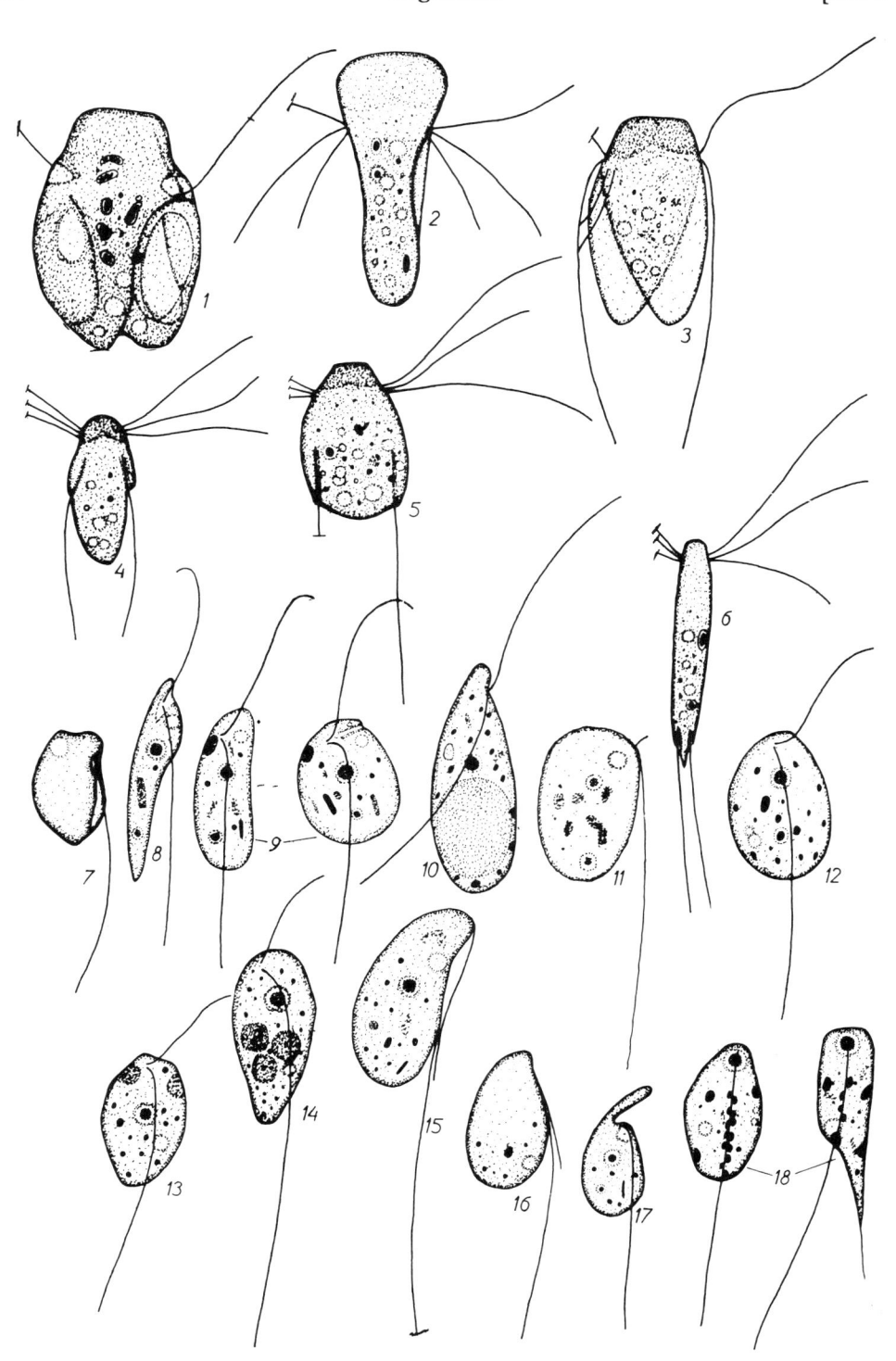

Fig. 3.3

Sec. 3.4]   **Protozoa (primitive animalcules)**   65

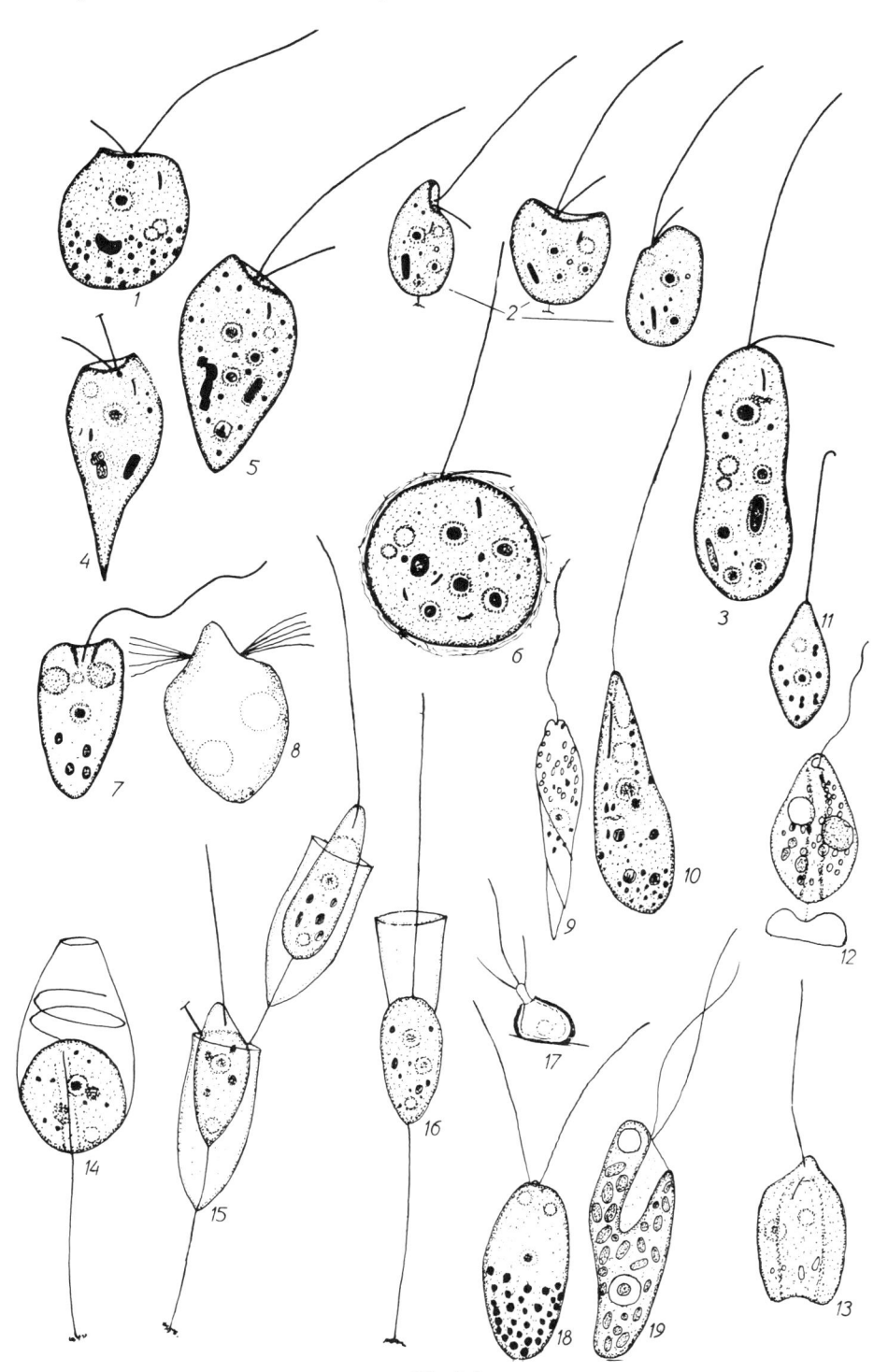

Fig. 3.4

66 Organisms [Ch. 3

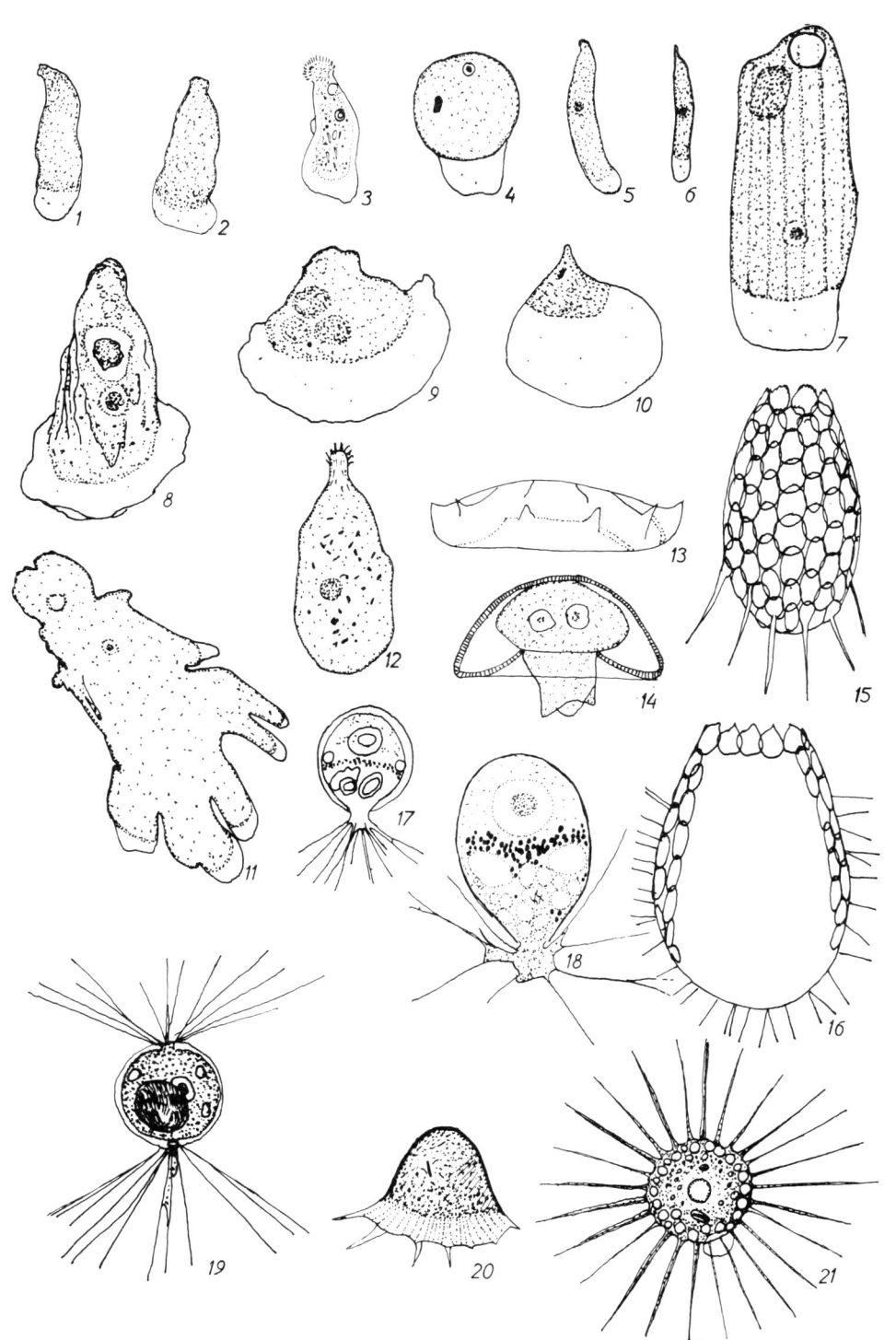

Fig. 3.5

swimming forms of the species shown which resemble *Astramoeba radiosa*, Dujardin, are not presented on account of the difficulty of distinguishing between them.

Order — Thecamoebae (shell-coated amoebae, Testaceae) Fig. 3.5(13–20).

Class — Heliozoa (sun-shaped animalcules) Fig. 3.5(21).

Pseudopodia arranged in ray-like configuration, sometimes supported by an axial fibre.

Ciliata

Highly developed primitive animal forms with cilia or similar organelles, dual nucleation, transverse membranes and conjunction. Around 160 species have been recorded in activated sludge.

Subclass — Infusoria

Cells enveloped in a pellicule. Movement occurs by the action of cilia or similar structures such as undulating membranes, cirri, etc. Nutrient uptake occurs through a mouth, the various shapes of which are of taxonomic significance.

Order — Holotricha (Figs 3.6 and 3.7(1–2))

Bodies usually completely covered with small hairs. The mouth opens at the body surface or in a groove in which free hairs, or rows of hairs, perhaps attached to membranes, can be found.

Order — Spirotricha (Fig. 3.7(3–15))

Mouth with an adoral membranelle zone wound in a clockwise spiral; often the sole member of the Hypotricha in activated sludge, for which the ventral hairs are exclusively composed of cirri and the dorsal ones almost always consist of short, stiff, barely flexible bristles.

Fig. 3.6 (p. 68) — *Holotrichal ciliates* From papers by Kahl (1935), Klimowicz (1974) Liebmann (1962), Uhlmann (1982). 1. *Chilodonella cucullula* Muller 75–300 μm. 2. *Chilodonella unciata* Ehrenberg 30–90 μm. 3. *Cinetochilum margaritaceum* Perty 15–45 μm. 4. *Coleps hirtus* Nitzoch 40–65 μm. 5. *Colpidium campylum* Stokes 50–120 μm. 6. *Colpidium colpoda* Ehrenberg 90–150 μm. 7. *Colpoda cucullus* Muller 50–120 μm. 8. *Cyclidium citrullus* Cohn 14–40 μm. 9. *Cyclidium glaucoma* Muller 15–32 μm. 10. *Cyclidium pellucidum* Kahl 40 μm. 11. *Drepanomonas revoluta* Penard 18–45 μm. 12. *Enchelymorpha vermicularis* Smith 30–45 μm. 13. *Glaucoma scintillans* Ehrenberg 40–86 μm. 14. *Hemiophry fusidens* Karl 130–200 μm. 15. *Hemiophrys pleurosigma* Stokes 200–300 μm. 16. *Lionotus carinatus* Stokes 80–100 μm. 17. *Lionotus hirunda* Penard 60–82 μm. 18. *Lionotus fasciola* Ehrenberg-Wrzesniowski 100 μm. 19. *Lionotus lamella* (Ehrenberg) Schewiakoff up to 200 μm. 20. *Paramecium caudatum* Ehrenberg 180–300 μm. 21. *Paramecium trichium* Stokes 50–120 μm. 22. *Tetrahymena pyriformis* Ehrenberg 40–80 μm. 23. *Trachelophyllum pusillum* Perty-Clap and Leachmann 40–50 μm.

Fig. 3.7 (p. 69) — Holotrichal (1–2) and spirotrichal (3–15) ciliates. From papers by Kahl (1935), Liebmann (1962), Hamm (1963), Matthes and Wenzel (1966). 1.*Trochilia minuta* Roux 20–30 μm. 2. *Uroriema marinum* Dujardin 30–50 μm. 3. *Aspidisca costata* Dujardin (ventral and caudal aspect) 25–40 μm. 4. *Aspidisca lynceus* Ehrenberg (ventral and caudal aspect) 30–50 μm. 5. *Aspidisca turrita* Ehrenberg (caudal aspect) 30–50 μm. 6. *Euplotes affines* Dujardin 40–70 μm. 7. *Euplotes carinatus* Stokes 70 μm. 8. *Euplotes eurystomas* Wrzesniowski 105–270 μm. 9. *Euplotes moebiusi* Kahl 50–70 μm. 10. *Euplotes patella* (Muller) Ehrenberg 80–155 μm. 11. *Oxytricha fallax* Stein 150 μm. 12. *Oxytricha ludibunda* Stokes 100 μm. 13. *Stylonychia mytilus* Ehrenberg (dorsal and lateral aspect) 100–300 μm. 14. *Stylonychia pustulata* Ehrenberg 150 μm. 15. *Urustyla weissei* Stein 300 μm.

68 **Organisms** [Ch. 3

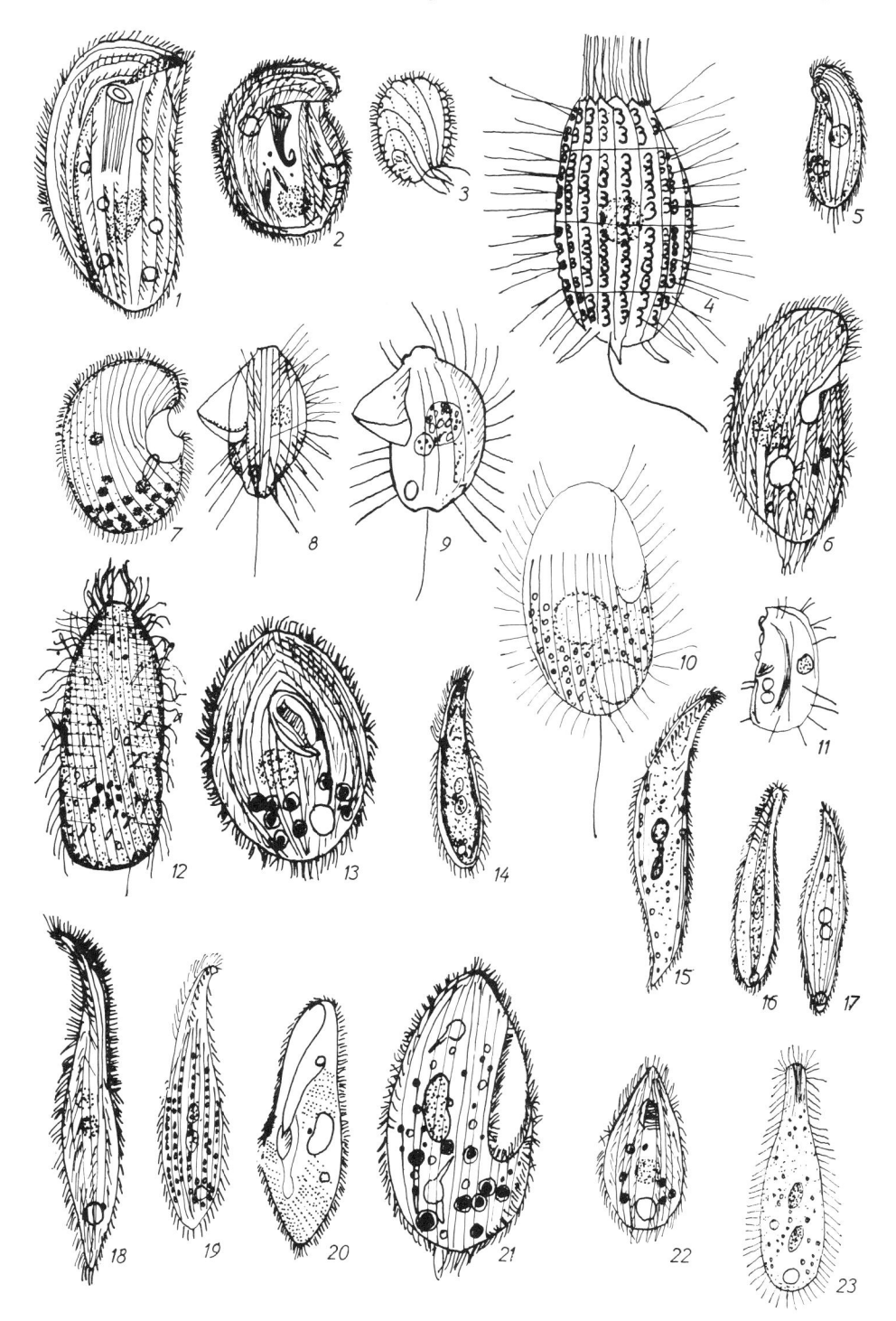

Fig. 3.6

Sec. 3.4]   **Protozoa (primitive animalcules)**   69

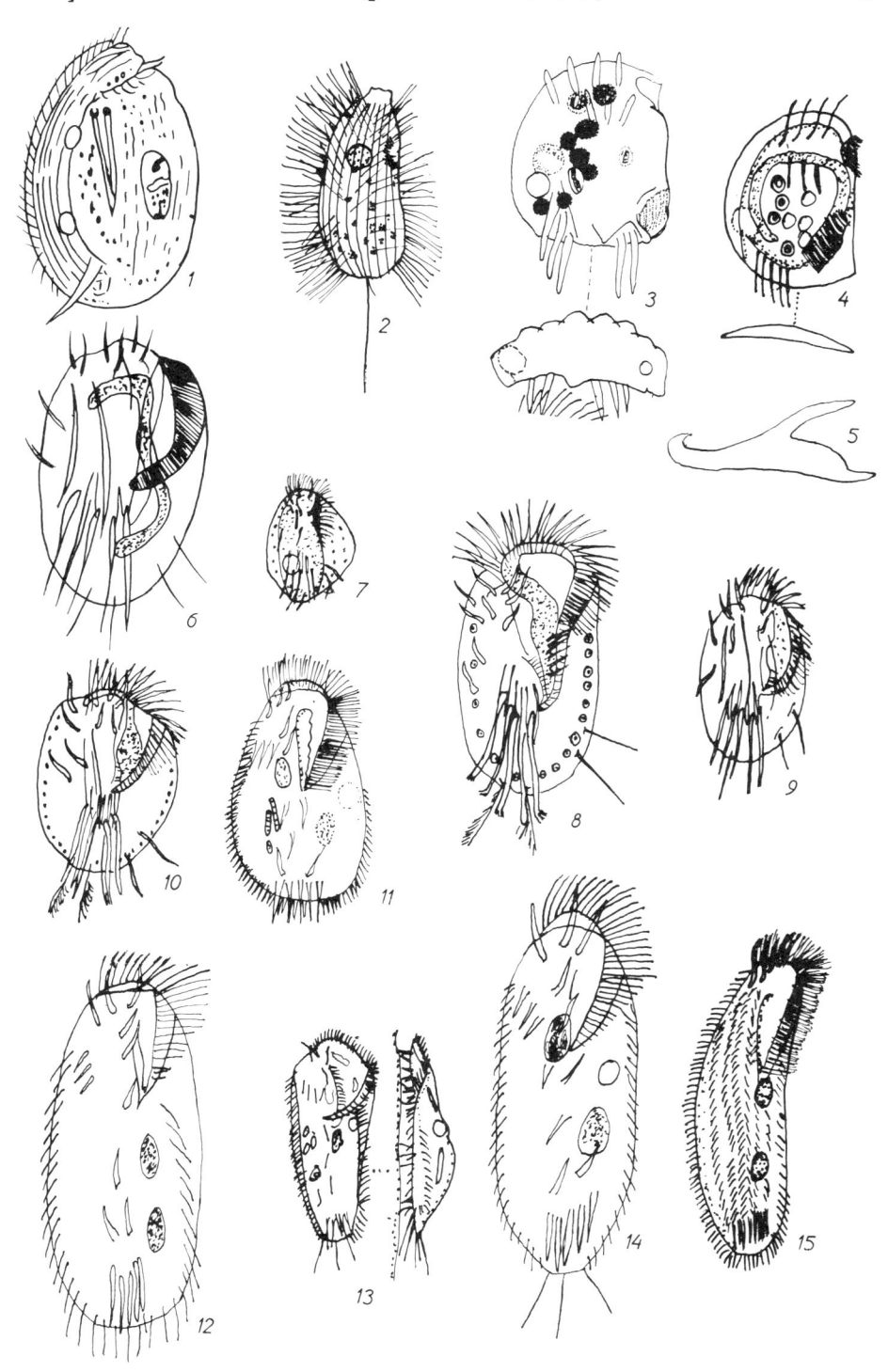

Fig. 3.7

Order — Peritricha Fig. 3.8
Bodies of the developed ciliates carry no other hairs besides those of the adoral zone. In activated sludges only attached species are present, which can form mobile fruiting bodies; these have a ring of hairs on the rear portion.

Subclass — Suctoria (Suction infusoria) Fig. 3.9(1–6)
Cells lack cytostomes. Possess capture and suction tentacles; juvenile forms have hairs but rarely possess suection tentacles; propagation mainly occurs by means of internal and external budding, and also by binary fission, while resting on the activated sludge flocs.

3.5 METAZOA (MULTICELLULAR ANIMALS)

Animals composed of many cells of widely varying type. The metazoa occurring in activated sludge are for the most part of only microscopic dimensions.

Rotatoria (rotifers) Fig. 3.10
Roundworms with a coronet-shaped array of hairs at the forward end. The sack or armorial shield-shaped body is divided into head, tail and trunk sections; both the head and the tail can be withdrawn into the trunk. The two hair coronets in the head section serve as propulsive and food gathering organs. The body surface is covered by rigid cuticles, to some extent like amour plating.

As the male forms of rotifers occur only rarely, or may even be unknown, only the female forms are shown. They invariably possess an ovipositor of single or double-pronged form divisible into ovary and egg-laying portions.

Fig. 3.8 (p. 71) — Peritrichal ciliates. From papers by Kahl (1935), Liebmann (1962), Pike and Curds (1971), Guhl (1979). 1. *Carchesium polypinum* Linne 45–140 µm. 2. *Epistylis lacustris* 50–70 µm. 3. *Epistylis plicatilis* Ehrenberg 100–120 µm. 4. *Epistylis rotans* Svec 70–100 µm. 5. *Opercularia coarctata* Clap and Lachmann (normal form, oxygen deficiency, cluster forming) 36–124 µm. 6. *Vorticella aequiliata* Kahl 40–50 µm. 7. *Vorticella alba* Fromentel 60–80 µm. 8. *Vorticella campanula* Ehrenberg 50–160 µm. 9. *Vorticella communis* Fromentel 30 µm. 10. *Vorticella convallaria* Linne 50–95 µm. 11. *Vorticella extansa* Kahl 50–100 µm. 12. *Voricella fromenteli* Kahl 90 µm. 13. *Vorticella hamata* Ehrenberg 40 µm. 14. *Vorticella microstoma* Ehrenberg 35–83 µm. 15.*Vorticella octava* Stokes 25–60 µm. 16. *Vorticella putrina* Müller-Kent 60–90 µm. 17. *Zoothamnion arbuscula* Ehrenberg 40–60 µm.

Fig. 3.9 (p. 72) — Suctoria (1–6) Insecta (7 and 9) Oligochaeta (8) Nematoda (10) and Helminth ova (11–17). From papers by Schoenichen (1927), Wilcke (1957), Schneider (1939), Lampert (1925) and [3]. 1. *Acineta cuspidata* Stokes 30–50 µm. 2. *Discophrya elongata* Clap and Lachm 85 µm. 3. *Podophrya collini* Root 40–50 µm. 4. *Podophrya maupasi* Butschli (with cluster formers) 30–50 µm. 5. *Podophrya fixa* Muller 10–50 µm. 6. *Tokophrya quadripartita* Clap and Lachm 50–100 µm. 7. *Culex pipiens* Linne (in final settling tank) up to 1200 µm. 8. *Aelosoma hemprichi* 1–5 mm. 9. *Chironomus thummi* Kieffer 1.3–2 mm. 10. *Diplogaster rivalis* Leydig 1.3–2 mm. 11. Egg of *Ascaris lumbricoides* Linne 50–75 µm. 12. Egg of *Enterobius vermicularis* (Linne) Leach 50–60 µm. 13. Egg of *Trichuris trichiura* Linne 45–75 µm. 14. Egg of *Strongylides* sp. 60 µm. 15. Egg of *Fasciola hepatica* Linne 130–150 µm. 16. Egg of *Hymenolepsis nana* (v. Subold) Blanchard 48–60 µm. 17. Egg of *Taenia solium* Linne 30–40 µm.

Fig. 3.10 (p. 73) — Rotatoria. From papers by Bartoš)1959), Klimowicz (1976), Voigt (1955). 1. *Brachionus rubens* Ehrenberg 150–280 µm. 2. *Cephalodella gibba* Ehrenberg 250–300 µm. 3. *Cephalodella gracilis* Ehrenberg 135–152 µm. 4. *Colurella colurus* Ehrenberg 60–120 µm. 5. *Encentrum lupus* Wulfert 150–200 µm. 6. *Epiphanes senta* Muller 400–500 µm. 7. *Lecane bulla* Gosse 170–180 µm. 8. *Lecane clara* Bryce 170–200 µm. 9. *Lecane clastocerca* Schmarda 50–90 µm. 10. *Lecane hamata* Stokes 120–135 µm. 11. *Lecane inermis* Bryce 92–154 µm. 12. *Lecane lunaris* Ehrenberg 135–145 µm. 13. *Lepadella patella* Muller 135–145 µm. 14. *Notommata glyphura* Wulfert 450 µm. 15. *Philodina roseola* Ehrenberg 320–540 µm. 16. *Pleurotricha petromyzon* Ehrenberg 220–280 µm. 17. *Rotaria rotatoria* Pallas 230–1090 µm. (Nos. 9–13 give ventral aspect).

Sec. 3.5] **Metazoa (multicellular animals)** 71

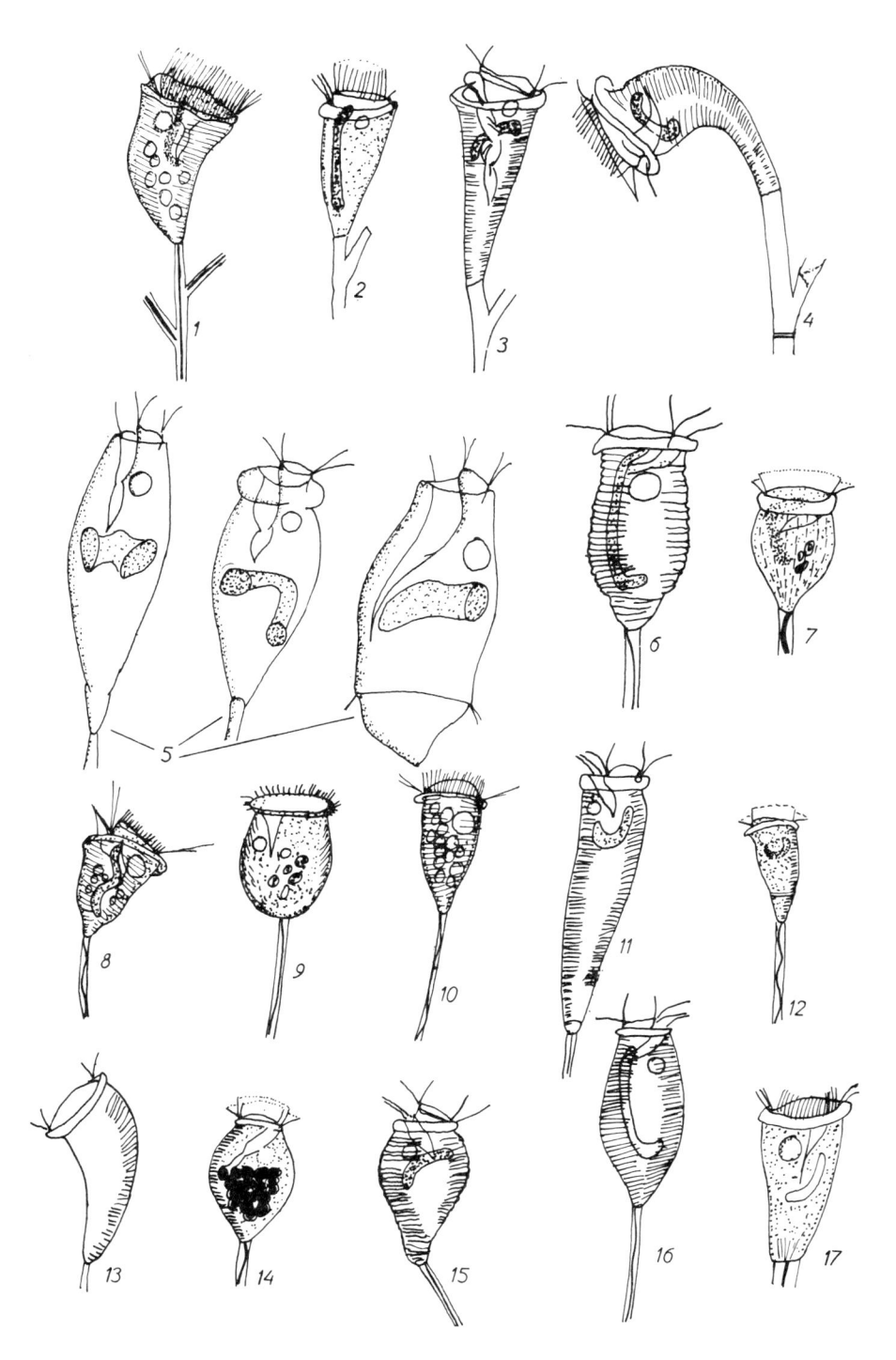

Fig. 3.8

72   **Organisms**   [Ch. 3

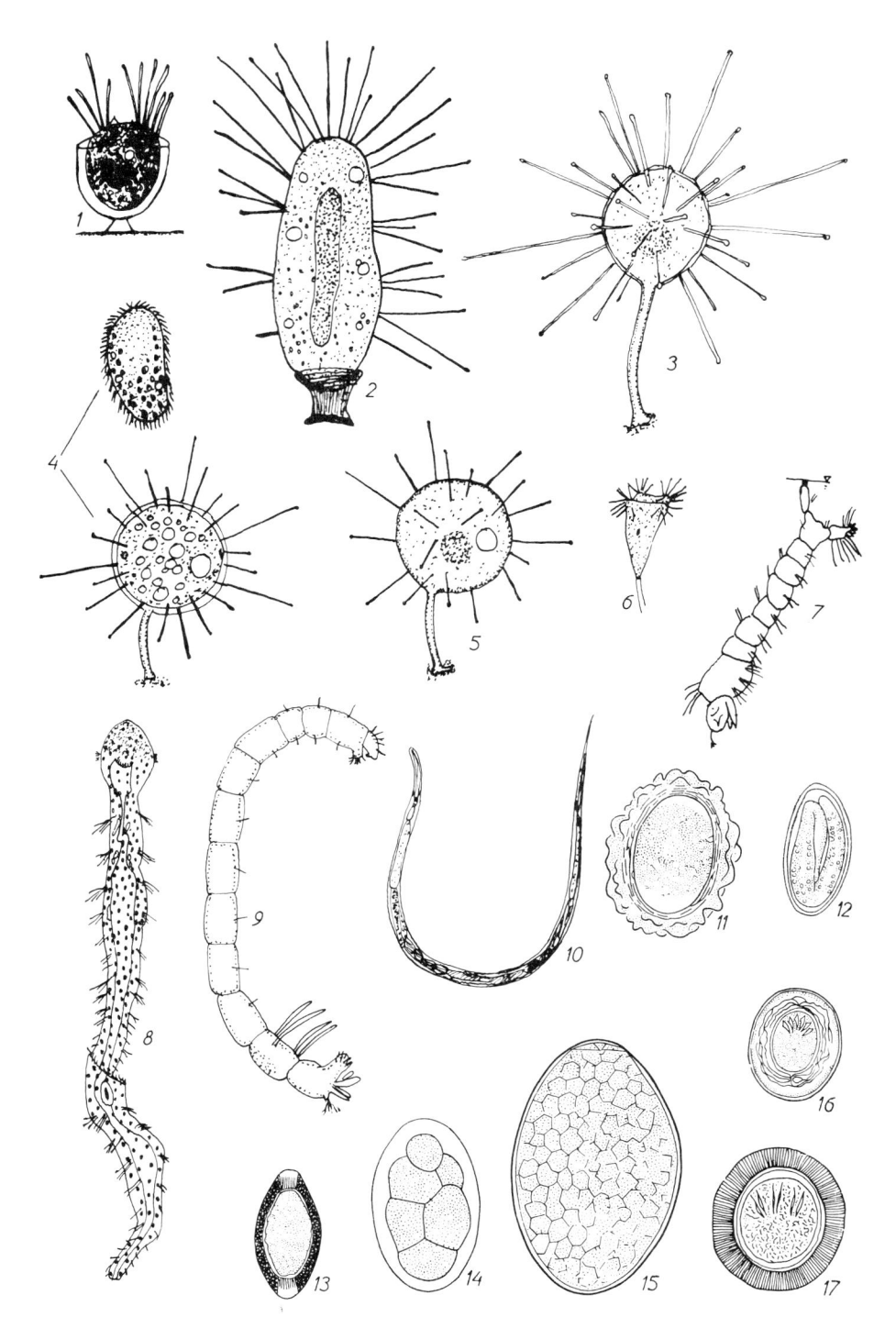

Fig. 3.9

Sec. 3.5] **Metazoa (multicellular animals)** 73

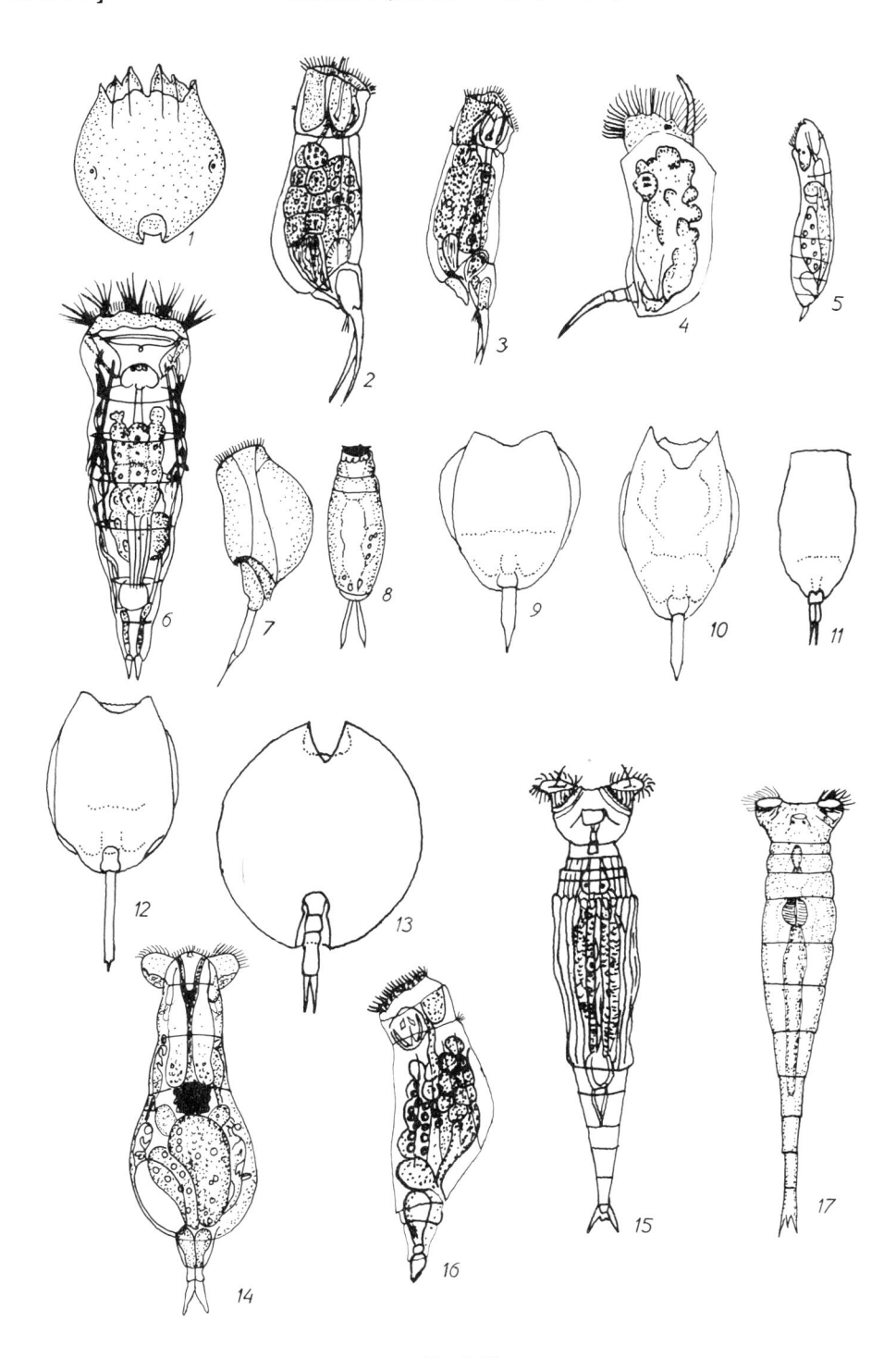

Fig. 3.10

Nematodes (threadworms) Fig. 3.9(10)
Roundworms with wriggling motion and round elongated bodies tapered at each end. Lacking external hairs. May be free-living or parasitic; the latter may occur in humans such as Trichina, maw-worms, maggot worms.

In sewage free-living aquatic, parasitic and even soil nematodes may occur. The identification of the species, which closely resemble each other in appearance, is usually a difficult task. In Fig. 3.9(10) the very widespread 1.3–2.0 mm long organism *Diplogaster rivalus* Leydig is illustrated.

Nematodes occur in small numbers in almost all activated sludge plants; they are often washed in with the incoming sewage. Chaudhuri *et al.* (1965) found members of the genera *Diplogaster*, *Diplogasterioides* and *Rhabditis* for which turbulence was considered to be the main growth-limiting factor in activated sludge plants, as it impeded bisexual propagation.

Oligochaeta (lightly bristled worms) Fig. 3.9(9)
Segmented worms (ringworms) with bristles.

Helminth ova (worm eggs) Fig. 3.9(11–17)
Sewage contains the eggs of human and animal-parasitising helminths, occasionally in large numbers. Table 3.2 and Fig. 3.9(11–17) show the ova of the most frequently encountered human parasitic organisms.

Insecta (Insects) Fig. 3.9(7 and 9)
Organisms divisable into head, thorax and abdominal portions with three pairs of legs. In activated sludge plants only the early development stages of two-winged insects occur regularly. In lightly-loaded activated sludge systems massive growths of red Chironomid larvae (biting midges) of various species may lead to operating failures, as on the one hand they consume the activated sludge flocs and on the other hand use them to build casings.

3.6 ALGAE

Green and blue-green algae of various taxonomic divisions almost invariably occur, often in considerable numbers, in activated sludge. It is not known whether these organisms are continually present as inhabitants of the sludge or whether they have only an incidental role as temporary occupants. Owing to the very low level of illumination they obviously cannot be exclusively autotrophic in their existence. Some algae, for instance, those of the genus *Chlamydomonas* can assimilate carbon from organic substrates in the same way as bacteria, so that their continuing presence in activated sludge cannot be ruled out. In addition, so-called algal-activated sludges are known which may exhibit very high degradative capacity in the course of biological sewage treatment.

Algae may be in part carried into the treatment plant in the incoming raw sewage; their origins may be multifarious, for example, from surface water, from waterworks, flower vases, aquaria, etc. As they may exhibit massive growths in the primary and secondary settling tanks, as well as on the wetted areas above the

Table 3.2 — Helminth ova found in sewage

	Species	Egg type	Egg size	Development pathway
1	*Ascaris lumbricoides* (human stomach worm)	Thick, nodulated outer case	50–75 × 40–60 μm	Sewage — vegetables/fruit — humans
2	*Enterobius vermicularis* (threadworm or pinworm)	Asymmetric–ovoid with worm-like embryo	50–60 × 20–32 μm	Sewage — (ffod chain) — humans
3	Strongylides, for example, *Ancylostoma duodenal* (pit — or hookworm)	Thin shell with segmentation stages or embryo	60 μm	Sewage — moist warm medium — humans
4	*Trichuris trichiura* (Trichina, whipworm)	Slender, lemon-shaped with pole protrusions and thickened skin	45–75 × 2–40 μm	Sewage — food chain — intermediate host (human or animal, e.g. rats, dogs, pigs) — terminal host (human or animal)
5	*Taenia solium* (pork tapeworm)	Roundish with thick brown radially striped shell	20–40 × 20–36 μm	Sewage — pigs — humans
6	*Taenia rhynchus saginatus* (beef tapeworm)	As for 5	As for 5	Sewage — cattle feed — cattle — humans

waterline in tanks and open channels, they are regularly transferred to the activated sludge in the course of cleaning and maintenance activities.

Pigmented blue algae of the type represented by the genus *Lyngbya* may in some cases give rise to bulking sludge, according to Eikelboom (1975).

3.7 LIVING COMMUNITIES

The branch of science known as ecology is concerned with the development of biological associations of living organisms. It studies the interactions between the organisms and their environment. The occurrence of a particular species under certain environmental conditions (autoecology) and the development of communal associations are based on certain rules, principles and even laws. The rules which are generally applicable and hence valid in the case of activated sludge are as follows.

— Action Law for environmental factors, the so called Minimum Law: the development of a community is governed by that particular environmental factor which differs most widely from the needs of the organisms.
— The more extreme the ambient conditions become, the more lacking in numbers of species but the more abundant in selected individuals are the actual communities.
— The more frequent and more detrimental the changes in the ambient conditions the more lacking in diversity and the more unstable are the communities themselves.
— *The abundance rule* In multi-faceted habitats those species with the greatest breadth of acclimatisation attain the largest numbers, while in singular or extreme biotopes those species with lower, but often specific, powers of acclimatisation achieve the greatest abundance of individuals.
— *Equilibrium Law* By the action and reaction of biocoenosis variables, a relative stability is achieved in the community. Where one species undergoes preferential development in response to temporarily optimal conditions, then their competitors begin to break down this excessive predominance.

3.7.1 Intake channel

The ambient variables consist of the fresh or partly putrefied raw sewage, with all its constituents, the high velocity of flow, and the markedly fluctuating water level. In municipal plants there is usually a greyish-brown surface growth of layer-forming and fibrous bacterial forms and in the case of effluents toxic to bacteria, a build-up of fungi on the walls of the channel, the so-called sewer-slime. As a consequence of the high velocity of flow, only those organisms will adhere which are firmly attached or can worm their way into the surface growth. Principally these comprise, in the case of the rhizopods, members of the genera *Vahlkampfia* and *Hartmaniella*, for flagellates the Distomatina *Boda caudatus* and *Oicomonas mutabilis*, and for the ciliates *Vorticella microstoma*, *Glaucoma scintillans* and *Enchelymorpha vermicularis*; the density of individuals is, however, low. From this growth there takes place a continual inoculation of the sewage with bacteria and protozoa, so that at the inlet to a municipal sewage treatment plant there will often be bacterial counts of 10^8–10^9 organisms/ml and protozoa in amounts of 50–300 organisms/ml.

As industrial effluents in particular may be of antisaprobic or cryptosaprobic character (see Table 2.2) the biocoenosis of the sewer wall has a high degree of indicator value in the case of operating failures of the succeeding treatment plant, and also for the planning of new installations. Sewer channels which convey toxic effluents do not exhibit any surface growths.

3.7.2 Aeration tanks
In the aeration tank there are three quite distinct biocoenoses, that of the activated sludge itself, the biocoenosis of the spray or wetted wall zone composed principally of blue aglae, bacteria and fungi, and the biocoensis of the surface-attached growth below the water line.

The growth on the walls of the tank is only visible when the water level is lowered. Its saprobic state may be widely different from that of the activated sludge, because it reflects the environmental conditions only at that particular part of the tank, and not those of the entire activated sludge system. Especially in the case of plug-flow reactors, various communities may be established at different distances from the point of entry of untreated sewage, which again may be quite different from those in the activated sludge biomass. For example in the final tank of a cascade system the surface growth consisted of a massive development of *Carchesium polypinum*, suggestive of good α-mesosaprobic conditions, while the activated sludge exhibited a pronounced polysaprobic state on account of anoxic zones in the initial compartments.

3.7.2.1 Structure of the activated sludge flocs
The activated sludge flocs are, as Plates 1–4 show, variously constituted. The heterogeneity of the floc structure extends both to flocs within a particular plant and also to flocs derived from different activated sludge plants, besides which temporal variations in floc structure are also a commonplace occurrence.

One can regard the floc structure as falling within several categories, namely:

— shape: spherical, flattish, star-shaped, interlaced, ragged and lobulated or loosely reticulated amorphous flocs;
— density: compact, slimy-gelatinous, or loose flocs;
— skeletal matter: detritus, mineral fibre and textile fibre and cellulose fibre-based flocs;
— dominating organisms: bacteria, fungi, Testaceae, *Zoogloea*, *Sphaerotilus*, *Nocardia*-based flocs;
— size: microflocs <50 μm: small flocs 50–250 μm, medium-sized flocs 250–500 μm, large flocs 500–1000 μm, giant flocs >1 mm.

The floc geometry is a consequence of various skeletal material and organisms, different species of organisms and disparate physiological groups, as well as of the various mechanical forces to which the flocs are subjected. It does, however, provide a clue to the ambient conditions in the activated sludge installation. It shows for example:

— slight oxygen deficiency in ragged lobulated, heavily disaggregated flocs;

— phosphorus deficiency in highly gelatinous, slimy flocs;
— high loading rates with dense, compact floc structure;
— low pH values where fungi are abundant;
— carbohydrate-rich media in the presence of *Sphaerotilus*;
— underloading in the case of microflocs.

The floc morphology on the one hand determines the available surface area of the flocs for contact with the surrounding liquid and hence also the relative numbers of organisms, and on the other hand the settling properties of the flocs themselves. Settling properties and exchangeable surface area are opposed with respect to their effect on treatment.

3.7.2.2 Colonisation of the activated sludge flocs

The growth of organisms in the activated sludge is governed by the combined action of all the environmental variables. The action of a given variable may indeed be well known in a particular case, but the combination of possibilities resulting from interaction with other variables is so large that generalised predictions are unrealistic. Some important variables are as follows.

Oxygen

Although the activated sludge process is by definition an aerobic decomposition process, and it is only by means of aerobic reactions that a practically complete elimination of organic substances can be achieved, the majority of plants suffer from localised, temporary (Fig. 5.3) or even permanent oxygen deficiency. Even in aerobically stabilised sludges, an oxygen-deficient condition can occur regularly in secondary settling tanks or in the recycled sludge as a consequence of the endogenous respiration of the sludge biomass, see section 7.4. The organisms in the biomass must be able to withstand severe fluctuations in the O_2-content with an occasional absence of oxygen. There are no natural biotopes exhibiting a similar rhythmic variation in oxygen supply.

The actual oxygen concentrations necessary in the aeration tanks are a subject of controversy. In the older literature a value of 2 mg/l O_2 was specified; at the same time it must be borne in mind that there were no oxygen electrodes available in the early days and consequently any exact determination of the oxygen balance of an activated sludge plant was out of the question. In the experiments with centrifugal aerators [19,20] it was shown that perfectly healthy biocoenoses could develop at oxygen concentrations of only 0.2–0.5 mg/l. In the course of comprehensive studies of the control of oxygen input in plants with centrifugal aerators, and by means of oxygen profile investigations involving numerous successive points in the aeration tank, distinct spatial and temporal variations in oxygen content have been demonstrated again and again. On the other hand neither the regular anaerobic storage conditions for sludge in the final settling tank nor the occasional oxygen deficiency indicated in Fig. 5.3, nor yet the purposely anaerobic or anoxic conditions in a denitrification tank, exert any serious effect on the activated sludge biocoenosis.

Owing to the inherent capability of many organisms to tolerate oxygen-free periods, there is no precise limiting value for oxygen concentration. That a spasmodic oxygen deficiency in the activated sludge, together with the BOD_5-sludge loading, restricts or excludes the occurrence of many organisms may, however, be assumed on the basis of the multitude of other ecological factors and the many other dominant species, but is nevertheless hard to prove, and of limited practical interest. The largely β-mesosaprobic rotifers illustrated in Fig. 3.10, as well as some other β-mesosaprobic organisms originated from oxidation ditches with retention times greater than three days, so that having regard to the retention behaviour in a completely mixed reactor illustrated in Fig. 6.1, many organisms only enter the final settling tank after periods of over 20 days. As the generation time of the organisms under these conditions is shorter than the aeration time and moreover development opportunities for an egg are provided in the attached growth on the ditch surfaces, such organisms may become dominant in the activated sludge, without having to survive the anaerobic phase in the final settling tank. Many protozoa are capable of surviving oxygen-free periods by means of resting stages, cysts and other forms.

In the case of permanent oxygen deficiency and very low redox potentials such as those in anaerobic contact processes, the precipitation of iron sulphide leads to the formation of black-looking sludges devoid of colonisation by protozoa or metazoa.

For an oxygen deficiency of long duration, such as 12 h/d and a concomitant BOD_5-sludge loading of more than 0.15 kg/kg, d, polysaprobic bacteria such as *Sarcina* and *Thiothrix* become dominant, together with polysaprobic flagellates, in particular *Trepomonas agilis*. With a less severe but still pronounced oxygen deficiency, the polysaprobic flagellates *Bodo caudatus*, *Monas ocellata* and *Helkesimastix faecicola* occur preferentially, as well as the rhizopods *Vahlkampfia* and *Hartmaniella* and the ciliates *Colpidium campylum* and *Vorticella microstroma*.

In activated sludges where a periodic absence of oxygen adversely affects the biocoensis, massive flagellate growths occur comprising *Bodo caudatus*, *Monos termo*, and *Helkesimastix faecicola*, plus the above-named ciliates with the addition of *Glaucoma scintillans* and *Tetrahymena pyriformis*. As such plants may also experience intermittent periods of elevated oxygen supply, for example at night, it may be possible for *Sphaerotilus natans* to utilise the low-molecular weight substrates (formed by decomposition of organic matter under anaerobic conditions) to such good effect that bulking sludge formation may occur.

For activated sludge biocoenoses which have been affected by oxygen deficiency, the frequent and sometimes massive development of free-swimming rod-shaped bacteria of *Pseudomonas* and *Spirillum* spp is a characteristic feature. Their massive growths can be seen with the naked eye after a settling time of two to five hours. The ciliated bacteria forsake the settled sludge and form a milky white layer about a centimetre thick at the sludge surface.

The saprobic classification system based on natural flowing waters still does not properly represent the biocoenosis of denitrification systems. Although, the BOD_5 sludge loadings of 0.10–0.15 kg/kg d are very low, oxygen-free conditions may occur several times every day, for example where the preferred method of a preliminary denitrification stage is adopted with a retention time under anoxic conditions of one to three hours. Nevertheless, the conspicuous ciliates *Aspidisca costata*, *Vorticella convallaria*, *Lionotus lamella* may become dominant. The population is certainly less rich in different species than systems without anoxic stages, but the saprobic index will point to α-mesosaprobic conditions.

BOD$_5$ sludge loading
The sludge loading is a lumped parameter indicative of the level of total nutrient supply for the organisms. It has a controlling effect on the water quality for the entire activated sludge plant and hence on the saprobic conditions. Nutrient supply, water quality and saprobic conditions together also determine the extent and manner of growth of the bacteria which represent the principal food source for the protozoa and metazoa in the sludge biomass. The numerical values cited in Table 3.3 for solitary bacteria substantiate the fact that protozoa and metazoa, which feed on solitary bacteria, occur principally in the more heavily loaded sludges. These include for example the flagellates *Bodo caudatus* and *Monas termo* as well as the ciliates, *Colpidium campylum*, *Glaucoma scintillans* and *Paramecium caudatum*. Where the bacterial flora is governed by oxygen-deficiency, the same organisms will also be dominant in lightly-loaded sludges. In sludges, however, which contain only very few solitary bacteria the dominant organisms are those which graze on the flocs, such as *Aspidisca* and *Euplotes*, or those which by reason of their mouth structure possess a high filter performance with respect to the smaller numbers of solitary bacteria, such as *Vorticella*, *Carchesium*, *Epistylis*.

Some typical biocoenoses of microorganisms are shown in Table 3.3 as a function of the sludge loading. In addition the bulking sludge forming thread-like organisms dealt with in section 3.7.2.3 and the bacteria growing within slimy gelatinous envelopes also develop to some extent with an obvious dependence on the sludge loading.

Reactor type/process variant
It is to be expected that different methods of operation for treatment of the same sewage will lead to the formation of different biocoenoses. In a completely-mixed reactor receiving septic raw sewage containing 300 mg/l BOD$_5$, 30 mg/l NH$_4^+$ and 20 mg/l H$_2$S, and assuming full treatment inclusive of nitrification and 100% sludge recycle, the organisms will experience a situation with only 10 mg/l BOD$_5$ and zero mg/l H$_2$S, while at the head of an ideal plug-flow reactor receiving the same sewage they would briefly be exposed to conditions of 155 mg/l BOD$_5$, 17 mg/l NH$_4^+$ and 10 mg/l H$_2$S. Concentration surges, sometimes involving toxic substances, may give rise to even more extreme conditions. In the case of a two-stage system, the second stage is quite unaffected by certain constituents of the original sewage. Despite these obvious differences there is little or no information in the literature on the development of distinct biocoenoses in response to such widely-differing saprobic states. Obviously the ecological adaptive capability of the organisms is so great that the diverse reactor conditions have little or no effect on the biocoenoses. As would be expected, the β-mesosaprobes occur preferentially in the lighly-loaded stirred reactors of small-scale activated sludge plants, in oxidation ditches and in the final stages of multistage treatment systems. The indicator value of these same β-mesosaprobes which are also dominant under plug-flow conditions, must be somewhat uncertain, or the sapro-biological limiting conditions require stricter

definition. Thus, for example, a massive development of *Coleps hirtus* may occur in a plug-flow reactor system treating effluent from a milk processing plant, simply because of the very low ammonium content in the medium.

The reactor types and process variants designed to avoid the formation of bulking sludges consisting of thread-forming organisms are dealt with under section 3.7.2.3.

Sludge age
In practice the sludge age in existing systems may range from 0.5 to 100 days, but as a rule lies between five and ten days. A particular species can only establish itself when its generation time is less than or equal to the sludge age. The following generation times are possible under favourable conditions.

Bacteria	*Pseudomonas* 0.3 h
	Nitrosomonas 15 h
	Acinetobacter lwoffi 3.6 h at 8°C, 1–1.5 h at 26°C
	(Fush and Chen 1975)
Flagellates	*Bodo caudatus* 3 h
	Monas termo 1.7 h (Hänel 1979)
Ciliates	Small species, like *Cyclidium*, *Glaucoma* 2–4 h
	(Fenchel 1980)
	Aspidisca costata 5 h
	Vorticella microstoma 10 h
Rotifers	*Brachionus calyciflorus* 20 h (Erman 1963).

This comparison indicates that the sludge age on its own cannot have any direct control over the nature of the most important components of the sludge biomass, as their generation times are appreciably shorter than the sludge age. The correlation evident from Table 3.3 concerning the link between sludge age and the biocoenoses is dependent on secondary effects, primarily a diminution in the saprobity along with increasing sludge age due to the elimination of organic substrates, organisms serving as prey and so on. Insect larvae, such as those of *Chironomus*, with generation times of 12 to 40 days, are likewise largely independent of the sludge age, as the eggs are deposited on the walls of the tank and also their multiplication takes place by parthenogenesis.

Turbulence
The energy input, which manifests itself as turbulence or mechanical force acting on the microorganisms, increases with increasing specific oxygen input into the aeration tank. For example, the energy input to an oxidation ditch amounts to 5–15 W/m^3, but with a cetrifugal aerator of type B2400 for a volumetric loading (B_R) of 3.6 kg/(m^3 d) is 46 W/m^3, and in intensively aerated tanks may be as high as 500 W/m^3. A high degree of turbulence interferes with the typical colony formation in *Anthophysa* and *Dendromonas* spp (Hänel 1979). Certain species cannot grow under highly turbulent conditions, for example, *Carchesium polypinum*. Among the peritrichal ciliates the sensitivity to turbulence increases in the order *Opercularia — Vorticella — Carchesium* (Buck 1968).

In the tower and deep-shaft processes [33] and sometimes also in the impeller

Table 3.3 — Common activated sludge organisms and their occurrence as a function of BOD_5 sludge loading

B_{TS} kg/(kg d)	SA d	S†	Schizomycetes	Flagellates	Rhizopods	Ciliates	Rotifers	Other metazoa
>2	<1	4	— solitary bacteria 10^7–10^9/ml — *Zoogloea uva* — *Streptococcus margaritaceus* — *Sarcina paludosa*	— *Trepomonas agilis*	—	—	—	—
~1	1	4	— solitary bacteria 10^7–10^9/ml — *Zoogloea uva* — *Sphaerotilus natans* — *Saprospira albida* — *Sarcina paludosa*	— *Bodo caudatus* — *Monas termo* — *Helkesimastix faecicola*	— *Hartmaniella* spec. div. — *Vahlkampfia* spec. div.	— *Vorticella microstoma* — *Colpiodium campylum* — *Uronema marinum*	—	—
~0.6	2	3.6	— solitary bacteria 10^7–10^8/ml — *Sphaerotilus natans* — *Microthrix parvicella* — *Leucothrix mucor*	— *Bodo caudatus* — *Bodo saltans* — *Monas termo* — *Pelatomonas pusilla*	— *Chlamydophrys* spec. div. — *Diplophrys archeri* — *Vanella* spec. div.	— *Opercularia coarctata* — *Podophrya* spec. div. — *Vorticella putrina* — *Trachelophyllum pusilum*	—	—

Sec. 3.7] Living communities

~0.3	4	3.0	— solitary bacteria 10⁶–10⁷/ml — Microthrix parvicella — Haliscomenobacter hydrossis — Leucothrix mucor	— Bodo saltans — Monas guttula — Petalomonas mediocannelata	— Diplophrys archeri — Chlamydophrys spec. div. — Vanella spec. div.	— Vorticella convallaria — Aspidisca costata — Euplotes affinis	— Philodina roseola — Rotaria rotatoria	—
~0.15	9	2.6	— solitary bacteria 10⁵–10⁷/ml — Microthrix parvicella — Haliscomenobacter hydrossis — Leucothrix mucor	— Bodo saltans — Petalomonas mediocannelata — Codonosiga botrytis	— Chlamydophrys spec. div. — Vanella spec. div.	— Vorticella convallaria — Aspidisca costata — Euplotes spec. div.	— Cephalodella gracilis — Lecane spec. div.	— Chironomus thummi
~0.05	25	2.6	— solitary bacteria 10⁵–10⁶/ml — Haliscomenobacter hydrossis — Peloploca fibrata	— Petalomonas mediocannelata — Bicoeca lacustris — Bodo angusta	— Arcella vulgaris — Thecamoeba verrucosa	— Zoothamnion arbuscula — Epistylis plicatilis — Vorticella campanula — Aspidisca costata — Euplotes spec. div.	— Lecane spec. div. — Cephalodella gracilis	— Chironomus thummi — Aelosoma hemprichi

† Saprobic index.

region of many types of sludge recycling pump, the sludge organisms are subjected to intermittently high pressures and pressure fluctuations. In an experimental deep-shaft system the activated sludge flocs were found to undergo pressure changes from 0 to 50 m water gauge every five to six minutes during continuous operation. The floc morphology and the associated biocoenosis did not exhibit any changes or injury as a result. The populations largely reflected the associations presented in Table 3.3, with even the large pertrichal ciliates such as *Vorticella convallaria* being present in appreciable numbers.

pH value
Although sewage may often exhibit fluctuating, and even toxic pH-values, which depart widely from the range of tolerance for the organisms concerned, nevertheless difficulties occur only infrequently, owing to the high buffer capacity of the bicarbonate ion. Many industrial effluents may, however, be so acidic that the pH in the aeration tank falls well below 6.0. In such extreme conditions protozoa and metazoa, as well as many bacteria, are unable to survive, but such a medium is well suited to the growth of fungi. Thus *Leptomitus lacteus*, *Oospora fragans*, *Endomyces lactis* and *Trichosporan* spp have been recorded at pH 3 and sometimes even down to pH 2.5. According to Sládeček (1955) some spiro- and peritrichal ciliates, and also *Cinetochilum margaritaceum*, are viable down to pH 5.

Temperature
Aeration tank temperatures for municipal activated sludge plants lie between 1°C and 22°C. There are undoubtedly auto-ecological and synecological effects as a result although the present state of knowledge is not sufficiently verified. Many species are highly adaptable towards variable temperatures; thus for example an α-mesosaprobic community with many peritrichal ciliates was found in an oxidation ditch at −1°C. According to Sládeček (1955) the ciliates *Stylonychia*, *Oxytricha* and *Podiphrya* as well as the rotifiers. *Encentrum* and *Habrotrocha* are mesophils, while the peritrichal ciliates *Chilodonella* and some flagellates are largely temperature-independent. Our own investigations indicated that massive growth of rotifers would only take place at water temperatures greater than 15°C.

Several types of industrial effluent may have the effect of raising the temperature in the aeration tank to 35–45°C. At these temperatures only those species indicated in Figs 3.1 to 3.10 occur while at the same time a trend to greater saprobity becomes apparent. Particularly heat-resistant forms comprise *Vorticella microstoma*, *Monas ocellata*, *Anthophysa steini* and *Trepomonas agilis*.

The temperature exerts a much greater secondary impact on the communities however. According to the Arrhenius Rule the respiration rate increases by a factor of 2–3 for every 10°C rise in temperature. As a result the oxygen supply is often no longer adequate at the higher temperatures to support the growth of nitrifying organisms, and may even be insufficient for the maintenance of aerobic conditions, so that polysaprobic forms can take over.

Predator–prey relationship
As almost all protozoa and metazoa are predatory organisms the relevant organisms coexist in a dynamic equilibrium, such that the Equilibrium Rule referred to at the

outset is repeatedly confirmed in text book style within a period of only a few days. In lightly-loaded plants the food chain illustrated in simplified form in Fig. 3.11 may become established. In reality the predator-prey relationships in a species-rich sludge biomass may be highly reticulated and very difficult to follow.

Industrial effluents
The effect of particular constitutents of industrial effluents on the development of the populations and their interactions on activated sludge is as a rule much less than might be expected from the variability of the effluents. The reason for this is that the protozoa and metazoa which occur in the biomass do not feed on the wastewater constituents, but on the bacteria which have developed from these substrates. As the bacterial community exhibits a fairly constant species composition, irrespective of the nature of the substrate, the food supply essential to the protozoa is mostly assured.

Some industrial wastewaters contain ingredients which inhibit the growth of protozoa and metazoa or at least of some species. Very little is known about the nature and limiting concentration of such toxic ingredients. Sludges which lack protozoa and metazoa are formed chiefly from effluents arising from coal carbonisation, chemical processes and flax-retting operations. Where biocoenoses containing protozoa and metazoa are formed, they often do not correspond with the interactions illustrated in Table 3.3. Such lightly-loaded sludges are often characterised by α- or β-mesosaprobic organisms, for example *Euplotes carinatus* in effluents from herbicide manufacture, *Cyclidium citrullus* in petroleum-containing effluents, without the other relevant members of the typical biocoenosis. The number of species is low, the number of individuals on the contrary is high. In effluents of one-sided composition the resistant polysaprobic species appear first as a rule; their powers of resistance towards toxic wastewater constituents are obviously greater than those of the mesosaprobic species.

The iron salts used in plants with simultaneous coagulation similarly bring about a reduction in the species diversity. According to Liebmann (1965) even 40 mg/l of iron results in inhibition of protozoan colonisation; relatively resistant species comprise *Paramecium caudatum*, *Podophrya fixa*, and especially *Anthophysa vegetans*.

Species selection also takes place as a result of fluctuating salt contents. With a relatively constant salt concentration, good treatment performance is possible at salinities equal to that of sea water and higher (Tables 4.14 and 4.17). In fish processing plants and in activated sludge plants on ocean-going vessels the salt concentration varies between that of fresh water and that of sea water, and for fish-packing operations sometimes even beyond. The protozoa as well as the bacteria require a period of about two weeks for adaptation to a new salt concentration. The salt water species of protozoa are not shown in the drawings in Figs 3.1–3.10.

When *running-in* activated sludge plants the organism-succession progresses through all the phase represented by the different sludge loadings in Table 3.3. In a plant with sludge stabilisation for example the sludge loading on start-up with screened primary sewage and a mixed-liquor suspended solids (MLSS) of 50 mg/l is about 3.6 kg/kg d and at 900 mg/l is only 0.2 kg/kg d. While initially only bacteria and protozoa washed in with the input are present, with increasing operating duration the polysaporobic protozoa commence their multiplication. Usually there is a flagellate

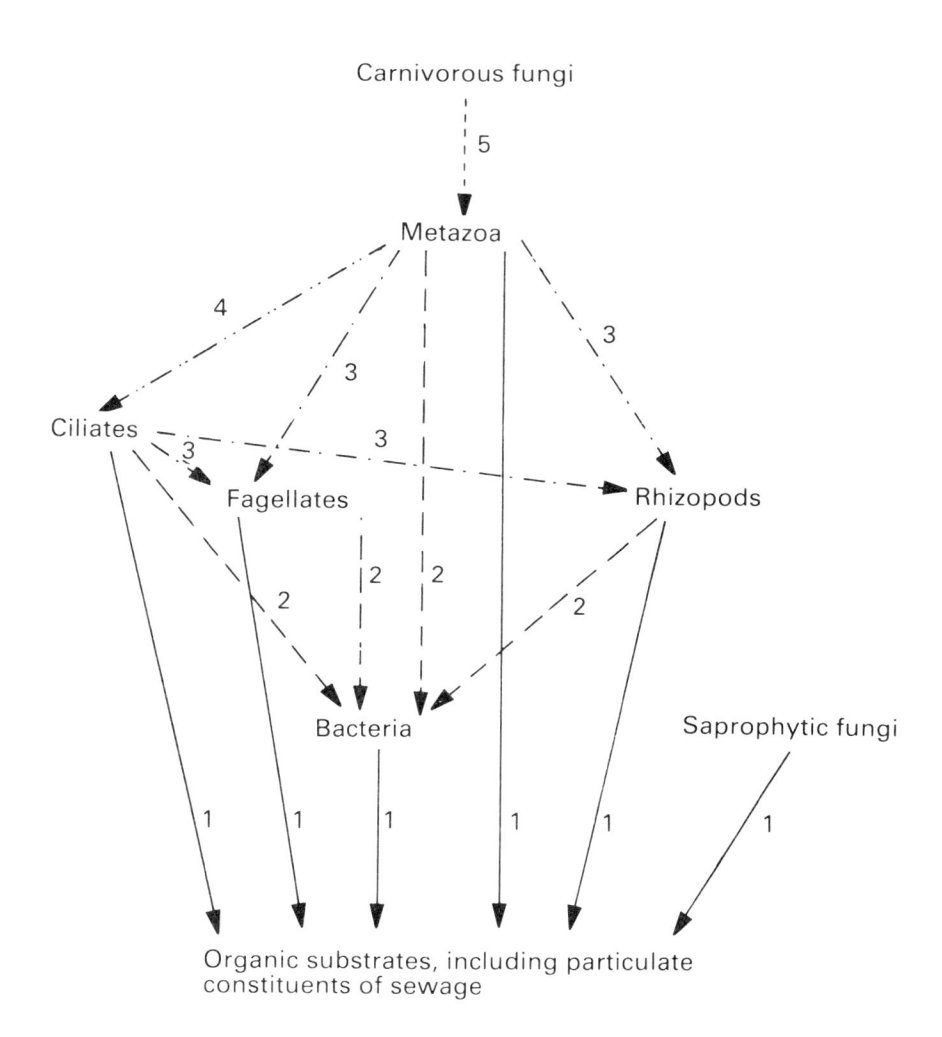

Fig. 3.11 — Simplified food chain in lightly-loaded activated sludge plants.

peak with *Bodo caudatus*, *Monas termo* and *Helkesimastix faecicola* and a subsequent peak of polysaprobic ciliates with *Vorticella microstoma*, *Glaucoma* and *Tetrahymena*. These massive growths are soon replaced by less numerous α-mesosaprobic species. As the development of these species requires a longer interval than the growth of the bacterial flocs the saprobic state of the biocoenosis during the running-in phase trails behind the diminution of the BOD_5-sludge loading. The ultimate biocoenosis becomes established after two to five weeks, depending on the sludge loading and other growth-limiting factors. Fig. 3.12 shows the rapid changeover of the dominant species during the start-up phase of an activated sludge plant.

3.7.2.3 Bulking sludge
Bulking sludge, a form of activated sludge with a sludge volume index of over 150 ml/g, occurs as a rule on account of large-scale development of *thread-forming bacteria* or *fungi*, or in exceptional cases due to the presence of bacteria with thick gelatinous envelopes of the *Zoogloea-uva* type. These organisms are characterised by a high level of bioactivity (Table 3.3) and — insofar as the thread-forming organisms are concerned — also by a high filtering action towards microflocs and solitary particles. Fibrous organisms are consequently inherently desirable as members of the activated sludge biocoenosis. The term originally employed and still often used colloquially of a 'sick sludge' is therefore totally inapposite, rather it is a question of healthy organisms with a high rate of multiplication and substrate metabolism. It is, however, unfortunate that the bacteria growing inside the gelatinous envelopes are of low density and hence exhibit poor settling properties, while the fibrous organisms possess ideal flotation characteristics and become entwined so that they can be separated from the surrounding liquid only with difficulty. Added to this, both types of organisms impede the thickening of the activated sludge in the final settling tank so that the recycled sludge cannot attain its normal solids content. The outcome is the accumulation of activated sludge within the settling tank until finally wash-out of the sludge solids with carryover into the final effluent occurs. Bulking sludge formation can also arise as a consequence of inanimate fibrous material, such as cellulose fibres, accumulating in the activated sludge as a result of too low a rate of turnover at low temperatures (Hurwitz *et al.* 1961). In the following only bulking sludge is considered which occurs as a result of thread-forming organisms and which exhibits a sludge volume index of over 150 ml/g.

Incipient bulking sludge formation manifests itself initially in a greater depth of visibility in the final settling tank as a consequence of the increased filtering action of the fibrous organisms, and also by means of increased production of surplus sludge (expressed as l/m^3 d, or m^3/d, rather less so if measured in terms of kg/m^3 d or kg/d) owing to the increased growth of the fibrous matter which effectively fluffs up the biomass. The increased sludge formation will be countered by the plant operators with an increased sludge wastage rate. As this increases the sludge loading and reduces the sludge age, the fibrous organisms no longer encounter suitable conditions for growth. If, however, the fluffing-up effect becomes greater, the final settling tank may even fill right up to the brim with sludge, and the sludge recycle rate must then be increased to the maximum. Should both these measures fail, then sludge wash-out and carryover of sludge into the final effluent occurs.

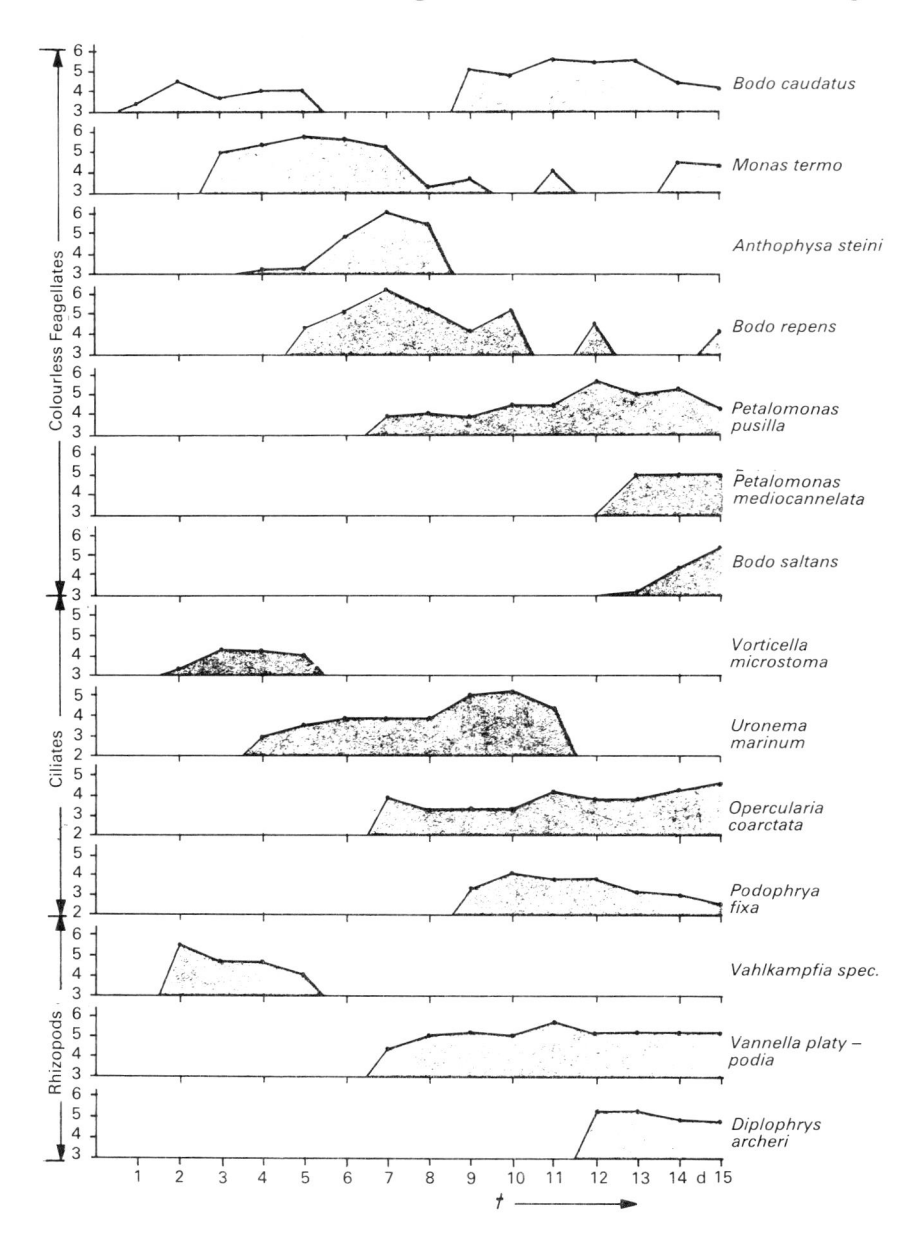

Fig. 3.12 — Microbial succession during start-up of an activated sludge plant. Municipal sewage: $B_R \sim 1$ kg/m^3 d, T $\sim$ 10°C. Ordinates: log 2 – log 6 = $10^2 - 10^6$ orgs/ml.

Roughly 50% of municipal sewage treatment plants will suffer from bulking sludge from time to time (or even continuously) with no obvious explanation in terms of sewage composition or operational changes; in the case of industrial wastewaters

Table 3.4 — Dependence of effluent quality on sludge volume index (I_{SV})

I_{SV} ml/g	BOD_5 mg/l	TS mg/l
50–105	25	15
105–135	13	8
135–145	8	5

Conditions: $B_R = 1.1$ kg/m³ d $B_{TS} = 0.62$ kg/kg d
$V_A = 0.8$ m/h   From Meissner et al. (1983)

containing a high percentage of low-molecular weight substances the percentage is even higher. Eikelboom (1975) investigated a total of roughly 200 bulking sludges and observed 26 microscopically distinguishable filiform bacteria, as well as several species of blue algae and fungi. Many of these fibrous organisms have not even received botanical names, to say nothing about our ignorance of their requirements for growth. The causes for the growth of fibrous organisms are manifold. As extreme cases the growth of *Endomyces lactis* in acidic pulp mill effluents, and *Sphaerotilus natans* in effluents from the food and beverage industries, may be cited. Both of these species are specific to the media concerned and neither can replace the other in the alternative medium. Hence a countermeasure which may be successful in suppressing bulking sludge in one case may fail entirely in another. The contradictory reports and hypotheses concerning bulking formation have been summarised by Schwägler (1980).

The projecting or free-swimming fibrous organisms associated with bulking sludge flocs possess the following physical characteristics in contrast to those occupying the denser cell aggregates.

— Each cell in the thread is in complete contact with the substrate, hence the substances can be asssimilated more readily than is the case for organisms adhering to or occluded by the flocs. In addition the surface/volume ratio of the fibrous organisms exceeds that of the organisms forming part of the flocs (Pipes 1967). Owing to these physiological and nutritional advantages the fibrous organisms can develop preferentially in lightly-loaded plants with a poor supply of nutrients and also in completely mixed continuous reactor systems, where the amount of available substrate is permanently at a low level. Fibrous organisms with a high substrate demand are favoured in a plug-flow reactor with localised oxygen deficiency, and those with low substrate demand in a stirred reactor, where they will outstrip the bacteria living within the cell aggregates.

The tendency to bulking sludge formation reported by Chudoba et al. (1974) in wastewaters containing easily degradable carbohydrates such as those from dairies, breweries, distilleries, has been confirmed in numerous cases. The starvation phases which occur near the outlet from plug-flow reactors, under conditions of discontinuous input and with segregated recycled sludge aeration facilities, restrict the development of many species of fibrous bacteria. As indicated by Schwägler (1980), bulking sludges always form with wastewaters of this kind in stirred reactors under sludge loading conditions of 0.02–1.10 kg/kg d; for a cascade system of several stirred tanks in series, if the sludge loading in the first tank was below 0.7 kg/kg d, bulking sludge formation could be safely prevented.

— The fibres have better access to dissolved oxygen, to the nutrient elements phosphorus and nitrogen, and to trace elements etc. Where these substances are in short supply or are growth-limiting, preferential conditions for growth of the fibrous organisms occur; according to Wood and Tchobanoglous (1975) for example, trace elements are of greater importance for bulking sludge formation in pulp mill effluents than carbohydrates. The phosphorus-nitrogen hypothesis for the formation of bulking sludge formulated by Wagner (1982) implies that the metal-phosphate salts formed in simultaneous coagulation systems lead to an improved nutrient supply for the floc-forming bacteria, especially for the essential phosphorus nutrient, with the result that the growth of these organisms is encouraged. Chudoba *et al.* (1974) reported that fibrous bacteria have a lower phosphorus requirement than other forms.
— The fibres are more severely exposed to toxic substances than bacteria which are protected by the flocs.
— The fibres which are anchored to flocs possessing anaerobic zones may make use of the easily metabolised substrate formed under anaerobic conditions. Many authors (e.g. Schwartz *et al.* 1980, Gassen 1980) have viewed the anaerobic interior portions of the flocs as being the principle cause of the growth of thread-forming organisms.

Owing to difficulties such as their species terminology, possible nutritional and physiological traits, and the often very uncontrollable or indeterminate substrate supply, further comprehensive investigations are required in order to be able to identify the causes of bulking sludge formation in any situation and to devise preventive measures.

Most of our experience to date has been concerned with *Sphaerotilus natans* although it is rarely responsible for bulking sludge formation in municipal sewage treatment plants. Its essential requirements have been fairly thoroughly investigated (Scheuring and Höhnl 1956). Various types of low-molecular organic compounds can act as substrates, such as organic acids, alcohols and carbohydrates, so that it occurs in activated sludge plants at dairies, breweries, fruit and vegetable processing plants, etc. In contrast to many other fibrous organisms it can withstand very low oxygen contents over long periods of time, but complete anaerobiosis leads to death.

Under severely septic conditions, such as in the final settling tank or in a preceding anaerobic treatment stage a *Thiothrix* bulking sludge may be formed. A quite different set of ambient conditions is occupied by the *Halisocomenobacter hydrossis* and *Microthrix parvicella* growths which are likewise anchored to or in the flocs. They occur particularly in lightly-loaded well-oxygenated municipal activated sludges. They find adequate substrate supplies in stirred reactors with a dissolved BOD_5 concentration of 5–10 mg/l, with production of marked bulking sludge growths. Free-swimming threads may be formed in lightly loaded activated sludge plants under polysaprobic conditions with occasionally elevated oxygen concentrations, and can give rise to extremely voluminous bulking sludges. Thread-forming fungi may occur especially in acidic pulp mill effluents, and *Thiothrix* in sulphide-containing and possibly septic effluents, leading to marked increases in the sludge volume index. According to Sarfert (1978), Type 021N is the most frequent, and *Microthrix parvicella* is the second most frequent bulking sludge-forming organism; Wagner (1982) reports that 90% of the bulking sludges are formed by only ten species.

Massive bulking sludge growths are less likely to occur in the following substances:

— in plants with separate reaeration of recycled sludge (Dautermann 1969) and discontinuous feed (Popp 1979);
— with a trickling filter plant connected upstream with no intermediate settling facilities;
— in simultaneous coagulation plants;
— in heavily-loaded activated sludge systems (B_{TS}) ≥ 0.5 kg/kg d);
— in oxygen-supplied aeration systems;
— with a high proportion of effluent derived from metal-working industries, textile processes, coal washing facilities, and blast furnaces;
— in anaerobic activated sludges;
— with the re-introduction of centrate from centrifugal sludge dewatering facilities (Schmidt 1977) or digester liquor (Kraus 1945);
— in plants without primary settling;
— in plants with preliminary denitrification as a consequence of alternating anaerobic and aerobic conditions.

Where there is a constant danger for bulking sludge formation, these possibilities should receive due consideration; on scientific and energetic grounds the last-named alternatives are the most promising.

3.7.3 Outlet from the final settling tank

The superficial layer of the attached growth is a mirror of the biological efficiency of an activated sludge plant. As the constituents of biologically treated sewage effluents serve as substrates for the growth of autotrophic and heterotrophic organisms, biofilms of a few millimetres in thickness are produced. As oxygen cannot reach the lower layers of this film, septic conditions may be re-established accompanied by a hypersaprobic state.

While the activated sludge organisms are exposed from time to time to raw sewage and also experience periodic oxygen deficiency so that only fairly resistant species are able to develop, the same does not apply in the outlet from the final settling tank. To this is added the fact that the previously oxygen-depleted or even anoxic liquid from the settling tank takes up oxygen as it cascades into the outlet duct. Owing to the effect of natural illumination, autotrophic species may proliferate, which are ordinarily indicative of good mesosaprobic conditions. In plants with sludge loadings of 0.03–0.6 kg/kg d saprobic indices of 2.3 to 3.5 are usually attained.

Despite these improved saprobic conditions the species occurring in the activated sludge are still fairly numerous as they are partially flushed out of the clarifier and exert a strong population pressure on the biocoenosis of the attached film.

The biological efficiency is usally apparent, for waters which are not strongly coloured, form the visual appearance of the growth. Where the growth is whitish, yellowish black, reddish or purplish, or possibly grey, then this is indicative of polysaprobic to hypersaprobic conditions (see Table 2.3). Mesosaprobic growths appear brownish or green; for relatively infrequent massive growths of peritrichal ciliates the colour may be bluish-white (Plate 14). Particularly characteristic are the green fibre bundles of α-mesosaprobic *Stigeoclonium tenue* and the medium-brown diatomaceous growths of *Navicula cryptocephala* and *Nitzschia palae*. The dark-green to blackish growths of the blue alga *Phormidium* in the spray zone of the ducts are of only slight indicator value, while the occurrence of several species of *Oscillatoria* in continuously wetted growths is usually indicative of good mesosaprobic conditions.

Owing to the light-dependence of the autotrophic organisms a distinct seasonal rhythm is noticeable in the attached film; in the spring and summer the autotrophicorganisms are at their peak. As a rule only those heterotrophic forms shown in Figs 3.1 to 3.10 occur in the outlet channel, while a multitude of autotrophic species may be present. New organisms or groups of organisms often appear within the submerged baffle region of the clarifier. If the water remains there for long periods, limnic biocoenoses may develop (autotrophic flagellates, algae, water fleas, water crickets, insects, etc.). The baffle itself may be used by many insects as a place for laying their eggs.

4

Biological, biochemical and biophysical processes

4.1 STRUCTURE AND FUNCTION OF IMPORTANT CELL COMPONENTS

The organisms controlling the sewage treatment process depend for their supply of carbon and energy on the substrates initially present in the sewage of those arising as the products of biochemical metabolism. They must therefore possess the capability for capturing such substances, transporting them into the interior of the cell, and subjecting them to biochemical action. Table 4.1 lists the most important cell

Table 4.1 — Bacterial cell organelles and their principal functions

Organelle	Function
Nucleus	Carrier of genetic information
Cell wall	Protective and supporting membrane
Cilia	Means of propulsion
Cell plasma	Carrier and transparent medium for cell contents
Ribosomes	Syntheses of proteins, including enzymes
Mesosomes	Membrane reservoir
Reserve substances	Energy storage
Gas vacuoles	Buoyancy control
Enzymes	Metabolic catalysts
Endoplasmic reticulum	Membrane structures for intracellular exchange processes

components, and Table 4.2 indicates the storage materials which may be produced by metabolic action.

4.1.1 The cell wall
4.1.1.1 Structure

The cell walls of the bacteria present in sewage and in activated sludge must protect the cells against adverse environmental effects, such as mechanical stress, differences in osmotic pressure within and outside the cell, harmful substances and so on, they must support the cell metabolism and as a special function in activated sludge, promote the formation of dense sludge flocs. The structure and function of the bacterial cell wall have been recently described very fully (Seltmann 1982) so that only a brief description will suffice at this point.

In principle there are three forms of cell wall construction which are distinguishable, namely: that of gram-negative bacteria (Fig. 4.1), that of the gram-positives, and that of the Archae bacteria. As the last-named have so far received little attention and are of only minor importance in sewage and activated sludge, only the first two types will be considered here.

Gram-negative and gram-positive cell walls retain their shape and stability against mechanical stress by reason of the component known as murein, also termed peptido-glycan. It consists of long polysaccharide chains which are cross-linked by means of peptide bridges. This results in a mesh-like structure of high mechanical strength. In the case of gram-negative bacterial only a few such mesh layers are superimposed on each other, while for gram-positive bacteria there may be up to forty.

Beneath the murein of both types of bacteria lies the cytoplasmic membrane. This is a rule not regarded as part of the cell wall, but is nevertheless of great importance in connection with materials transport and biosynthesis.

The principal difference between the gram-negative and gram-positive bacteria lies in the fact that the former possess an outer membrane. This is constructed very differently from normal biological membranes and contains phospholipids only in its innermost monolayer. The outer monolayer by contrast contains lipopolysaccharides as characteristic components, which make a substantial contribution to the resistance of these bacteria towards detergents, certain antibiotics and other materials. They are positioned with their lipid portion within the outer membrane, while the polysaccharide portion projects outwards. Both of these monolayers contain proteins in addition, among which are the so-called poreins, that is pore-forming proteins which are of importance for the transfer of materials through the membrane.

Between the cytoplasmic membrane and the outer membrane there is the periplasmic zone, in which the materials passing through the outer membrane are prepared for transport through the cytoplasmic membrane.

The gram-positive cell wall lacks the outer membrane and is therefore much more permeable than the gram-negative cell wall. The function of the lipopolysaccharides is assumed by the teichoic acids. Two different forms of teichoic-acids can be distinguished — those of the cell wall and those of the membrane. Both forms consist of acidic polysaccharides in which the acidic character is due to the presence of phosphate groups. The cell-wall teichoic acids are bound with one end in the murein,

Sec. 4.1]   **Structure and function of important cell components**   95

Table 4.2 — Storage components of microorganisms

Component	Main constituent
Lipid granules	Poly-B-hydroxybutyric acid
Volutin	Polyphosphate, besides RNA, lipids protein and Mg^{2+}
Sulphur granules	Polysulphides
Glycogen	Polysaccharide
Parasporal bodies	Protein
Myclin-like forms	Lipids
Starch	Polysaccharides
Metachromatic granules (Babes-Ernst granules)	Polyphosphate

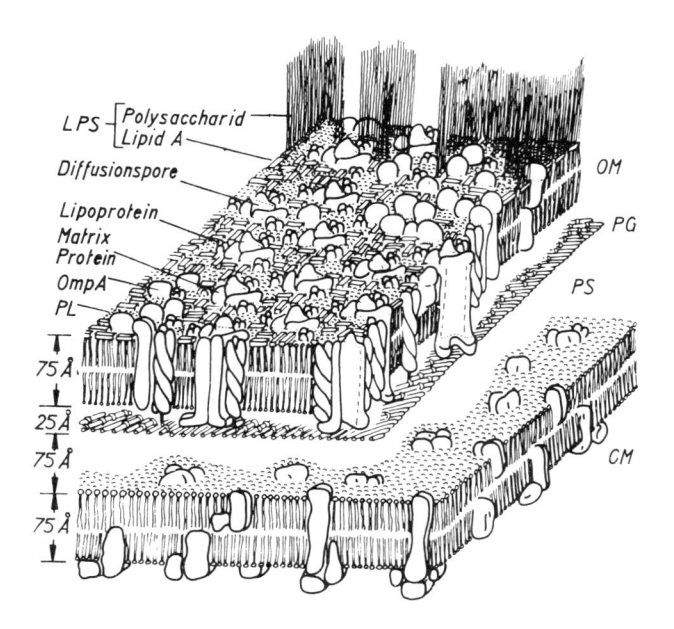

Fig. 4.1 — Model of the gram-negative cell wall. After Di Rienzo et al. (1978). LPS: lipopolysaccharide; OmpA: pore-forming membrane; PL: lipoprotein; OM: outer membrane; PG: peptidoglycan; PS: polysachharide; CM: cytoplasmic membrane. Fimbria not shown.

with the other projecting outwards. The membrane teichoic acids contain a lipid residue at one end and are thus known as lipo-teichoic acid (LTA); the lipid residue is the means by which the LTA is fixed to the cytoplasmic membrane. The LTA penetrates right through the murein and its opposite end projects outwards. The teichoic acids are evidently of great importance for the gram-positive bacteria as they can be formed even under phosphate-limiting conditions. Where phosphate is completely absent then teicho-uronic acids are formed instead, in which the acidic character is determined by the uronic acid residues. In addition to the teichoic acids, the gram-positive cell wall contains a range of proteins which are in part linked to the acidic groups of the teichoic acids.

The fimbria and the capsular material do not belong to the cell wall. The fimbria consist of thread-like structures attached to the cell wall and projecting outwards, being made up of protein subunits and responsible among other things for the floc-forming properties of the activated sludge biomass. The capsular material comprises layers of varying thickness superimposed on the cell wall and consisting usually of acidic polysaccharides. They constitute an additional barrier a a protection for the cell itself.

4.1.1.2 Materials transport through the cell wall

For the transport of water soluble materials through the cell wall there are three principal mechanisms available:

— simple diffusion;
— passive transport;
— active transport.

Simple diffusion is possible only in those cases in which the external concentration exceeds the internal concentration. The same applies to passive transport, which can only take place in the presence of a concentration gradient. The difference between it and simple diffusion lies in the fact that the substance being transported is bound to a carrier. Active transport takes place against a concentration gradient, the necessary energy being obtained from, for example, the cleavage of ATP. Carriers are also involved in the process of active transport.

As the gram-negative and gram-positive cell walls are constructed very differently, transport through them also occurs by different means.

The permeability of the gram-negative cell wall is limited by the outer membrane in the case of *E. coli* for free diffusion to uncharged molecules of molecular weight not exceeding 600–800, and for *Pseudomonas aeruginosa* up to 6000±3000. It occurs in this case by means of about 10^5 water-filled pores in each cell for *E. coli*, with a diameter of about 0.6 nm (Nikaido and Rosenberg 1981). For charged substances there are selection mechanisms in these pores which may be either inhibitory or augmentary in their effect. In the latter case it is suspected that there are conjugating proteins which assist the passage of the material through the pores. For certain substances, such as phosphate, specific pores exist or may be induced as a consequence of a deficiency state. Similarly the transport of molecules of greater size than that stated above may be effected through specific pores.

The activated sludge bacteria often have at their disposal only very low concentrations of the substances necessary for their growth, and for this reason free diffusion and passive transport are of very little significance for them. They require selective pump mechanisms in order to accumulate the respective substances within their internal organs, and hence depend on active transport.

After the penetration of substances through the outer membrane, they enter the periplasm, a space between the two membranes which can be readily visualised by means of the electron microscope. Here they are prepared for onward transport through the cytoplasmic membrane; for example, molecules which are too large may be broken down into smaller fragments.

The gram-positive cell wall, owing to the absence of the outer membrane is highly permeable, even for large molecules with a molecular weight of 10^5, so that the molecules arrive quite rapidly at the cytoplasmic membrane. Gram-positive bacteria continually secrete hydrolytic enzymes with the aid of which these large molecules (see Fig. 4.3) can be broken down far enough for passage through the cytoplasmic membrane into the interior of the cell. This transport process proceeds in a similar fashion to that across the cytoplasmic membrane of the gram-negative bacteria. The lack of an outer membrane in the gram-positive bacteria confers the benefit of being able to utilise high-molecular substances, but has the disadvantage that they are very much more sensitive to harmful substances, for example certain antibiotics, than the gram-negative forms.

For the passage of fat-soluble substances across the outer membrane of gram-negative bacteria there is a second route, besides that of transport through the water-filled pores, which is designated the hydrophobic path (Nikaido 1979). In this case the transfer takes place directly through the lipid-double layer of the membrane.

A single example will show that the conditions obtaining in a particular case are often very much more complex than has been suggested so far. For the transport of maltose in *E. coli* cells the following gene products are required (Richarme 1982):

— mal B: responsible for transport through the outer membrane;
— mal E: periplasmic conjugating protein;
— mal F and Mal K — responsible for transport through the cytoplasmic membrane;
— mal G: function unknown.

Finally there takes place, very probably due to the action of the mal E gene product and a range of chemotactic proteins, an activation of the chemotactic machinery, which controls the direction of rotation of the cilia, by means of which the bacteria propel themselves in the direction of the highest maltose concentration.

According to the foregoing, it would appear that passage of large molecules in particular through the gram-negative cell wall is virtually impossible. In opposition to this, however, is the fact that the entry of the deoxyribonucleic acids of phages and plasmids, as well as that of other high-molecular substances such as colchicine, takes place readily. However, in this case mechanisms are involved which have no connection with those of normal transport processes. Although these mechanisms may differ widely from one case to another, it can be stated that in general the substances infiltrating the cell are bound to a specific receptor and then can give rise by a variety of methods to defects in the structure and hence the permeability of the cell wall, as a result of which their entry into the cell becomes possible. A second possibility, which is also of importance for the exfiltration of macromolecules from within the gram-negative bacterial cells, is the temporary rupture and reclosure of the membrane in the course of the normal life cycle of the bacteria. Finally, a special transport mechanism from the interior to the exterior of the outer membrane has been described for the *E. coli* haemolytic agent.

4.1.1.3 Floc formation

A prerequisite of floc formation is that the surfaces of the bacterial cell walls have the necessary adherent properties. Irrespective of their special habitat, the bacteria must be in a position to cling to surfaces; a capability which they have acquired in the course of their evolution. The bond with the support must on the one hand be strong enough to prevent the bacteria from becoming dislodged, for example by mechanical forces; on the other hand the bond must only be just strong enough for it to be relaxed by the bacteria at any time without an undue expenditure of energy. The adhesion of the bacteria to the underlying surface, for example to an activated sludge floc, may be of a specific nature, via a receptor, or non-specific in nature, and it is also possible that an initial non-specific means of attachment may undergo a transition to a specific mode. The adhesion may take place as follows:

— mechanically by means of the 'sticky' acidic polysaccharides, for example capsular substances;
— electrostatic forces; the bacteria and the 'host' surface are oppositely charged;
— by chelate-forming bonds; bacteria and host surfaces having similar negative charges may be connected through the action of bivalent metal ions;
— by hydrophobic linkages.

The last three modes of linkage occur in cases of specific adhesion, but with the occurrence of sterically modified multiple bonding centres (key-in-slot situation) the attachment become very much stronger and more specific. Linkages of this kind occur frequently in nature, for example the enzyme-substrate (see Fig. 4.21) and the lectin-polysaccharide combinations.

All externally situated components of the bacterial cell may function as organs of attachment. A special role seems to be played by the fimbria, however, as they may enmesh or entwine each other. Owing to the adhering properties of the bacteria a floc may be formed by colony-like growth from a single cell, or by the superposition of bacteria of different species. The majority of activated sludge flocs in municipal sewage treatment plants consist of a mixture of species. The growth of the flocs may take place by cell-division, or by means of the aggregation of solitary bacteria, detritus, etc.

As a consequence of the relatively weak bonding energies, the adhesion of the bacteria is easily impaired by large changes in pH or by the addition of anticoagulants such as urea or detergents. Similarly substances capable of attacking the biosynthesis of cell wall components, for example sublethal doses of many antibiotics, also lead to changes in the attachment capabilities of the bacteria.

As the cell membranes of fungi in addition to hemicelluloses are largely composed of chitin, no such stable attachments like those of bacteria can be formed. Besides, the fungal mycelium has a specific growth habit, so that the pure fungal flocs tend to bulking sludge formation (Plate 2). Usually the fungal hyphae are surrounded by superimposed or enveloping bacteria. Probably this coating is facilitated by the presence of an outermost layer of polysaccharides, which has been demonstrated for some fungi.

Almost all toxins present in sewage exert their initial impact on the cell wall structures, which encourage floc formation. By the addition of floc-forming metal

4.1.2 Enzymes

Enzymes are the means of regulating the entire metabolic function of microorganisms. They constitute a form of biological catalyst which speeds up chemical reactions, or makes them possible in the first place, without actually appearing in the nett balance for the reaction. Their mode of action lies in reducing the activation energy for those chemical transformations which take place only very slowly or not at all under ordinary conditions of temperature, pressure or ionic strength.

According to Ruttloff *et al.* (1978) there exists a total of ca 12 000 different enzymes, of which from 2000 to 10 000 may be present in a single cell. The enzymes are subdivided according to their site of action as follows:

— *intracellular enzymes*: dissolved in the cellular fluid or bound to specific cell structures;
— *ectoenzymes*: attached to the cell wall and only capable of acting externally;
— *extracellular enzymes*: excreted into the surrounding medium and only active outside the cell. In activated sludge these enzymes are needed for the decomposition of particular sewage constituents (see Fig. 4.3).

The science of enzymes, or 'enzymology' has developed over the last 20–30 years into an extensive and economically important branch of biochemistry. Industrially produced enzymes are currently employed for catalysing biochemical reactions in many manufacturing processes. In addition sewage technology, especially in the area of sludge treatment [36] and cleavage of resistant organic substances, appears to offer possibilities for enzyme application.

Enzyme structure
Enzymes consist of three principal component types:

Apoenzymes — generally now referred to simply as enzymes;
coenzymes;
metallic activators.

They are composed of proteins of complex composition, and four structural levels may be distinguished.

— *Primary structure*: this is based on about 20 different amino acids which are linked together by means of peptide bonds (—CO—NH—) in a characteristic sequence which is specific for each enzyme. The total number of amino acids in a polypeptide chain of this kind can vary from 50 to several thousand.
— *Secondary structure*: the polypeptide chain which forms the primary structure is

wound in a spiral or aranged in a series of folds. There are also examples of an unordered structure.
— *Tertiary structure*: the arrangement of the molecule in space involves a spatial cross-linking of the polypeptide chain by bridging links, such as disulphide bridges, hydrogen bridges, etc. A breakdown of this tertiary structure due to, for example, inappropriate shifts in pH or heavy metal ions, may cause irreversible damage to the enzyme.
— *Quaternary structure*: the final composition of the enzyme molecule from a number of peptide chains or subunits.

One or more functional groups of certain amino acids in the enzyme molecule are mainly responsible for the catalysis of an entire reaction process; they are termed 'active centres'. At these sites the substrate becomes superimposed by means of hydrogen bridges, hydrophobic interactions, ionic bonds and/or covalent bonds. A special active group, the so-called coenzyme, is also attached to the active centre and helps to promote the catalytic activity. The structural formulae of some of the important coenzymes in sewage treatment have been reported by Uhlmann (1982). While the apoenzyme is highly specific towards certain reactions, the same coenzyme may be combined with different apoenzymes, in order to catalyse the same chemical reaction for a number of different substrates. Table 4.3 shows the metabolic

Table 4.3 — Metabolic roles of some selected coenzymes (From Reinbothe 1975)

	Coenzyme	Symbol	Substance transferred
1.	*Oxidoreductase coenzymes*		
	Nicotinamide adenine-dinucleotide	NAD	Hydrogen
	Nicotinamide adenine-dinucleotide phosphate	NADP	Hydrogen
	Flavine mononucleotide	FMN	Hydrogen
	Diphosphopyridine nucleotide	DPN	Hydrogen
2.	*Radical-transferring coenzymes*		
	Adenosine disphosphate	ADP	Phosphate (PO_4^{3-})
	Adenosine triphosphate	ATP	Phosphate (PO_4^{3-})
	Phospho-adenosine phosphosulphate	PAPS	Sulphonyl group ($-SO_3H_2$)
	Pyridoxal phosphate	PAL	Amino group $-NH$)
	S-Adenosyl-l-methionine	S-AM	Methyl group ($-CH_3$)
	Tetrahydrofolate	COF, THF, FH_4	formyl group ($-CHO$)
	Biotin	—	Carboxyl group ($-COOH$)
	Coenzyme A	CoA	Acetyle group (CH_3CO-)
	Thiamine pyrophosphate	TPP	Aldehyde group $-CHO$)

functions of some typical coenzymes. Furthermore the activity of some enzymes may be governed by the presence of certain metal salts or ions (for example Na, Ca, Mg, Co, Cu, Fe, Zn, Mn, Mo) as for example occurs with magnesium ions in the enzymatic transfer of phosphate.

Many enzyme molecules exhibit a certain flexibility in their tertiary structure so that different so-called conformation states may be adopted. By this means the arrangement of the functional groups relative to a particular substrate may be optimised. This form of optimisation is of importance in the field of sewage treatment and is known as 'induced adaptation' (see Fig. 2.5, Curve 3).

Enzyme function
Each enzyme acts in a substrate-specific and reaction-specific role. The *substrate-specificity* for some enzymes is absolute, that is only one substrate is acted on by the relevant enzyme. Most enzymes, however, exhibit a relative specificity, that is a number of related substrates of more or less structurally similar form may be acted on by one enzyme. On account of this substrate-specificity, Fischer (1894) compared the relationship between substrate and enzyme to that between a key and a lock (Fig. 4.21). However, in view of the known flexibility of the enzyme, this long-held analogy is now regarded as somewhat too rigid. The enzymes themselves seek out those substrates suited to them; the turbulence in the aeration tank has a supporting role, in that it promotes the possibilities for contact between enzyme and substrate. The *reaction specificity* consists in the fact that the substrate undergoes a specific transformation due to the action of the enzyme.

The enzyme complement of a bacterial species only provides for certain specific reactions, hence an individual species is to some extent 'pre-programmed' for the metabolism of certain substrates. With increasing numbers of species, the number of programmes increases. The reaction itself takes place by means of electron displacements at the coupling point between the substrate and the enzyme. An enzymatic reaction can be subdivided into the following partial stages:

— Formation of an enzyme-substrate complex;
— Structural repositioning to allow optimal activity;
— Conversion of the substrate to the product;
— Separation of the enzyme-product complex.

The enzymatic decomposition of a given substrate to its various end products as a rule involves the action of several different enzymes.

Enzyme synthesis
The information required for the formation of the species-specific enzymes of a cell is stored in the deoxyribonucleic acid (DNA), which is located in the cell nucleus or in the nucleus equivalent of the bacteria. From this DNA it is transferred to a transport-ribonucleic acid and carried into the cell plasma. Under the action of the ribosomes the relevant peptide chains can then be synthesised.

Each species of organisms has a specific DNA, by which its range of possible metabolic reactions is pre-programmed. The more diverse the species composition of the sludge biomass, the greater the variety of its complement of enzymes, and hence also its flexibility for the decomposition of different substrates.

Enzyme regulation
The biological control circuit of the enzymatic metabolism is constructed in a similar manner to a technical control circuit. The enzymatic control serves to maintain the dynamic situation, according to which the metabolic performance of the individual

enzymes is effectively and economically co-ordinated, surplus output of intermediates and end products is avoided and the substrate is sparingly utilised. The enzymatic control system is a biological barrier to the intensification of sewage treatment. In this context it should be noted that some microorganisms only begin to produce cetain enzymes when a particular substrate is present in the medium. One therefore differentiates between inducible and constitutive or continuously available enzymes.

Classification and nomenclature
The large number of enzymes requires a systematic organisation. Originally trivial names were employed such as pepsin, trypsin, etc., but later the practice evolved of adding the suffix '-ase' onto the name of the substrate as a means of characterising enzyme function, hence peptidase for peptide-splitting enzymes, or else involving the specific reaction type, for example transferase for enzymes concerned with the transfer of certain chemical groups. The system of classification and nomenclature currently in use divides the enzymes into six classes, which are further subdivided into subgroups, namely:

— Oxidoreductases: hydrogen and electron transporting enzymes, responsible for oxidation and reduction processes;
— transferases: transfer of a particular chemical group from one substrate to another, such as nitrate, phosphate or aldehyde groups;
— hydrolases: catalysis of hydrolytic cleavage and condensation reactions for example for C—C, C—N, ester and peptide linkages;
— lyases: coupling or decoupling of certain groups to or from a substrate with a double bond remaining, or the addition of groups to double bonds;
— isomerases: catalysis of reversible transformations between isomeric compounds;
— ligases: uniting of two molecules with simultaneous cleavage of an energy rich bond, such as formation of C—O, C—S, C—N, C—C compounds.

4.2 FUNDAMENTALS OF METABOLIC PROCESSES

All organisms are involved in a continuous exchange of materials and energy with their environment. Materials uptake occurs with the object of the maintenance and multiplication of the biomass and the securing of energy for vital processes. The endproducts of metabolism are excreted into the surrounding medium or deposited at certain sites within the cell.

Metabolic processes can be classified as:

— anabolic processes: that is the processes designed to synthesise the characteristic structures essential to life;
— catabolic processes: that is dissimilatory processes for the breakdown of organic compounds by hydrolytic changes and oxidative reactions;
— reforming processes: that is materials transformations in the direction of synthesis or breakdown. Since processes of this kind are usually masked in the sewage treatment processes they are not treated further here.

Fundamentals of metabolic processes

All three processes impinge on each other. An organic compound may act as a building block for cell synthesis, and also as the starting point for decomposition reactions. Food and body mass frequently have the same chemical structure. The energy-giving decomposition processes are a prerequisite for synthetic and reforming processes. Where the cells are furnished with an excess of nutrients, then the synthetic or anabolic processed predominate, hence for example surplus activated sludge production occurs; where nutrients are available in only limited quantities, then they must be used for the production of vital energy supplies by oxidative decomposition. According to Fig. 4.2 the utilisation of organic substrates for

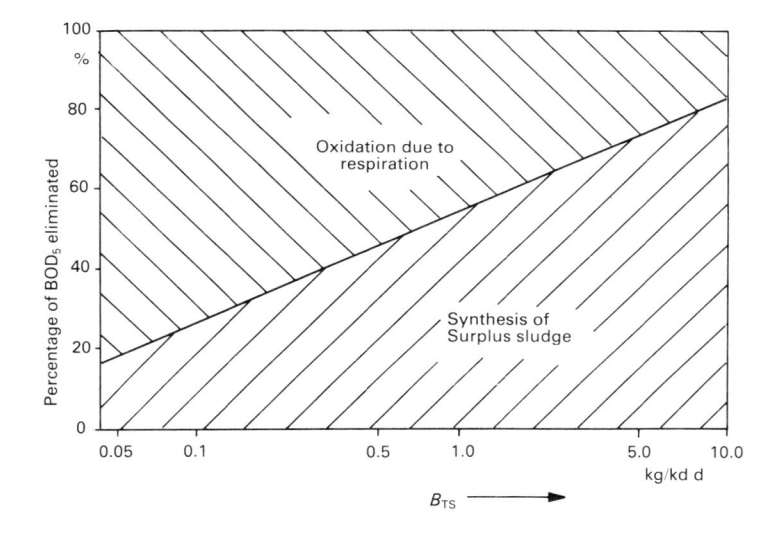

Fig. 4.2 — Ratio of synthesis to oxidation rates for BOD_5 eliminated, as a function of BOD_5-sludge loading. From Sawyer (1964), redrawn.

synthetic or decomposition reactions is bound up with the BOD_5-sludge loading, and so also is the oxygen demand for elimination of a given BOD_5 load.

The distinctions which were formerly made between constructive and up-keep materials turnover, or between energy and subsistence metabolism, can no longer be justified because they cannot be adequately distinguished from each other biochemically nor are they necessarily opposed. The widely but incorrectly used term 'decomposition' in sewage treatment technology concerns not only the the mechanical but also the biological process; in the following pages the following preferred terms will therefore be adopted:

— elimination or turnover rate instead of rate of decomposition;
— metabolisable substrate in place of degradable substrate.

4.2.1 Kinetics of enzyme-catalysed reactions

Any conversion reactions to which the organic substrates present in sewage are subjected takes place by an enzyme-catalysed pathway. This may be expressed by means of the following equation.

$$E + S \underset{K_{-1}}{\overset{K_1}{\rightleftharpoons}} ES \overset{K_2}{\rightarrow} E + P$$

where the terms have the following meanings: E = enzyme; S = substrate; ES = enzyme–substrate complex; K_1, K_{-1}, K_2 are rate constants; the rate constant K_{-1} shows that the formation of the enzyme substrate complex may be reversible; P = product.

The reaction rate is determined by the constants K_1, K_{-1}, and K_2. Michaelis and Menten (1913) established for the enzymatic hydrolyses of cane sugar that K_2 was appreciably smaller then K_1 or K_{-1} and hence was rate-determining for the complete reaction. The ration $\dfrac{K_{-1} + K_2}{K_1}$ was termed the substrate constant K_S K_S has a characteristic value for each enzyme-substrate complex (see Table 5.5). Under the conditions where $K_2 \ll K_1$ and K_{-1}, then K_S is termed the *Michaelis–Menten constant*, K_m.

The maximum reaction velocity is achieved when all the enzymes present are loaded with substrate; this is the case where the substrate availability is appreciably in excess of the number of enzyme-bonding sites. Mathematically this can be expressed by the Michaelis–Menten equation:

$$V = V_{max} \frac{S}{K_m + S} = -\frac{dS}{dt} \frac{1}{TS_{BB}}$$

where the symbols have the following meanings:

V = reaction velocity (mg/g h);
V_{max} = maximum reaction velocity (mg/g h) V_{max} is reached when $S \gg K_m$; for $S = 100\ K_m$ ther difference from V_{max} is less than 1% (see Fig. 4.4B).
K_m = Michaelis–Menten constant (mg/l). It corresponds to the substrate concentration at $V = 0.5\ V_{max}$ K_m is the most prominent symbol in enzymology; the smaller value of K_m, the more readily metabolisable the substrate;
S = substrate concentration;
TS_{BB} = activated sludge concentration in the aeration tank (mg/l or g/l).

The modifications of the Michaelis–Menten equation applicable to the activated sludge process inclusive of sludge recycle and washout rates, have been given by Uhlmann (1982) but are only valid under constant conditions.

From the ratio between substrate concentration and K_m it is possible to distinguish reactions of different orders:

— reactions of zero order: substrate concentration $\geq 100\ K_m$, $V \approx V_{max}$;

— reactions of mixed order: substrate concentration = $0.01 - 100\ K_m$;
— reactions of first order: substrate concentration $\leq 0.01\ K_m$.

The K_m or K_S values for the individual substances present in sewage are as a rule between 0.1 and 1.0 mg/l, from which it can be inferred that the reactions taking place in the aeration tank are primarily reactions of first or mixed order, since the concentrations of the individual substances rarely attan levels of 10–100 mg/l. More difficult metabolisable compounds, however, have high K_m-values, for example phenol, 2 mg/l, pentaerythritol 47 mg/l (Ilič 1977). In estimating the time for decomposition of a substance with a high K_m-value, it is to be noted that for a substrate bonded to an enzyme it is the sludge age and not the hydraulic retention time in the aeration tank that counts. The effect of particle size of the metabolisable substrate on the rate of decomposition can be seen in Fig. 4.3.

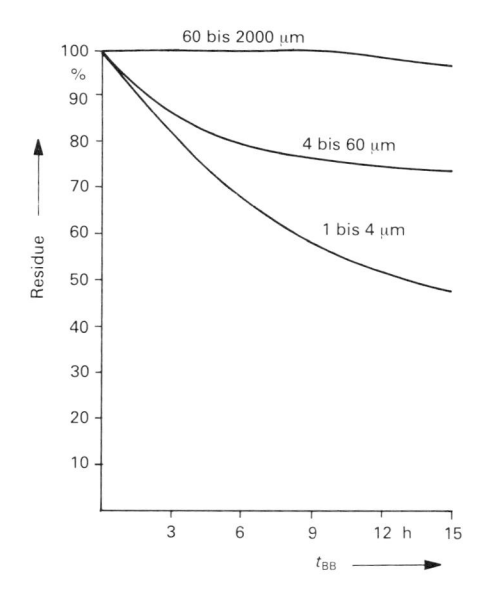

Fig. 4.3 — Dependence of metabolism of proteinaceous substances on the particle size. Substrate 150 mg/l, $TS = 3$ g/l, $T = 30°C$. From Takahasi *et al.* (1968) with modifications.

For determination of the reaction constants V_{max} and K_m one employs linearised diagrams with transformed co-ordinates, for example the method of Lineweaver and Burk (1934). For this purpose one measures the rate of turnover with varying levels of substrate concentration and plots the results using the reciprocal values as co-ordinates, as shown in Fig. 4.4C and Fig. 4.5. From the resulting straight lines, values of V_{max} and K_m can be obtained. From the concrete example of Curve 1 in Fig. 4.5 it is apparent that for municipal sewage a high conversion rate is linked to a K_m value such that low values of effluent BOD_5 are not achieved; a comparison of reaction rate

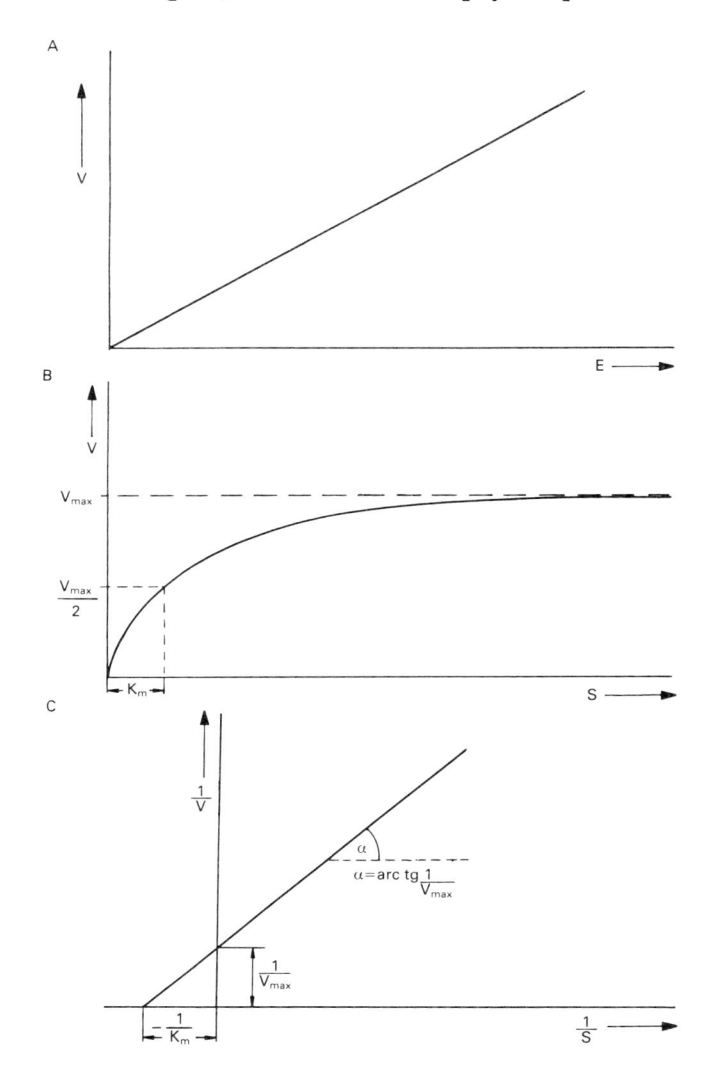

Fig. 4.4 — Kinetics of enzyme-catalysed processes. A: Dependence of reaction velocity on the enzyme concentration, that is on the active biomass; B: Graphical form of the Michaelis–Menten relationship; C: Determination of K_m and V_{max} by the method described by Lineweaver and Burk (1934).

constants for Curves 1 and 2 proves on the basis of reaction kinetics, the advantages of a two-stage activated sludge process.

From Table 5.5 and Fig. 4.6 it is apparent that different substrates have different reaction or Michaelis–Menten constants and hence also different *conversion rates*. If the mixed bacterial population of an activated sludge biomass containing several hundred bacterial species is simultaneously supplied with several substrates then all the free enzymes can combine with the relevant substrates and set about their

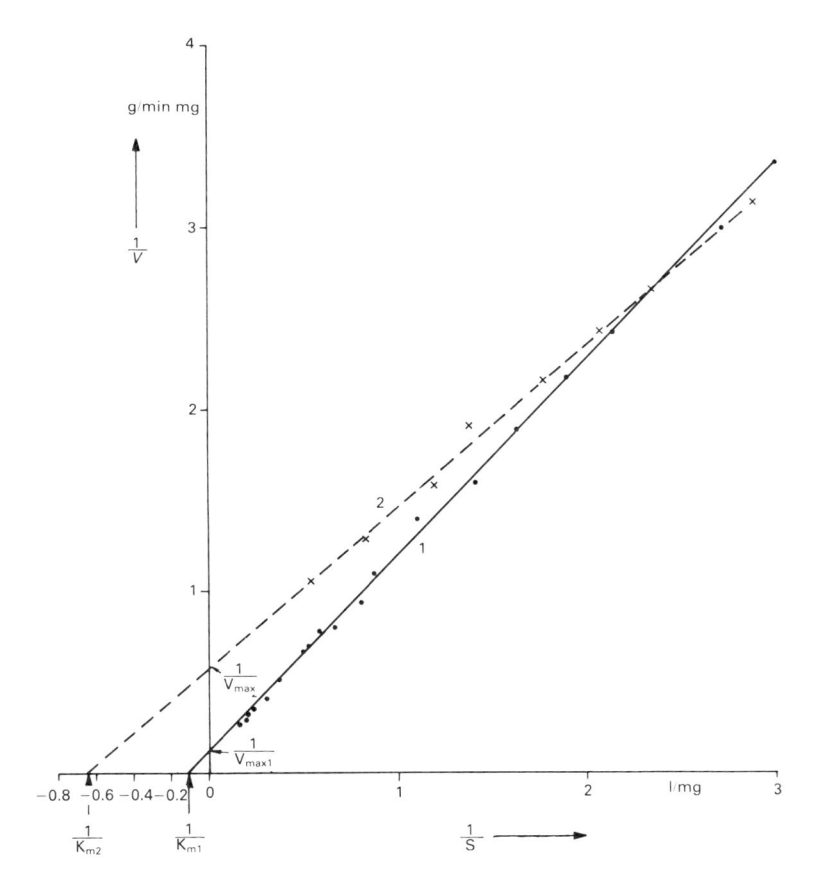

Fig. 4.5—Determination of reaction constants K_m and V_{max} from the rate of BOD_5 elimination by activated sludges in a two-stage activated sludge plant. Curve 1: B_{TS} = 8. kg/kg d; K_m 9.1 mg/l; V_{max} = 8.33 mg/g min, =12 kg/kg d; Curve 2: B_{TS} = 0.5 kg/kg d; K_m 1.52 mg/l; V_{max} = 1.67 mg/g min, = 2.4 kg/kg d. A comparison between Curves 1 and 2 demonstrates the effect of BOD_5 sludge loading on the reaction constants.

metabolic breakdown, provided no other factors such as diauxy or polyauxy are at work. The immediate combination between substrate and non-occupied enzyme sites results in the very rapid, almost 90%, elimination of organic substrates which occurs in the first few minutes in lightly-loaded stirred reactor systems. This process, often erroneously attributed to adsorption, is thus largely of a biochemical nature. The diagrammatic sketches (Fig. 6.3(A) and (B)) showing the oxygen demand and BOD_5-reduction in a plug-flow reactor demonstrates the distinction between the BOD_5-elimination and decomposition rate.

The simultaneous metabolism of the different substrates is independent of the K_m value and independent of the turnover rate of the individual substrate. The self-regulating overall conversion rate is not given by the sum of the individual conversion rates for particular substrates, but once again reflects the form of the typical

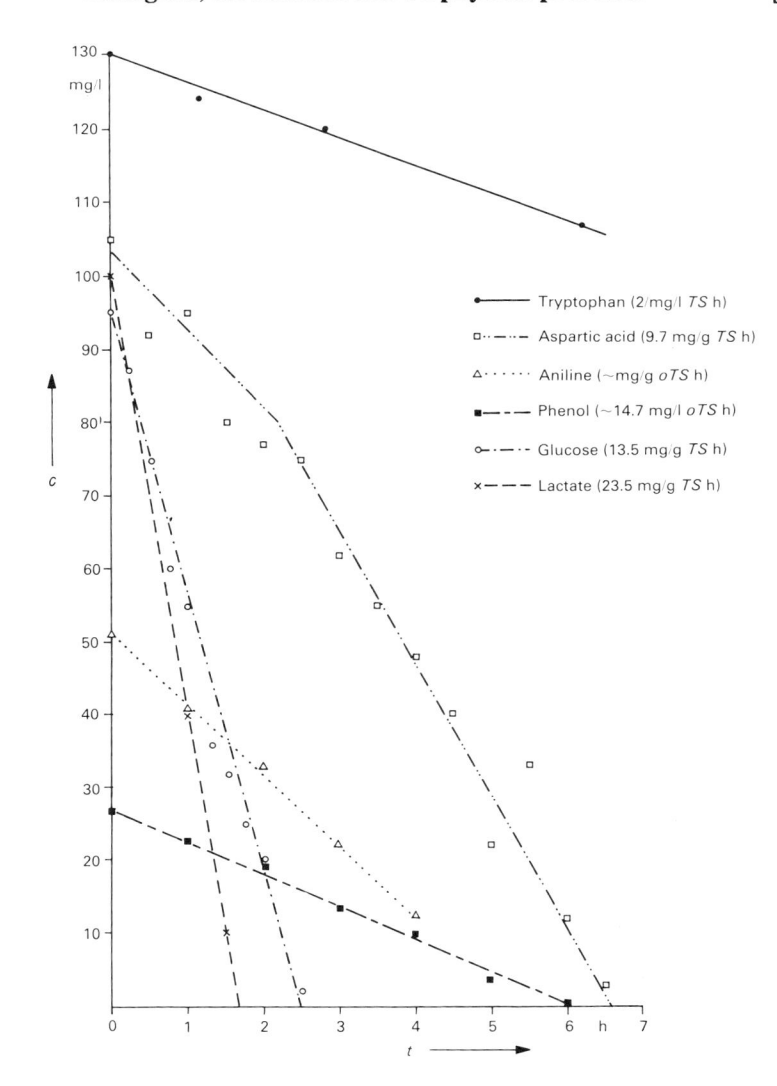

Biological, biochemical and biophysical processes [Ch. 4

Fig. 4.6 — Elimination of particular substrates following zero-order reaction kinetics. Aniline and phenol from Tischler and Eckenfelder (1968). Other substrates from Wuhrmann *et al.* (1958).

Michaelis–Menten relationship illustrated in Fig. 4.4(B). For heavily-loaded activated sludges in which the adsorption phase does not appreciably suppress the subprocesses, the curves for conversion rate may be used to indicate reactions of zero, mixed and first order (Fig. 4.7).

If the conversion rate is related to the active microorganisms then there is a direct proportionality between the enzyme concentration and conversion rate (Fig. 4.4(A)) where acclimatised organisms at a constant temperature are concerned. Where the

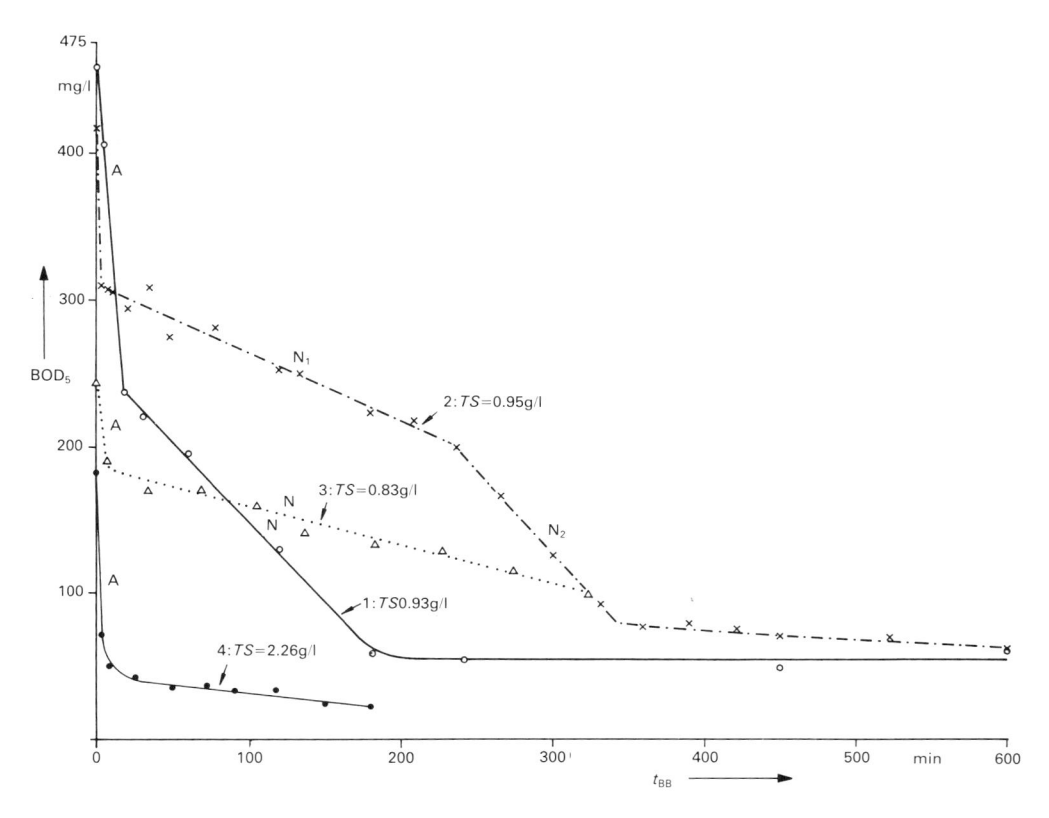

Sec. 4.2] **Fundamentals of metabolic processes** 109

Fig. 4.7 — Elimination of BOD_5 in batch tests by means of adsorptive processes (A) and also reactions of zero (N, N_1, N_2) and mixed orders.
Curve 1: sewage from a gelatin factory with sludge previously acclimatised to it. Curve 2: the same effluent as 1 but with a municipal activated sludge $V_{N_2} > V_{N_1}$ owing to induced acclimatisation. Curve 3: sewage from the exhaust air scrubbers at an animal by-products establishment (ABP) with non-acclimatised sewage sludge from an ABP. Curve 4: Municipal sewage with an appropriately acclimatised sludge biomass.

organisms are acclimatised incompletely or not at all however, then various rates of conversion may be observed (Fig. 4.8).

The substrate-enzyme ratio corresponds roughly to the *BOD_5-sludge loading* of sewage treatment technology. As will be seen from Fig. 4.9 and also in part from Table 10.1, if the BOD_5-sludge loading is increased severalfold, the BOD_5 in the treated effluent does not increase proportionately, while with a decrease in the BOD_5 sludge loading the BOD_5-elimination rate falls only relatively slowly. The causes of this behaviour lie chiefly in the change in the relative synthesis/decomposition conditions (Fig. 4.2 and 4.10). Owing to the proportionality between the enzyme or biomass content and the conversion rate (Fig. 4.4(A)) an increase in the sludge biomass reduces the time for treatment of the sewage. In sewage technology this is expressed by the *sludge age* as the product of the sludge biomass content and the aeration time. However, the sludge biomass content cannot be increased

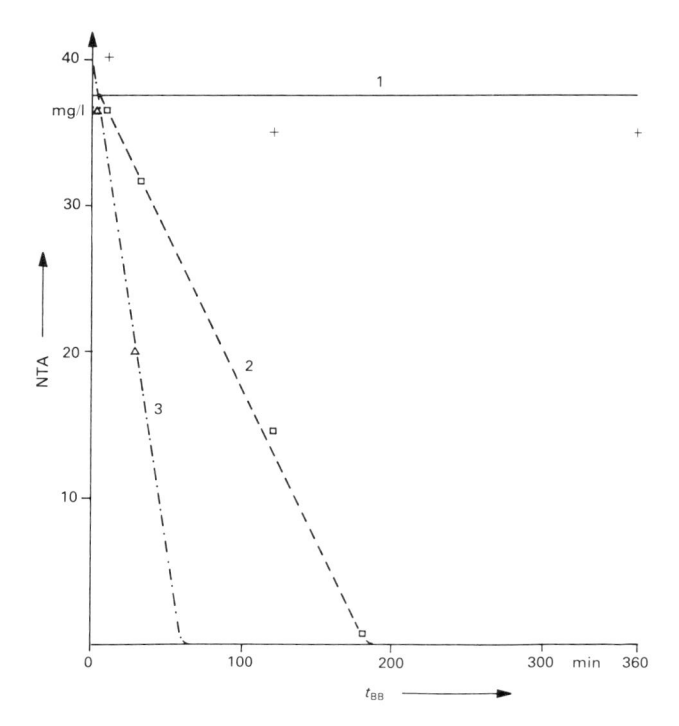

Fig. 4.8 — Effect of acclimatisation on the metabolism of nitrilotriacetic acid (NTA) a builder used in household detergents. Curve 1: non-acclimatised activated sludge $V = 0.0$ mg/g h; Curve 2: acclimatised activated sludge $V = 25.7$ mg/g h; Curve 3: acclimatised activated sludge $V = 58.8$ mg/g h; From Gudernatsch (1970) with modifications.

indefinitely on account of hindered settling at the higher sludge concentrations (see Figs 7.1–7.3). The optimum of 3.2 g/l indicated in Fig. 7.2 may fall to only 1 g/l where there is no primary settling stage. Furthermore the theoretical connection between the substrate–enzyme ratio and the sludge loading is usually not quite so pronounced in practice. For exxample, under otherwise identical conditions, an activated sludge plant with an aeration time of 4 h and a sludge biomass content (mixed-liquor suspended solids — MLSS) of 2 g/l will usually perform better, with a superior effluent quality, compared with one having an aeration time of only two hours and an MLSS content of 4 g/l. Apart from the operational reasons, this behaviour may be ascribed to the free floc surfaces, as described in the following section.

An approximate design calculation for activated sludge tanks used on experimentally-determined reaction constants can be derived from the Michaelis–Menten equation for first-order reaction kinetics:
— for plug flow reactors ($\simeq$ so-called Henri-equation)

$$t_{BB} = \frac{K_m}{V_{max}} \ln \frac{S_0}{S_t} + \frac{S_0 - S_t}{V_{max}} \text{ (h);}$$

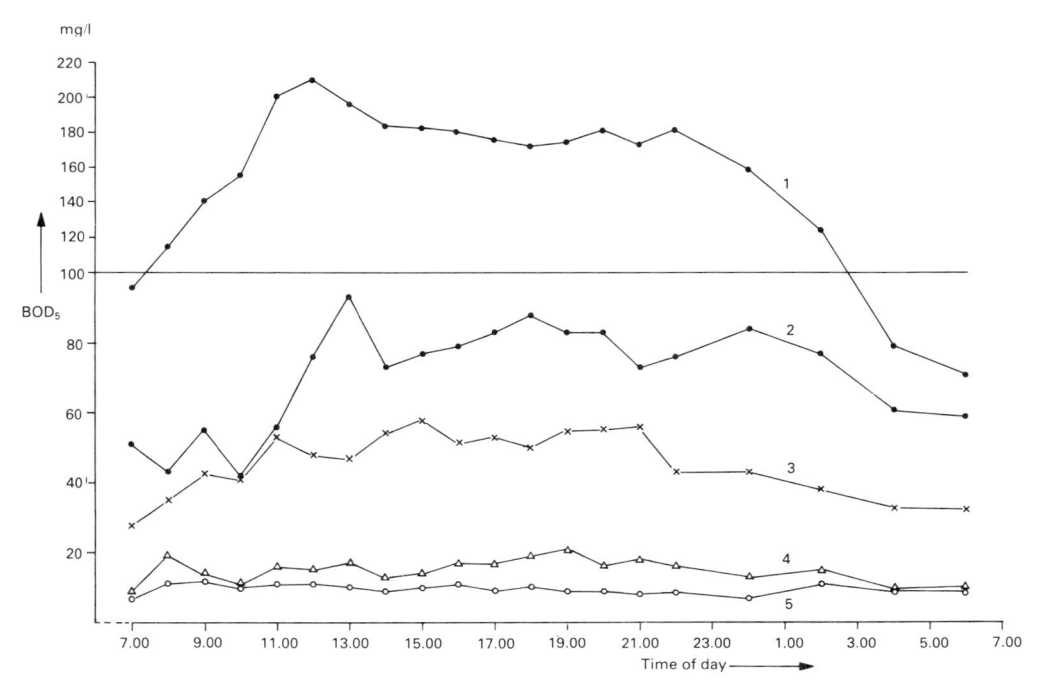

Sec. 4.2]   **Fundamentals of metabolic processes**   111

Fig 4.9 — BOD$_5$ profiles for influent and effluent at various different sludge loadings in an experimental sludge plant of pilot or full-scale dimension. Curve 1: influent x = 153 mg/l calculated mean from twelve 24-h samples at constant hydraulic loading.

Curve No.	B	Effluent concentration
2	7.4 kg/kg d;	65.5 mg/l. S = 15.1
3	2.9 kg/kg d;	44.2 mg/l. S = 9.5
4	0.8 kg/kg d;	14.4 mg/l. S = 3.4
5	0.3 kg/kg d;	9.5 mg/l. S = 1.4

Data from [2] and [31].

— for stirred reactors (see Eckenfelder 1966)

$$S_t = \frac{S_0}{1 + K \, oTS_{BB} \, t_{BB}} \quad (\text{mg/l});$$

— for plug flow reactors, as given by Hunken (1960)

$$S_t = S_K + (S_0 - S_k) \, 10^{-K \, TS_{BB} \, t_{BB}} \quad (\text{mg/l});$$

where:
S_K = non-removable fraction of substrate (mg/l), ~ about 4–8% of total substrate content of the influent;
S_0 = substrate concentration in the sewage-activated sludge mixture at the time $T \equiv 0$ (mg/l);

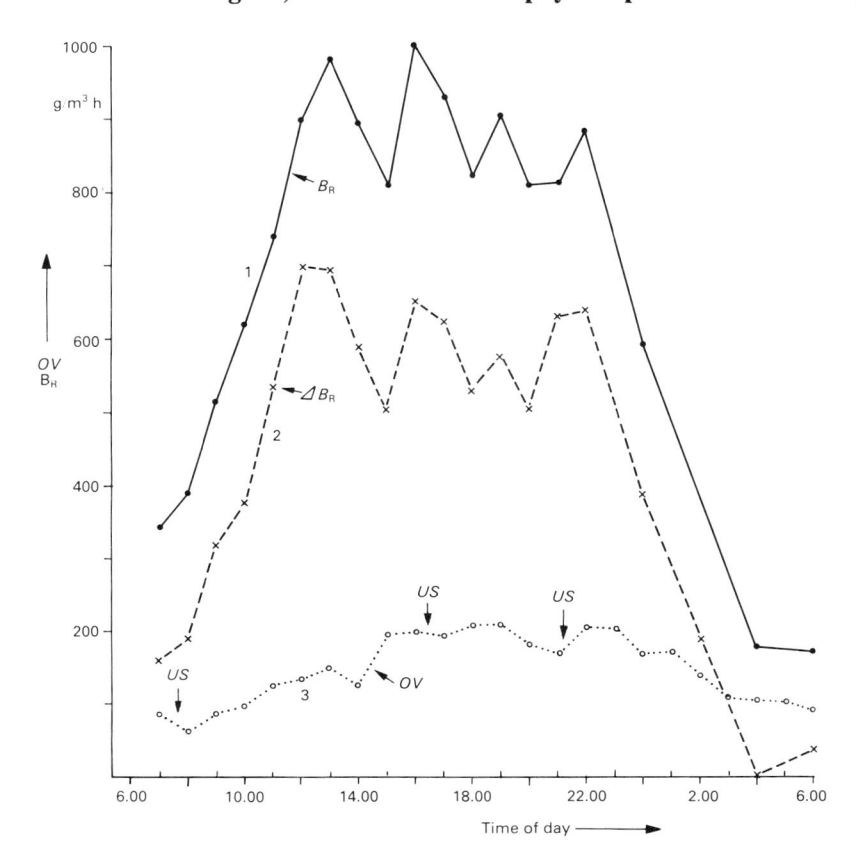

Biological, biochemical and biophysical processes [Ch. 4

Fig. 4.10 — BOD$_5$ loading and elimination and corresponding oxygen consumption figures at constant hydraulic loading. Test conditions $T = 18.5$ to $.4°C$; $TS = 2.65$ g/l. Curve 1: BOD$_5$ of influent = 685 g/m^3 h; 16.4 kg/m^3 d. Curve 2: BOD$_5$ eliminated = 448 g/m^3 h; 65.4% of input, of which 68.1% was employed for synthesis or was adsorbed. Curve 3: Respiration = 143 g/m^3 h; 31.9% of total BOD$_5$ eliminated

S_t = substrate concentration at time t;
$S_0 - S_K$ = eliminable portion of substrate (mg/l);
K = reaction constant; 0.7–0.9 l/g h in the second equation and 0.4–0.7 l/g h in the Hunken equation;
t_{BB} = aeration time (h).

Vavilin (1982) gives 14 different equations for calculation purposes. Uncertainties in the use of such equations arise from the temporal variability of the reaction constants, and from the change in the actual fraction of the flocs contributing to aerobic metabolism as a function of time and under the influence of changing levels of turbulence in the aeration tanks. Consequently more accurate knowledge based on research into these aspects is desirable.

4.2.2 Metabolic activity of the sludge flocs

The activated sludge is only partly composed of active organisms (Fig. 4.11).

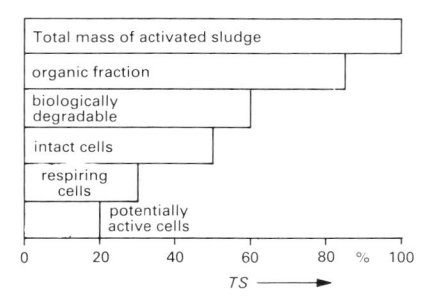

Fig. 4.11 — Diagram of fractions in an activated sludge of sludge age ten days. From Brunsfield and Phillips (1978).

According to figures given by Pike (1975) raw sewage contains 5.6×10^8 bacteria/ml and the activated sludge 1.4×10^{10}/ml or 6.2×10^8/mg TS, and hence only about 25 times the numbers present in raw sewage. If the moisture content of a bacterial cell is taken as 80%, the density as 1 g/cm^3 and the diameter 0.8 μm, then the value quoted of 6.2×10^8 bacteria/mg TS represents a weight fraction of only 3.3%.

From the oxygen demand determined in a respirometer for 3×10^7 aerobic, solitary bacteria per millilitre of municipal sewage, an OV_{max} for each bacterium of 8×10^{-11} mg/h can be calculated. With 3.5 g/l of activated sludge and a BOD_5 volume loading of 0.5 kg/m^3 d, a maximum oxygen uptake of 48 g/m^3 d can be realised. When related to the oxygen demand of a single bacterium it can be deduced that the numbers of bacteria actively respiring amounted to only 6×10^{11}/ml and hence around 1% of the total solids.

For similar respirometric studies with sodium acetate as the substrate, a maximum respiration rate for activated sludge of 32.6 mg/g h was recorded and for an effluent sample derived by paper filtration and containing 1.1×10^7 solitary bacteria a value of 8.5×10^{-11} mg/h. bacterium was obtained. Using the same values for bacterial properties as those given above, this works out at an active fraction equal to 2.1% of the total solids.

The specific respiration rates at the optimal temperature for growth of 28–30°C were stated by Schlegel (1969), for the most frequently-occurring kinds of bacteria present in activated sludge, to be:

pseudomonas: 1200 μl/mg h $\simeq$ 1715 mg/g h;
acetobacter: 1800 μl/mg h $\simeq$ 2572 mg/g h;
azotobacter: 2000 μl/mg h $\simeq$ 2858 mg/g h.

With the aid of the Arrhenius rule, respiration rates of temperatures of 10°C and 20°C may be calculated as around 430–1430 mg/g h. These values when compared with those of Fig. 4.19 and with those recorded for activated sludge flocs in the presence of acetate, confirm that the active fraction of the bacteria in the flocs under normal loading conditions (Table 10.1) only amounts to 1–3%.

Fig. 4.12 shows the decline in oxygen content in a respirometer as a consequence

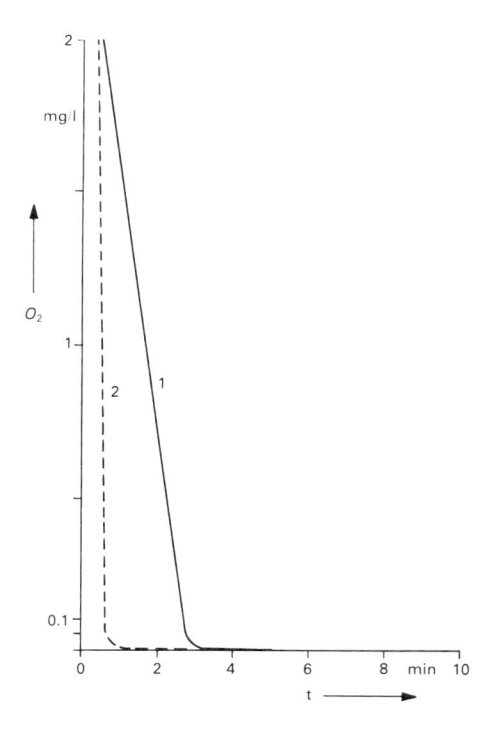

Fig. 4.12 — Oxygen profiles for respiration and chemical oxygen demand obtained using an oxygen sensor Model AM 220 with a range of 3 mg/l. Curve 1: Respiration of an activated sludge, $OV=53.1$ mg/l h; Curve 2: Reaction of oxygen with sodium sulphite in the presence of a catalysts.

of the respiration of the activated sludge, and, so that the accuracy of the O_2-selective electrode could be assessed, over the range of values resulting from the addition of sodium sulphate at concentrations up to 3 mg/l. The respiration rate curve reflects a constant rate of respiration independent of the level of dissolved oxygen down to about 0.05 mg/l; below 0.05 mg/l the inertia of the electrode interferes. From this constant rate of respiration (see also Figs 2.5 and 2.8) it may be deduced that the proportion of active biomass remains constant to well below 0.05 mg/l of O_2. Were this not so, the proportion of oxygen-consuming cells would diminish as a consequence of the decreasing extent of penetration of oxygen into the flocs in response to the reduction in the dissolved oxygen content, so that the rate of respiration would diminish and the graph would exhibit a curvilinear profile. The shape of Fig. 4.12 thus demonstrates that the oxygen hardly diffuses at all into the interior of the flocs,

Fundamentals of metabolic processes

but is consumed only by those bacteria situated on the floc surfaces or freely suspended in the liquid.

It follows that the entire inner region of the flocs must be oxygen-free and the bacteria situated there must be either inactive or undergoing some form of anaerobic metabolism. The aerobically active portion of the flocs of significance for sewage treatment can accordingly be correlated with the floc surface. For flocs of ideal spherical shape, the oxygenated portion thus would be about 60% at 10 μm, 6% at 100 μm and only 2% at 300 μm floc diameter. From 90 to 95% of the sludge volume consists of flocs of >100 μm, as long as there are no toxic effects or operating failures. The actual floc morphology, however, deviates considerably from that of a smooth sphere so that the surface area is greatly increased, for example for slab or star-shaped flocs.

Fig. 2.5, Curve 1, illustrates the dependence of respiration rate on the energy input and hence on the turbulence. Under the experimental conditions employed, there were noticeable effects up to 84 W/m^3 but hardly any further change thereafter up to 1970 W/m^3. Obviously beneath a certain size the floc becomes so stable with respect to increased turbulence that any further incease in energy input causes no further increase in activity and hence in treatment performance of the activated sludge. Facultative anaerobic activated sludge bacteria, for which the Pasteur effect applies, exhibit a higher respiration rate at low levels of turbulence than at high levels.

Hence for the sewage treatment process the mass of the activated sludge and thus also the BOD$_5$ sludge loading do not exert a controlling influence, but rather the free floc surface area and, to some extent, the solitary bacteria appear to be the decisive factors (Sladká 1975, Fig. 4.13). However, since for a constant level of turbulence the floc surface area is largely correlated with the activated sludge mass, the empirically-determined relationships between sludge loading and treatment performance are understandable. On the other hand, however, as a result of changes in morphology, the aerobic fraction of the flocs may fluctuate considerably and thus the oxygen demand, treatment effect, settling behaviour, and so on, will also be affected. Also, owing to the dependence of treatment performance on the free surface area of the bacterial flocs it is also possible to explain why, despite an adequate supply of oxygen, the treatment effect remains constant above a certain activated sludge content, or even falls somewhat: the free surface area decreases on account of the greater floc density associated with high sludge biomass contents. In the Soviet operating recommendations [15] for example, with an initial BOD$_5$ of 150 mg/l, a 13% greater aeration time is specified at 4 g/l MLSS, and a 55% longer time at 6 g/l, compared with the customary value of 3% total solids in the sludge suspension.

In elucidation of the oxygen uptake measurements cited above, a conceptual model of the oxygen transfer at the boundary of an activated sludge floc was devised (Fig. 4.14). The picture shows that the bacteria situated at the boundary layers between the flocs and the liquid experience the same oxygen and substrate contents as that in the activated sludge–sewage mixture. In addition they receive breakdown products from processes taking place in the floc interior. The underlying layer only receives oxygen and substrate from the outside at those points at which it is not concealed. This proportion of the total surface varies according to the morphology of the bacteria and their disposition; for round cells it may be from 9% to 22% of the

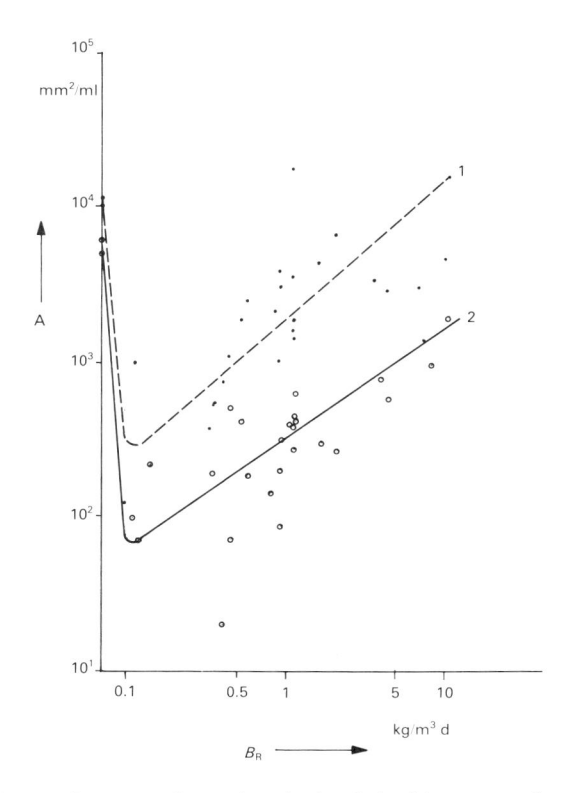

Fig 4.13 — Free surface area of organisms in the sludge biomass as a function of the BOD_5-volumetric loading. Curve 1: total surface area; Curve 2: surface area for solitary bacteria. From Sladka (1975) with modification.

total surface. The majority of these bacteria will accordingly still be capable of an aerobic metabolism, possibly utilising either nitrite or nitrate. The substrate supply is abundant as the major portion of the products of fermentation occurring in the interior becomes available. The cells in the floc interior, however, in so far as they are capable of adopting anaerobic metabolism, only carry out anaerobic fermentation processes with their cell storage components (Table 4.2). With increasing cell age however, these reserves are very soon exhausted and the production of anaerobic breakdown products ceases; the flocs are then stabilised. More heavily-loaded sludges contain high levels of breakdown products in their interiors. Some rhizopods, fungi, and obviously also some thread-forming bacteria, have become adept in the use of these breakdown products as nutrients. Furthermore it may be assumed that certain injured or dead bacteria situated within the flocs may serve as nutrients for certain specialised organisms.

There may be departures from the basic principles of the metabolism of activated sludge flocs that have been enunciated, which are determined by the various species and their particular metabolic functions. For example Krul (1976) reported that anoxically maintained pure cultures of floc-forming bacteria of the *Alcaligenes* group

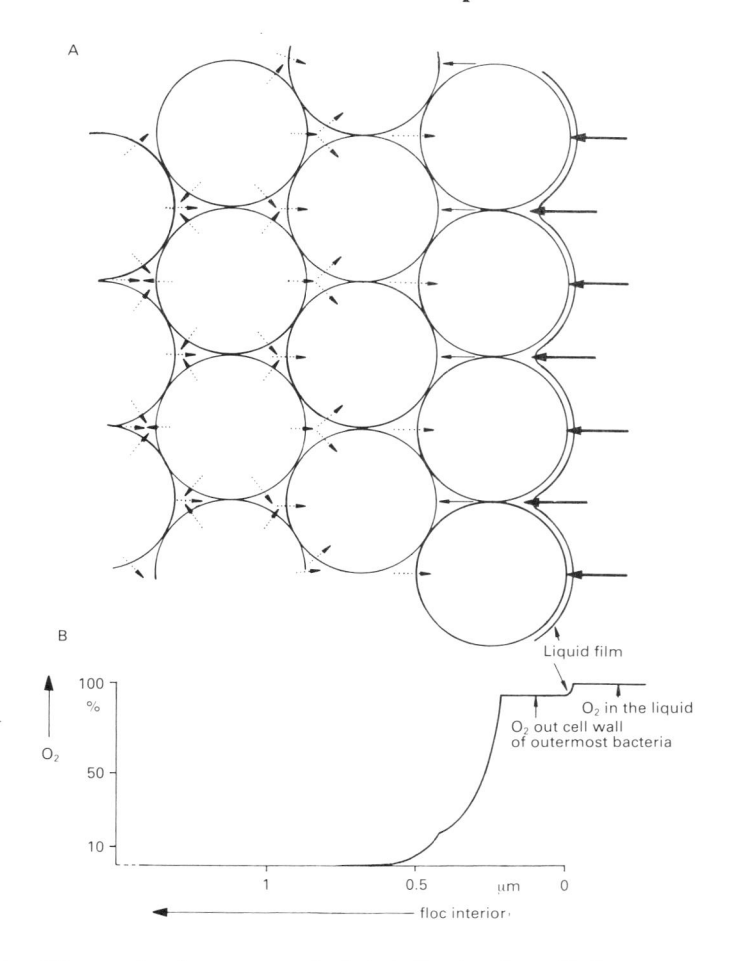

Sec. 4.2] **Fundamentals of metabolic processes** 117

Fig. 14.4 — Diagram showing oxygen and substrate flow together with the oxygen content in and around a sludge floc composed of spherical bacteria. A: Materials flow — heavy arrows: oxygen and substrate flow at the floc margin; light arows: oxygen and substrate flow following passage through the outermost layer of bacteria; dotted arrows: products of anaerobic metabolism. B: Oxygen saturation values in the vicinity of the floc margin.

show a marked decline in the respiration rate in the range from 2.5 to 0.5 mg O_2/l, in the presence of nitrite, which is explained by the inhibitory action of NO_2-reductase.

Zoogloea-like bacteria may form hollow cavities within the flocs, so that the supply of substrate and oxygen to the interior of these cavities becomes responsible. Several floc-forming bacterial species, such as *Pelodictyon luteolum*, possess the ability of opening up the structure of over-compact flocs in a sponge-like manner (Häussler 1982). Whether many activated sludge flocs are susceptible to this kind of treatment is, however, questionable. It can be reliably assumed, however, that the variable floc morphology is at least in part a direct response to the prevailing ambient conditions; thus star-shaped or branched or extensively fragmented flocs are formed in response to critical levels of dissolved oxygen. Finally, those bacteria should be mentioned which disentangle themselves from the flocs if the surrounding conditions no longer suit them.

4.2.3 Growth and surplus sludge production

As a result of the anabolic metabolism of the microorganisms, a growth of biomass and multiplication of the organisms occurs. Under favourable environmental conditions a bacterial cell can double its volume every 15 to 20 minutes and split into two. As this involves a geometrical progression for the numbers of organisms ($2^0 \rightarrow 2^1 \rightarrow 2^2 \rightarrow 2^3 \rightarrow 2^n$) theoretically 2^7 new cells can be formed from one cell in a single day. If this growth potential could be exploited in the sewage treatment context, then the length of the treatment process could be reduced to only a few minutes. This growth potential stands in contrast to the sludge ages indicated in Table 10.1.

On the basis of enzyme kinetics the growth of organisms can be represented by the Monod equation (as for the Michaelis–Menten equation, according to Peters (1976) the half-life constants for substrate decomposition and growth, K_m and K_S are identical), namely:

$$\mu = \frac{\mu_{max} S}{K_S + S} = \frac{dTS_{BB}}{dt} \frac{1}{TS_{BB}} \; (d^{-1});$$

where: μ = growth rate, per day (d^{-1});
μ_{max} = max possible growth rate (s^{-1});
S = substrate concentration (e.g. mg/l BOD_5);
K_S = substrate concentration (mg/l), at which $\mu = 0.5 \, \mu_{max}$;
TS_{BB} = solids content in the aeration tank (= MLSS) (g/l)

When applying this equation to the growth of activated sludge, certain problems arise due to the multicomponent nature of the sewage substrate and the fluctuating load S, while K_S also varies continually (Fig. 4.6), and also μ_{max} does not remain constant owing to induced acclimatisation. The equation therefore contains three variable terms. For batch experiments a good agreement with the equation can often be obtained as a result of the well-defined conditions. Fig. 4.15 shows the growth curves of the sludge volume when starting up a municipal activated sludge plant under various BOD_5 volume loading conditions. Although the sludge volume is a very inadequate measure of the amount and growth of the organisms concerned, the curves demonstrate the exponential nature of the sludge growth. Nevertheless no evidence of a 2^{72}-fold growth per day, but only a daily $2^{3.2}$-fold volume increase in the sludge biomass is apparent. The reason for this is that only those bacteria situated in the aerobic boundary layers of the flocs take part in the aerobic metabolic activity and, consequently, only those bacteria are able to exhibit a sustained high rate of growth. The maximum growth rate for activated sludge of around 40 ml/l d which can be inferred from Curves 2 and 3 may be exceeded under similar loading conditions by well-developed activated sludges if a large quantity of surplus sludge is suddenly bled off. This is accounted for by the fact that during the start-up phase an activated sludge

Fundamentals of metabolic processes

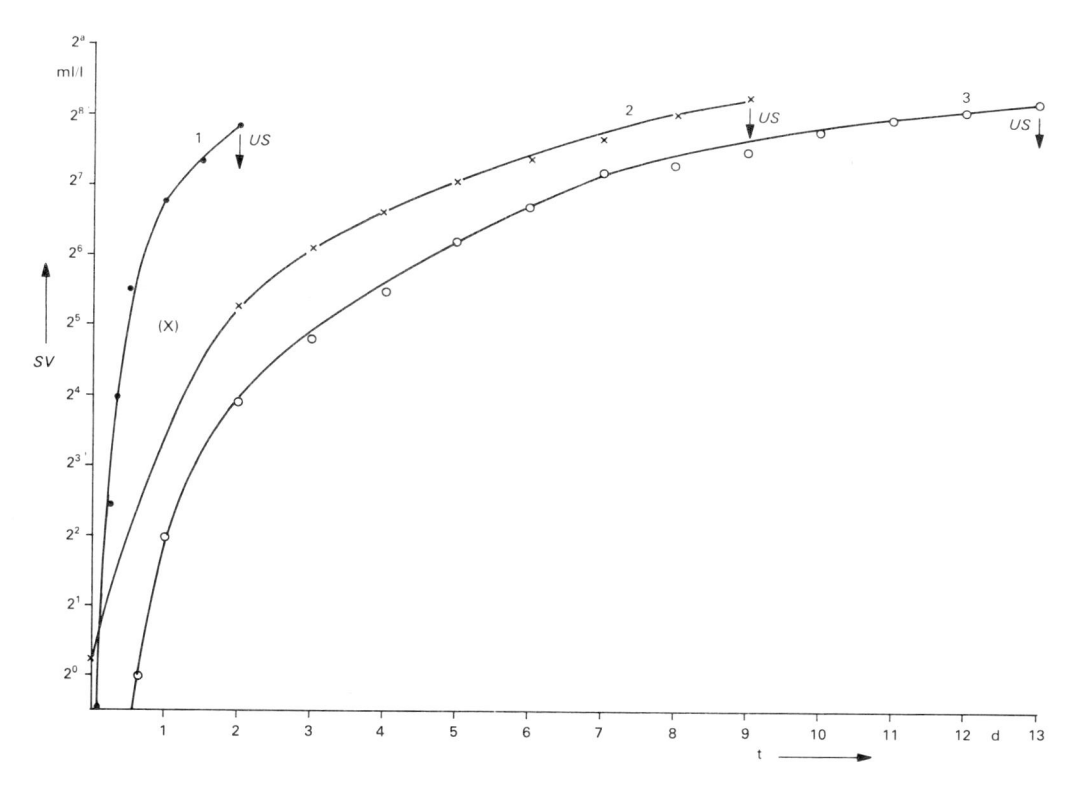

Fig. 4.15 — Growth curves of activated sludge during start-up of a municipal treatment plant with primary treated sewage. Curve 1: $B_R = 28.8$ kg/m^3 d, $\mu_{max} = 11.4$ d; Curve 2: $B_R = 1.7$ kg/m^3 d $\mu_{max} = 3.2$ d; Curve 3: $B_R = 0.7$ kg/m^3 d $\mu_{max} = 2.2$ d.

is not well adapted to the surrounding medium either in respect of its species composition (synecology) or in the respect of specific organisms (autoecology).

The theoretical growth curve of a bacterial culture and the working range of the activated sludge process are shown in Fig. 4.16 The lag-phase, that is the delay period, in the case of non-toxic municipal sewage inputs, is virtually absent. For floc formation, however, periods from several hours to a few days may be required, during which the flocculation of the organisms in the final settling tank is of crucial importance. For industrial effluents, however, the lag phase may last from a few days to several weeks. The working range of activated sludge plants lies in the region of diminishing growth, in which the organisms are obliged, as far as possible, to incorporate the available substrates into their metabolism. The lower the substrate supply is, the smaller will be the growth of the activated sludge, until finally the endogenous growth phase is entered. In this latter phase the organisms consume their cell reserve materials (Table 4.2). From there on there occurs a further phase in which the sludge volume and dry matter generally undergo a logarithmic decline (Table 4.4). As a result of fluctuating loads in activated sludge plants in general, this endogenous growth phase may occur briefly in any plant. Under prolonged con-

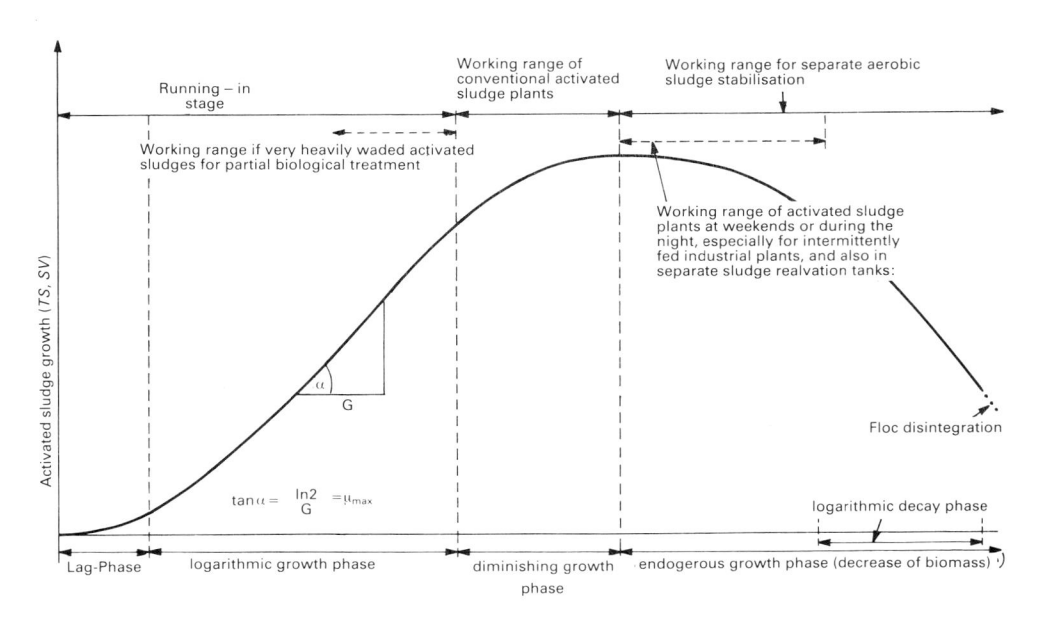

Fig. 4.16 — Working ranges of activated sludge plants as a function of the growth phases of the organisms.

Table 4.4 — Changes in sludge content in three series-connected aeration tanks. Conditions B_R 0.51 kg/m³ d) C_z=155 mg/l BOD_5 and 98 mg TS, T~16°C

	1st tank anoxic or anaerobic	2nd tank aerobic	3rd tank aerobic	Decrease tank 1 to tank 3
t_{BB} (h)	4.2	1.6	1.6	—
TS (g/l)	3.42	3.35	3.16	0.26
SV (ml/l)	335	326	288	47

ditions, however, the operation of an activated sludge plant in the endogenous growth phase is impossible, because the activated sludge flocs are broken down by self-oxidation and are washed out of the aeration tank.

As a rough estimate of the growth of the sludge biomass (≙USP in Table 1.1) a figure of 0.8 kg/kg is often assumed for municipal sewage. For industrial effluents which do not contain any ingredients favouring bulking sludge formation the value may often be as low as (0.05 or) 0.3–0.4 kg/kg. It is much more difficult to calculate the rate of waste activated sludge production in the case of those effluents which contain a high proportion of non-settleable coarse particulate solids of organic composition, such as animal slurry (Gruz 1981). Fig. 4.17 shows how the waste activated sludge production is influenced by the BOD_5 sludge loading and the

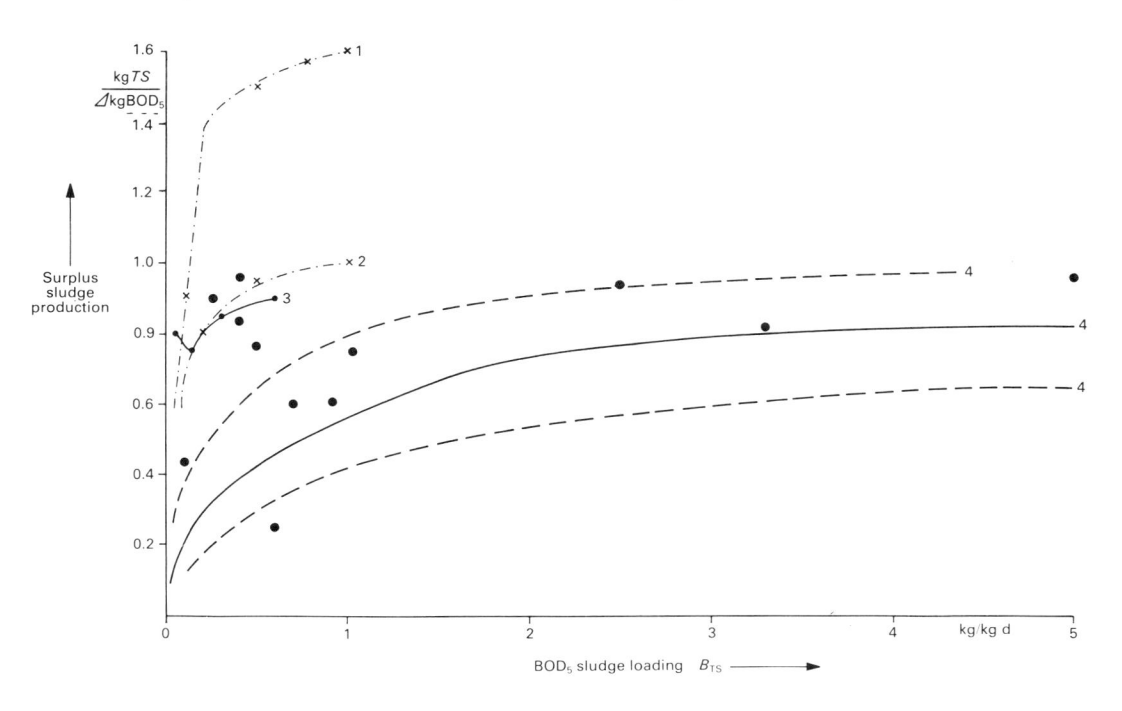

Sec. 4.2]　　　　　　　**Fundamentals of metabolic processes**　　　　　　　121

Fig. 4.17 — Surplus sludge production as a function of BOD_5 sludge loading or BOD_5 sludge degradative performance. (1) From Veits (1977) for non-settled raw sewage; (2) For settled raw sewage; (3) From [14]; (4) From Röske, Hackenberger and Uhlmann (1982).

primary settling of municipal sewage. A noticeable effect on sludge growth is also exerted by the temperature, as with increasing temperature more of the substrate will be consumed by respiration, so that less substrate is available for growth; the organic fraction of the sludge during the winter months may therefore be greater than that in the summer months by 6%–8%.

Despite a multitude of formulae for calculation purposes, an exact advance forecast of sludge production from industrial activated sludge plants is often problematical. If no comparable plants exist, then the sludge growth rate must be determined from experiments on a pilot-scale flow-through installation.

4.2.4 Respiration and oxygen consumption

The general equation for aerobic sewage treatment by biological methods, quoted in section 1.2, shows that carbon dioxide and water are formed as reaction products. The hydrogen withdrawn from the organic substrates in the course of the citric acid cycle and other metabolic reaction sequences (Fig. 4.18) is combined with pyridine nucleotide or flavine enzymes and is transferred by the process of respiration to the terminal hydrogen acceptor — oxygen or very often either nitrite or nitrate oxygen — so that the energy essential to vital processes is obtained. The carbon dioxide split off during a round of the citric acid cycle is of no significance for the energy gain of the organisms, neither does the oxygen, as generally assumed, originate from the oxygen

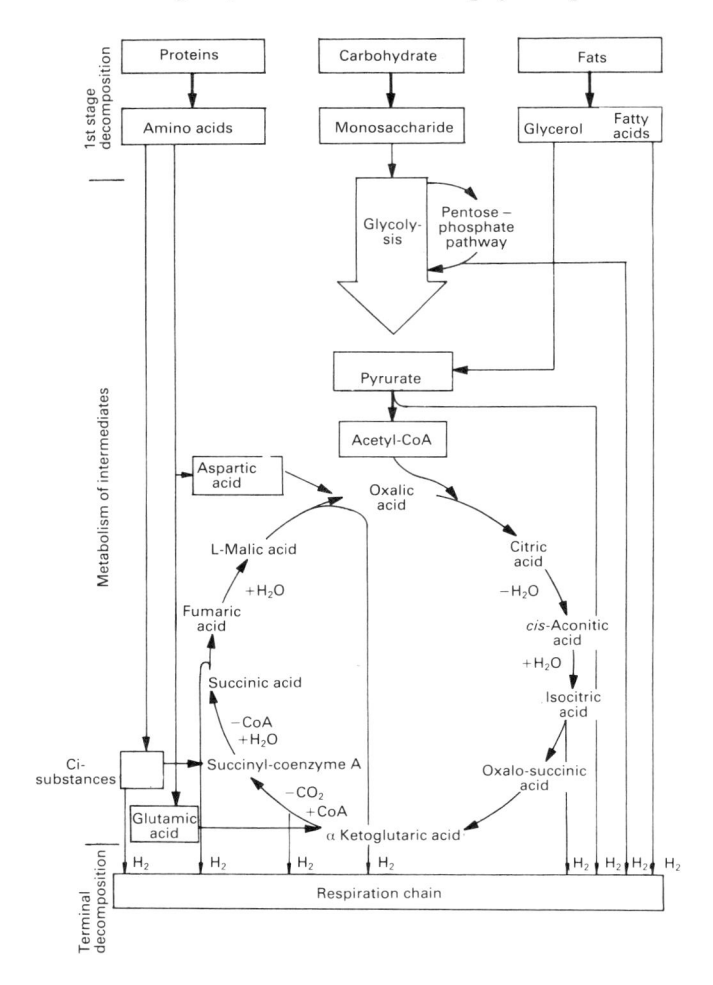

122 **Biological, biochemical and biophysical processes** [Ch. 4

Fig. 4.18 — Diagram showing the aerobic degradation pathways for important sewage constituents via the citric acid cycle and cleavage of hydrogen for production of energy in the respiration chain. From Vogel and Angermann (1972), modified.

input from the aeration system but, as will be seen in Fig. 4.18, from the water molecules. The microbial cell only uses hydrogen as an energy source. Since during one lap of the citric acid cycle two molecules of CO_2 are split off and eight hydrogen atoms prepared for respiratory use in which they react with four oxygen atoms, the overall balance gives the appearance of a carbon oxidation process. As moreover the citric acid cycle and the subsequent respiration reaction constitute the major pathway of aerobic biological elimination reactions, measurements of the ratio of carbon dioxide produced to oxygen consumed often give a molar ratio of unity. As will be discerned from Table 4.25, the CO_2 liberated largely remains in the aqueous phase and represents an almost inexhaustible carbon source for succeeding autotrophic biocoenoses.

Depending on the oxygen regime, aerobic, microaerophilic, facultatively anaerobic and anaerobic bacteria may be distinguished. The former possess a particular enzyme complement for respiratory purposes. They comprise dehydrogenases, oxidoreductases, electron-bearing proteins and an endo-oxidase. Copper and, more especially, iron are essential metallic components of the respiratory enzymes; those which incorporate iron are referred to as cytochromes.

The respiratory enzymes are arranged like pearls on a string. This arrangement provides a check on the reaction, which is ultimately a form of explosive gas reaction:

$$2H_2 + O_2 \rightarrow 2H_2O + 592 \text{ kJ}.$$

For each mole of hydrogen oxidised, three moles of ATP are formed; this so-called P/Q ratio of 3 is realised during the oxidation of $NADH_2$ (Table 4.3) but for microorganisms the ratio is usually only about 1.0. The formation of high-energy ATP — other high-energy but less important phosphate compounds are also formed — takes place according to the equation:

$$ADP + PO_4^{3-} + 74.2 \text{ kJ} \rightarrow ATP.$$

During respiration of say one mole of glucose according to the equation

$$C_6H_{12}O_6 + 6O_2 \rightarrow 6CO_2 + 6H_2O$$

the energy produced amounts to 2872 kJ.

This amount of energy, which is equivalent to 0.78 kWh, is stored in 38 moles of ATP at the rate of 75 kJ/mole. The energetic efficiency at around 40% is relatively high in comparison with technical process efficiencies. From the equation for the formation of ATP from ADP it is inferred that where phosphate is deficient only a low energy gain and hence only a limited, or much retarded, sewage treatment process can proceed. Iron deficiency, which gives rise to an inhibition of respiratory enzyme formation, has exactly the same effect.

The oxygen required for respiratory purposes is introduced into the aeration tank by the aeration equipment. However, the securing of oxygen from nitrite, nitrate and other oxygen-containing anions also belongs to the respiratory activity, as this oxygen is also utilised in the respiratory chain.

The enzymatic control of the respiration rate permits a high *utilisation rate* for the liberated energy. With an optimal substrate supply there is an upper limit to the respiration rate which lies in the range 25–60 mg/g h at temperatures below 20°C. The value of OV_{max} is temperature dependent, but is also governed by the species concerned and the BOD_5-sludge loading. The OV_{max} is achieved under conditions of excess substrate availability, when the organisms consume large amounts of substrate per unit of time in order to use it for the synthesis of biomass or to lay down cell reserves for future use in emergencies. The elimination of 1 kg BOD_5 under high sludge loading rates accordingly requires only about 0.2 kg O_2 according to Fig. 4.2, or 0.32 kg according to Fig. 4.10. Maximal respiration rates thus signify minimal oxygen consumption per unit of BOD_5 eliminated — a crucial factor in the context of the energy requirements of an activated sludge treatment plant.

In order to achieve a thorough treatment of the sewage the activated sludge biomass must, at least by the time the outlet from the aeration tank is reached, have consumed the substrate to the extent that a deficiency is created, and only very little hydrogen remains to be oxidised. In an extreme case, endogenous respiration proceeds, according to which the cell reserve materials are metabolised at an enzymatically controlled rate (Table 4.2). The reduced oxygen consumption rate under conditions of substrate deficiency permits the organisms to survive long periods of starvation (for example, during the weekend shutdown of inflow at industrial treatment facilities). The endogenous respiration is independent of the BOD_5-sludge loading and hence of the sludge age (Table 4.5). Extremely heavily-

Table 4.5 — Dependence of endogenous respiration on sludge age. From Eikhoff (1969)

Sludge age (days)	Oxygen uptake, Ov_{end} (mg/g h)
0.5	15–22
10	7–14
40	4

loaded municipal activated sludges may commence endogenous respiration at BOD_5-values of 50–100 mg/l in the aeration tank, so that there is thus no further BOD_5-reduction during the nightly input of dilute sewage (Figs 4.9 and 4.10, curve 2 between 04.00 and 06.00). Of course there may also be a respiration of the products of fermentation generated in considerable amounts under such conditions in the floc interior, these being more accessible and more readily utilised than the substrates present in the free liquid phase.

The pronounced transition from endogenous to assimilatory respiratory metabolism as a consequence of acetate addition, and to renewed endogenous respiration as a consequence of substrate deficiency, is apparent from the batch experiments illustrated in Fig. 2.7; these upsurges have so far only been observed with acetate and they are certainly occasioned by its role in the citric acid cycle. The oxygen inputs required to cover the oxygen uptake of the organisms for operational purposes are represented in Fig. 4.19.

The effects of anaerobic holding conditions for activated sludge on the respiration in a subsequent aerobic stage are indicated in Fig. 4.20(A). The increased respiratory activity arises from the breakdown of the fermentation products generated according to Fig. 4.25. As Figs 4.20(B) and (C) show, these substrates are reincorporated into the metabolic cycle in only three to five minutes under aerobic conditions; the somewhat enhanced oxygen demand levels of the recycled sludge

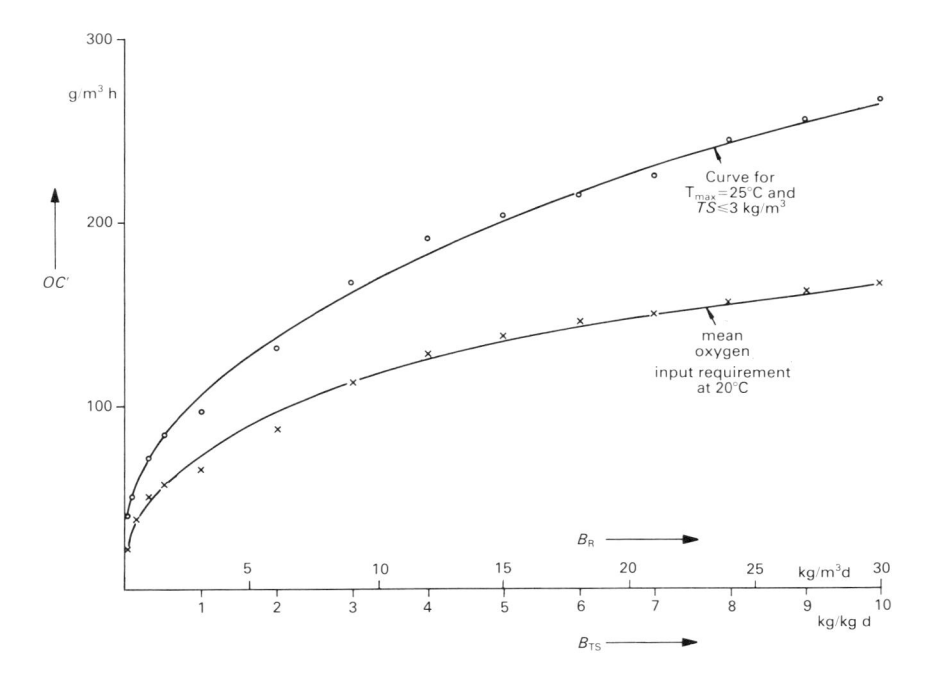

Fig. 4.19 — Oxygen requirements as a function of BOD$_5$ volumetric loading and sludge loading rates.

need not be given special consideration in calculating the oxygen uptake in the aeration tank.

The temperature has a decisive effect on the respiration of microorganisms. According to the Arrhenius Rule referred to in section 4.2.6, a rise in temperature of 10°C causes a doubling or trebling of the respiration rate. The effects of this rule on the energy demand for oxygen input are reflected in Fig. 4.24. Increasing the respiration rate under the same sludge loading conditions also results in a shift of the ratio of respired to biomass-producing substrate, so that less surplus sludge is processed.

The enzymatic control of respiration (and hence the maximal respiration rate) may be upset by injury to the enzymes involved in the respiration chain. Depending on the frequency of upsets in the operation of activated sludge plants, different effects may occur, namely:

Pasteur effect

Where a distinct oxygen deficiency occurs over a prolonged period in an activated sludge plant, then a facultatively anaerobic biocoenosis is formed. If an adequate oxygen supply is then restored, the OV_{max} is increased to two or three times the normal value. The metabolic control is probably raised as a result of competition between the respiration and fermentation processes for the available phosphate and

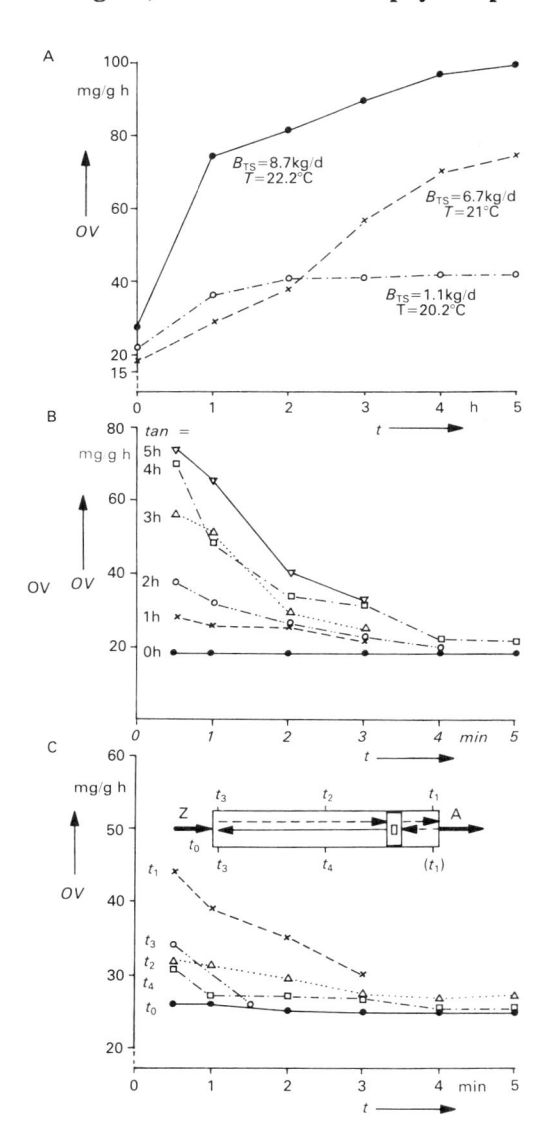

Fig. 4.20 — Changes in oxygen demand as a function of anaerobic holding. A: Effect of holding time for variable sludge loading and temperature conditions; B: Reduction of oxygen demand on re-establishment of aerobic conditions as a function of the anaerobic holding time. Curve for $B_{TS}=6.7$ kg/kg d taken from Fig. A; C: Oxygen demand and its decline in an activated sludge from a full-size final settling tank as a function of the position of the sludge removal carriage. Time of day for: $t_0=09.45$, $t_1=09.57$, $t_2=10.10$, $t_3=10.23$, $t_4=10.36$ h.

the ATP. With very heavily-loaded activated sludges, or in cases of interrupted sludge recycle, it is possible for facultatively anaerobic bacteria to develop so strongly despite an adequate supply of oxygen to the aeration tank, that this competitive respiration becomes a normal state of affairs. Under these conditions respiration rates of over 100 mg/g h may be observed at temperatures of only 20°C (Figs 4.20(A) and (B)).

Effect of toxic substances
Electron transport and ATP formation in the respiration chain may be decoupled in the presence of toxic substances so that the cell without ATP formation, that is without energy gain, continues to respire either at normal or increased rates. Toxic substances may exert various different effects, for example both 2,4-dinitrophenol and arsenate are true decoupling agents, while ATP formation and also electron transport may be actually inhibited by other toxins, for example sodium azide. Cyanide, among other substances, only inhibits electron transport. Decoupling may be either reversible or irreversible. The respiration rates may from time to time be increased by two- to -three-fold as a result. It is assumed that in the phase of acclimatisation to the toxins a non-inhibited substrate respiration may occur as a consequence of the formation of a new enzyme system.

Crabtree effect
Regression of respiration due to the introduction of glucose into the aeration tank: under certain conditions glucose may inhibit the enzymes of the citric acid cycle and the respiration chain.

4.2.5 Toxicity
From time to time, or even continuously, industrial discharges may convey substances which can interfere with the enzymatic processes in the activated sludge. Although over 500 different sewage constituents capable of adverse effects (toxic substance inventory [7]: 550 substrates) are known, it is nevertheless difficult to lay down limiting concentrations for activated sludges, because biological decomposition, acclimatisation, synergism/antagonism and mixing ratio in the aeration tank usually invalidate the toxic threshold determined under quite different conditions.

Concentrations of degradable toxic materials in the influent may frequently be allowed which are a few powers of ten greater than those in the respirometer, provided sufficient dilution occurs in the aeration tank, accompanied by adaptation and biological decomposition. Similar considerations apply to fluctuating pH values: dilution, bicarbonate buffering action and carbon dioxide production all contribute to a high buffering capacity. On the other hand they may often be a magnification of the toxic effect of a particular substance due to the presence of heavy metals, formaldehyde with benzaldehyde, phenols and thiocyanate, etc.

Reversible inhibitors
These combine with the enzymes until an equilibrium is reached, characterised by the equilibrium constant K_i. Depending on the type of reaction between the inhibitor and the enzyme, four types of inhibition may be distinguished.

Competitive inhibition
This occurs with structurally-related substrates, and hence also as a product inhibition. The enzyme becomes attached to the 'false' substrate, whereby a partial

inhibition of the catalytic function of the enzyme occurs. In Fig. 4.21 a pictorial representation of the blocking of succinic acid dehydrogenase by malonic acid in this way is presented. If in such circumstances an excess of the 'correct' substrate is introduced, the toxic action can be diminished; consequently heavily-loaded activated sludges are able to counteract such effects more rapidly than lightly loaded sludges.

Non-competitive inhibition
The inhibitor, for example heavy metal ions, attaches itself to certain groups in the enzyme which are concerned with substrate activation, even though the substrate itself is correctly positioned on the enzyme surface. In this context, complexes composed of enzyme- substrate, enzyme-inhibitor and enzyme-substrate-inhibitor may be formed.

Uncompetitive inhibition
The inhibitor can only react with the enzyme-substrate complex and not with the free enzyme.

Substrate inhibition
In cases of substrate excess, two substrate molecules may be bound by the enzyme simultaneously. A glance at Fig. 4.21 will show that the key-in-the-lock situation is also destroyed when this occurs. Fig. 4.22 shows the Lineweaver–Burke diagrams for these four types of inhibition. In the case of competitive inhibition, the maximum decomposition rate is maintained, but the K_m value is increased. For non-competitive and uncompetitive modes the V_{max} value is decreased, while the K_m value is also decreased for uncompetitive inhibition.

It may be assumed that the reversible and irreversible inhibition processes just described occur very much more often in activated sludge plants than would be apparent from a consideration of the quality of the treated effluent. Reasons for such non-recognisable toxic effects are the enzyme potential and enzyme regulation capacity by which the knocked-out enzymes essential to the metabolism of the organisms can be reformed. The blocking of enzymes by toxins also appears to be an important mechanism for the detoxification of effluent.

Toxic effects on activated sludge manifest themselves in different ways according to the strength of the toxin, its nature and duration of action, usually in the following sequence:

— a rise in the turbidity of the treated effluent;
— a reduction of nitrification;
— a reduction in the elimination of COD;
— a reduction in nitrite oxidation;
— a reduction in the elimination of BOD_5.

4.2.6 Effect of temperature

The biochemical reactions in activated sludge are affected by temperature changes in accordance with the Arrhenius equation (Reaction velocity — temperature Rule, Arrhenius or Van't Hoff Rule, RGT Rule) according to which the reaction velocity is

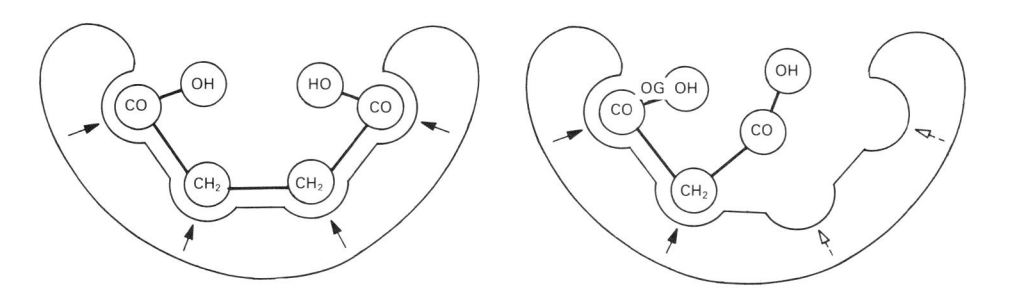

Fig. 4.21 — Diagram of mode of competitive inhibition of succinic acid dehydrogenase. The substrate (succinic acid) matches the active centres completely (left) while the malonic acid molecule (right) can only partially attach itself to the bonding centres, with the result that the enzyme is blocked. After Straub (1972), revised.

increased by from two to three times for every 10°C rise in temperature. In many cases, however, counter effects lead to a decrease in the reaction rate at high temperatures.

For the biocoenosis of activated sludsge the empirical relationships of Streeter and Phelps (1925) are usually regarded as valid:

$$\frac{t_2}{t_1} = \frac{K_2}{K_1} = e^{c(\theta_2 - \theta_1)} \text{ and } e^c = \frac{K_2}{K_1} \text{ for } \theta_2 - \theta_1 = 1°C;$$

where: K_1, K_2 = reaction rates at T_1 and T_2 (mg/g h);
t_1, t_2 = reaction time (h);
θ_1, θ_2 = temperature (°C);
c = constant.

In practice c is usually assigned the value 4.7% per °C (Fair and Geyer 1961); for a temperature rise of 10°C the factor after compound reckoning works out at 8.3% per °C. Peters (1978) quotes published values of 2.6%–13.5% per °C so that the value of 4.7% per °C usually adopted for sewage treatment appears to be purely arbitrary. The same author reports that temperature also affects the K_m value. For *Escherichia coli* she quotes the following changes in K_m value for a rise in temperature of 10°C: glucose 1.4, fructose, 1.3, pyruvate 1.6, acetate 1.2.

For nitrification Ilič (1977) sites the following K_m values: 2.0 mg/l at 20°C and 1.5 mg/l at 8°C expressed as NH_4^+–N. The temperature coefficients are accordingly substrate-dependent and not constants peculiar to the organisms. The substrate dependence arises as a result of the temperature dependence of the enzyme activity, which may range from a 1.2 to a four-fold increase for a rise of 10°C (Ruttlof *et al.* 1978).

The bacteria may be classified into three groups according to their temperature tolerance:

psychrophilic bacteria   0–20°C;

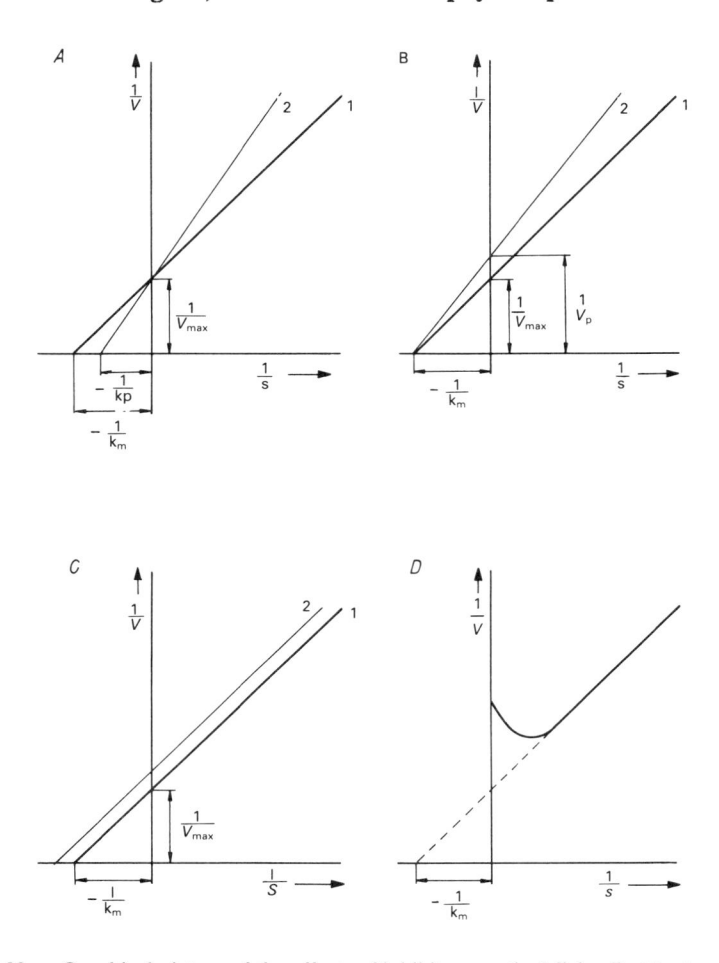

Fig. 4.22 — Graphical picture of the effects of inhibitors on the Michaelis–Menten constants and the reaction velocity. Curve 1: no inhibitor; Curve 2: with inhibitor. A: competitive inhibition; B: non-competitive inhibition; C: uncompetitive inhibition; D: substrate inhibition. From Ruttloff et al. (1978) revised.

mesophilic bacteria 13–46°C;
thermophilic bacteria 42–69°C.

Under the climatic conditions prevailing in central Europe municipal activated sludge plants almost always operate in the psychrophilic range, but many industrial systems are in the mesophilic or possibly even the thermophilic range. Physiological adaptation in the psychrophilic range normally only takes a few days but the acclimatisation of psychrophilic biocoenoses to mesophilic conditions may take weeks or even months. At temperatures of about 35°C the vital processes of many organisms come to a halt, for example the nitrifying bacteria. In municipal treatment plants the temperature changes normally follow a seasonal pattern so that there is

Sec. 4.2] **Fundamentals of metabolic processes** 131

almost always sufficient time for acclimatisation to take place. For relatively warm wastewaters and combined sewerage networks, there may be temperature-related drops in treatment performance during the onset of the snowmelt period, and in the case of industrial mesophilic sludges, when production ceases at weekends during frosty weather.

Comments in the literature on the temperature dependence of reaction rates for special substrates are relatively infrequent, probably because there are only a few substrates for which the enzyme system of cold water bacteria is unsuited. About 96% of the Earth's water reserves have a temperature of about 4°C, hence low temperature-adapted enzyme systems predominate. Eden et al. (1972) reported that for nitrilotriacetic acid (NTA) decomposition under otherwise identical conditions was complete at 20°C, incomplete at 10°C and practically nil at 5°C. Fig. 4.23 shows

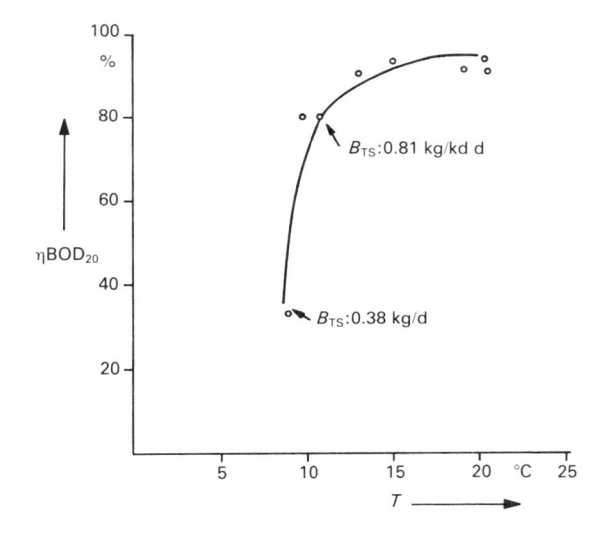

Fig. 4.23 — Dependence of the BO_{20} elimination on temperature for pharmaceutical wastewater. After Stikute et al. (1979) revised.

the marked effect of temperature on the elimination of BOD_{20} from effluents discharged by the pharmaceutical industry. Temperature effects on the oxygen demand, denitrification, aerobic sludge stabilisation and energy requirement can also be seen in Tables 4.10 and 11.1, as well as in Fig. 4.24.

Published reports almost invariably indicate that BOD_5 elimination, particularly for municipal sewage, does not follow the rule. The comprehensive set of data presented by Pöpel (1971) for example cites the following effluent BOD_5 values for treated municipal sewage:

Dec–Feb   17.5 mg/l BOD_5;
Mar–May   18.6 mg/l BOD_5;
Jun–Aug   16.6 mg/l BOD_5;
Sep–Nov   17.4 mg/l BOD_5.

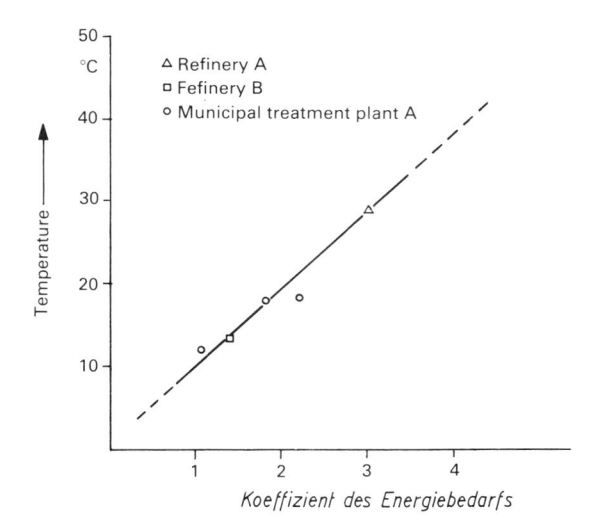

132   Biological, biochemical and biophysical processes   [Ch. 4

Fig. 4.4 — Dependence of energy requirement on water temperature as a consequence of the temperature-dependent respiration rate of the biomass. From Grutsch (1980) revised.

Where there are no secondary influences such as oxygen deficiency at elevated temperatures, or bulking sludge, our own observations confirm the absence of any pronounced temperature-dependent fluctuations in the BOD or COD of treated municipal effluents. The causes undoubtedly lie in the fact that besides the physiological and ecological adaptation of the organisms, the reserves existing in the BOD_5 sludge loading can be better utilised in the formation of surplus sludge biomass. (see Fig. 4.8, curves 4 and 5, where raising the sludge loading by 266% only increases the effluent BOD_5 by 5 mg/l).

As recent research shows however, it is also not beyond the bounds of possibility that higher temperatures prevail within the cells of mesophilic microorganisms than in the surrounding liquid.

In *oxidation ditches* relatively large temperature fluctuations may occur by reason of the long retention times and the large surface:volume ratio for the liquid. Kienzle (1974) for temperatures of < 10°C quotes effluent BOD values of 10–25 mg/l, and for temperatures > 10°C, below 10 mg/l BOD_5. The reduction in BOD_5 elimination at low temperatures occurs as a consequence of carryover of microorganisms from the settling compartment due to impaired floc-forming behaviour in the biomass at low temperatures. Also temperature shocks over 5°C, as a result of the entry of snow or meltwater in combined sewer systems, can cause deflocculation. At very low temperatures it should also be remembered that the settling rate of the sludge biomass in the final settling compartment diminishes on account of the increased kinematic viscosity of the water. With hydraulically fully-loaded settling tanks, the sludge content must then be reduced somewhat.

Despite the enzymatic control of energy gain, about 40% of the energy acquired from respiration is lost as heat (Uhlmann 1982). From the respiration equation given in section 4.2.4, it is possible to calculate the heating of the liquid as a consequence of the heat loss, based on the exothermic biochemical processes occurring during BOD_5 decomposition (not elimination!) and from respiration of the sludge biomass.

For ΔBOD_5:   100 mg/l → 0.17°C/l;
                      10 000 mg/l → 17°C/l.
For OV: 15 mg/l h → 0.0255°C/l h;
                      150 mg/l h → 0.255°C/l h.

For warm, highly-concentrated industrial effluents an inhibition of the metabolic performance of the organisms may occur as a result of this heating effect, especially in the transition range from mesophilic to thermophilic conditions, so that the treatment performance of the plant is reduced.

A programme for the pocket calculator Model K 1002 was devised by Bernsen (1983) for heat balance calculations in small activated sludge plants affected by low night-time flows, and also for industrial plants subject to the effect of severe frosts during weekend shut-down periods.

The heat transfer number K_1 which denotes the rate of heat loss or heat gain by the liquid for temperature differences between air and water is usually in the range 50–75 kJ/m³°C h, for concrete tanks sunk into the ground. The effect of the aeration system on the heat balance is closely linked to the geometry, the material of construction, and the insulation of the aeration tank, as well as the length and the embedding of compressed air pipes in the ground. Plants especially at risk during periods of severe frosts are those such as oxidation ditches where the surface/volume relationships impair the heat balance, and those employing compressed air systems in which high heat losses may occur between the blowers and the aeration tank, such that the heat of compression and friction is dissipated before entry into the liquid.

4.3 MATERIALS CONVERSION PROCESSES

At the present time over four million, chiefly organic, chemical compounds are known. These comprise on the one hand natural substances which are used in industry, trade and domestic premises, and are understandably metabolised by human and animal organisms, and are thus transformed into new compounds. These organic substrates, if one overlooks the deposits of coal and petroleum, have been almost completely mineralised for periods of thousands of millions of years. Nature has provided us with a system of organisms which are capable of decomposing organic compounds into their mineral constituents as a basis for the further synthesis of biomass. These substances of natural origin are by no means confined to substrates of simple molecular structure such as carbohydrates and fats. In fact all those substances which contribute to 'life' can find their way into sewage, including those with the most complex organic structures. On the other hand there are the synthetic organic compounds which are often derivatives of natural products, but also include others of complicated structure formed under extreme conditions foreign to those in nature. Every year about 1 000 000 new compounds are developed by the chemical industry.

Table 4.6 — Bacterial families frequently encountered in activated sludge, together with some of their preferred nutrients and specific metabolic functions. Numerical classification as given in [16]

	Bacterial order	Principal genera in activated sludge	Preferred nutrients and specific metabolic transformations
1	Phototrophic bacteria	*Rhodospirillum*	fatty acids, intermediate products of the citric acid cycle, ethanol, amino acids, fructose
2	Gliding bacteria	*Leucothrix* *Thiothrix* *Beggiatoa*	Many simple organic substrates Sulphide Sulphide, sulphide, acetate
3	Sheathed bacteria	*Sphaerotilus* *Streptothrix*	Alcohols, fatty acids, sugar, aminio-acids, gelatin, peptone Carbohydrate, gelatin, peptone
4	Budding bacteria		
5	Spirochaebes		
6	Spiral bacteria	*Spirillum*	Organic acids, amino acids, alcohols, peptone
7	Gram-negative aerobic rods and cocci	*Psuedomonas*	>100 substrates: carbohydrates, amino acids, fatty acids: denitrification, petroleum products
		Zoogloea	Carbohydrate, ethanol, amino acids
		Azotobacter	Carbohydrate alcohols, organic acids, N_2 fixation from the atmosphere
8	Gram-negative facultatively anaerobic rods	*Escherichia*	Carbohydrate, acetate
		Salmonella	Carbohydrate, citrate, gluconate, cyanide; denitrification
		Enterobacter	Carbohydrate, gelatin, organic acids
		Flavobacterium	Carbohydrate, organic acids, cellulose, clutin, starch casem

9	Gram-negative anaerobic bacteria	*Desulfovibrio*	Lactate, pyruvate, malate; reduction of sulphate and other reducible S-compounds to H_S
10	Gram-negative cocci and coccoid bacteria	*Acinetobacter*	Carbohydrate, cyanide, acetate; storage of phosphate
11	Gram-negative anaerobic cocci		
12	Gram-negative chemo-lithotrophic bacteria	*Nitrobacter* *Nitrosomonas* *Thiobacillus*	Oxidation of nitrite to nitrate with CO_2 binding Oxidation of ammonium to nitrite with CO_2 binding Oxidation of sulphides, sulphur, thiosulphates, polythiomates sulphiles to sulphate with CO_2 binding
13	Methane bacteria	*Methanobacterium*	Acetate, formate, methanol; reduction of CO_2 to methane
14	Gram-positive cocci	*Micrococcus* *Staphylococcus*	Pyruvate acetate, lactate, carbohydrate Many carbohydrates; formation of extracellular enzymes and toxins
		Streptococcus	Carbohydrate, amino acids, proteins, fatty acids
		Sarcina	Carbohydrate, amino acids
15	Endospore-forming rods and cocci	*Bacillus*	Carbohydrate, polysaccharide, pectin, gelatin, poly-peptides; denitrification
		Clostridium	Carbohydrate, polyalcohols, organic acids, proteins, amino-acids
16	Gram-positive		
17	Actinomycetes and related organisms	*Arthrobacter*	Carbohydrate, peptone, fatty acids, aromatics (inc. hetero-cyclics) amino acids; denitrification
		Nocardia	Carbohydrate, phenols, paraffins, urea, amino-acids, cellulose, keratin, alcohols, benzidine denitrification, N_2 fixation from the atmosphere

The biological metabolism of organic substrates is rendered possible by an immense variety of enzymatic decomposition and conversion processes. It is usually only a question of time before a population of organisms emerges capable of effecting a particular decomposition process. The human being concerned with rationality and impelled by economic forces would, however, like to be able to 'decompose' the constituents of sewage in a few hours or even minutes. The organisms are, however, not designed for this purpose, because their metabolic reaction systems have been developed in response to the conditions to which they have been subjected for billions of years. The pre-programmed decomposition pathway for the metabolism of organic substances embodied in the deoxyribonucleic acid of the organisms imposes certain conditions which must be fulfilled, but which man is nevertheless capable of optimising.

Chemists generally subdivide the vast range of organic compounds into four principal groups.

— *Aliphatic or acyclic compounds*: sugars, organic acids, ketones, aldehydes, alcohols, etc.
— *Carbocyclic compounds*: the basic structural unit consists of the benzene ring.
— *Heterocyclic compounds*: rings which contain C or N atoms.
— *Proteins*: basic units are the amino acids.

Biological studies of metabolism usually distinguish between the metabolism of carbon and hydrogen, and of nitrogen and sulphur. On grounds of expediency the same system of classification will be adopted here.

Concerning the metabolic processes applicable to activated sludge, there are a number of penetrating accounts such as those of Reinbothe (1975), Aurich and Weide (1975), and for some special classes of materials, Fush (1961). Table 4.6 lists some of the bacterial forms common to activated sludge and their metabolic capabilities.

4.3.1 Aerobic metabolism of carbon and hydrogen

The basic scheme of aerobic degradation of organic substrates can be seen from Fig. 4.18. In this the high-molecular organic substrates are broken down by hydrolysis without energy gain, into low molecular compounds, with the help of enzymes belonging to the category of hydrolases, which add or subtract water from the substrate.

The central metabolic feature of aerobic organisms is the *citric acid cycle*, also sometimes referred to as the tricarboxylic acid cycle (TCC), or the Krebs cycle. In this cycle acetic acid bound to coenzyme A, the so-called acetyl-CoA, is broken down into two molecules of carbon dioxide and four atoms of hydrogen in the sequence of reactions indicated in Fig. 4.18. The hydrogen is then oxidised in the respiratory chain with a gain of energy; from one mole of acetic acid, 800 kJ or 12 moles of ATP are obtained. Besides acetyl-CoA, other products, such as glutamine and aspartic acid formed during the decomposition of proteins, may enter the cycle. On the other hand intermediates may be released from the cycle and used for the synthesis of biomass.

For sewage treatment it is important that the enzymatic decomposition of organic compounds occurs as a rule within the living cell; the resulting intermediate compounds do not appear in the free liquid. Where, however, the organic substances are metabolised by *exoenzymes*, either excreted into the surrounding liquid or bound to the cell wall, then intermediate products or even biologically stable residues may occur in the free liquid phase. Thus Behrens and Hannes (1984) were able to measure, under special culture conditions for detoxification of about 460 mg/l formaldehyde, amino-acid concentrations in the free liquid of over 200 mg/l as a result of the activity of exoenzymes bound to the cells. Owing to the mixed population of the activated sludge biocoenosis and the mixing effect in the aeration tank, such processes usually proceed undetected in activated sludge plants.

In recent years extensive studies have been performed on surface-active products of the metabolism of microorganisms (Cooper and Zajic 1980). These so-called *biosurfactants* comprise a broad spectrum of organic compounds of chemically different structures, which lower the surface tension of water from 0.73 to less than 0.30 N/m, and may thus give rise to foam formation. They chiefly consist of lipids with hydrophilic and hydrophobic groups which can break up organic particles and emulsify fats and oils and may also possess anti-microbial properties. These biosurfactants are as a rule formed only when the metabolic rate of the organisms is depressed on account of nutrient deficiency. In activated sludge plants such surface-active substances may be produced, especially by lightly loaded activated sludges, where the rate of inflow ceases, or in the last tank of a cascade system or in sludge reaeration tanks, etc. Although the biosurfactants should be more readily biodegradable than synthetic detergents they are probably largely protected from decomposition owing to their accumulation in the foam, since the metabolic activity of organisms occurring profusely in the foam is restricted on account of oxygen deficiency.

The metabolism of organisms is characterised by growth and respiration. In so-called co-metabolism, however, substrates are utilised from which the energy gain is employed only for the maintenance of vital processes and not for the purpose of cell growth, so that other, more readily metabolisable substrates must be available for growth. By means of co-metabolism a whole range of poorly degradable substrates, such as chlorinated hydrocarbons, plant protection chemicals, halogenated aromatics and alkylbenzenesulphonate, can be broken down. According to a literature review by Horvath (1972) many bacteria are capable of this process of co-metabolism; the intermediates listed by him also probably appear in the free liquid.

4.3.2 Anaerobic metabolism of carbon and hydrogen

Although aerobic conditions are necessary for complete biological treatment, technical reasons (such as failure of the aeration system, insufficient circulation and mixing, or merely inadequate oxygen input) may give the opportunity for anaerobic processes to develop in the final settling tank as well as in the sludge recycle lines. For these 'anaerobic sludges' strictly anaerobic decomposition and conversion processes may be occasioned by methanogenic fermentation.

As has been already explained in section 4.2.2 it is assumed that passive diffusion of oxygen into the flocs only exerts any immediate influence in the boundary region. The bacteria which are trapped within the flocs can accordingly undergo only such fermentation reactions as are described below.

As Fig. 5.3, Curve 3a, shows, oxygen-free or anoxic spells may occur from time to time in many parts of the aeration tank. As the oxygen contained in nitrite and nitrate is respired during such anoxic periods, the organic substrates are eliminated oxidatively, that is with the involvement of the citric acid cycle and the respiration chain, and there is no adverse effect on the quality of the treated effluent. Even in the absence of nitrite and nitrate, and of physically dissolved oxygen, anaerobic processes may not necessarily become dominant. Curve 3a shows this progression, when the oxygen demand stands at 60.1 g (m^3/h) and the oxygen input is only 60 g/m^3 h. The amount of non-removable and oxidisable hydrogen under such conditions is so small, that the enzymes remaining in reserve can store the excess of organic substrates for periods of several hours without there being any noticeable effect on the quality of the treated effluent. As the determination of oxygen level provides no clue as to the magnitude of the deficit, an oxygen concentration of 0.0 mg/l does not enable one to deduce what biochemical processes will be initiated as a consequence of the deficiency. The list ranges from denitrification, via concealed but still dominant aerobic processes, to the formation of organic acids and alcohols, to desulphurisation and on to methanogenic fermentation. Criteria for evaluation of the oxygen deficit comprise the redox potential, the formation of organic acids, hydrogen sulphide and methane, a rise in BOD$_5$, an increase in saprobity, and so on. Table 4.7 shows the course of BOD elimination and formation under anaerobic conditions.

Table 4.7 — BOD$_5$-elimination and liberation under rigorous anaerobic (not anoxic) conditions of a continuously-fed compartment in a cascade-type activated sludge system. Condition 1 — adsorptive binding of organic substrate with oxidation or fermentative decomposition. Condition 2 — Liberation of BOD$_5$-forming fermentation products from the cell reserve materials of microorganisms

Time of day	Condition 1		Condition 2	
	BOD$_5$ (mg/l)		BOD$_5$ (mg/l)	
	influent	effluent	influent	effluent
7.00–12.00	90	13	11	22
13.00–18.00	93	16	17	31
19.00–24.00	79	12	16	45
1.00– 6.00	36	9	11	35

For condition 1: $t=2.0$ h; $T=16.8°C$; $TS=3.3$ kg/m^3.
$B_{TS}=0.3$ kg/kg d; $OC'<0.5$ g;m^{i3} h.
For condition 2: $t=4.2$ h; $T=15.0°C$; $TS=1.1$ kg/m^3.
$B_{TS}=0.7$ kg/kg d; $OC'<0.5$ g/m^3 h.
For condition 2 the feed consists of recycled sludge only, with no fresh sewage input.

For relatively short-lived oxygen deficiency and lightly-loaded activated sludge a high rate of BOD$_5$ elimination, unaffected by the oxygen deficit, continues to occur; the substrates are bound by the enzymes, forming enzyme-substrate complexes

Materials conversion processes

without the ability to release hydrogen for use in the respiration chain. For relatively longer-lasting anaerobic retention times and higher sludge loadings, digestion products are formed which can cause a rise of about 20 mg/l in the BOD_5 in the recycled sludge.

The anaerobic processes which proceed under such conditions may be subdivided into two completely different metabolic pathways, both with the same objective, which is that of energy gain:

obtaining oxygen from already oxidised that is oxygen-containing compounds like NO_3^-, NO_2^-, SO_4^{2-}, ClO_3^-, ClO_4^-, CrO_4^{2-}, CrO_7^{2-} which thus serve as oxidising agents in the respiration of aerobic organisms, with release of energy;
obtaining energy from organic substrates by means of facultative or anaerobic bacteria.

The anaerobic forms of energy production from organic substrates are termed anaerobic fermentation, and thus constitute a substitute for respiration under anaerobic conditions. The liberation of energy takes place as a result of the esterification of alcohols by phosphoric acid, while the energy uptake occurs as for respiration by the transformation of ADP into ATP. As the end products of anaerobic fermentation listed in Fig. 4.25 are still very rich in energy, the energy

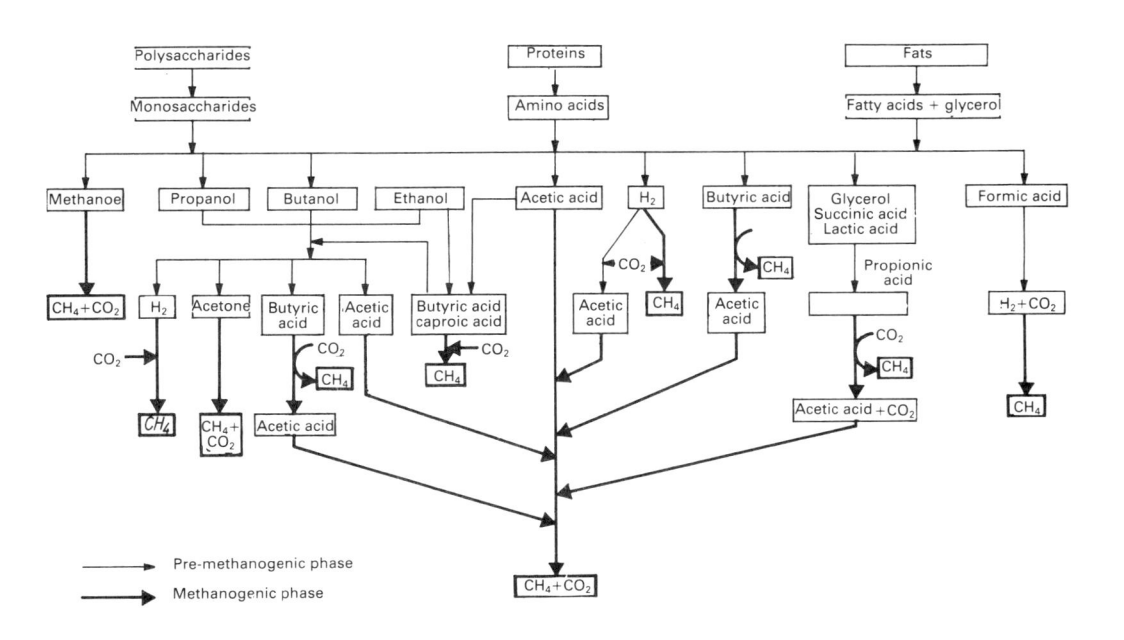

Fig. 4.25 — Diagram showing the anaerobic decomposition pathways for important sewage constituents. After Dohányos et al. (1982) with modifications.

yield from anaerobic processes only amounts to around 8% of that from respiration. In addition the turnover rates of organic substrates and the surplus sludge production are reduced in equal proportion. The high-energy products of anaerobiosis give rise, as far as they do not escape (like methane) to the atmosphere, to BOD_5 values in the effluent which are several times those indicated in Fig. 4.9. Should sulphate-reduction occur, then the BOD_5 values in the effluent from the final settling tank may even exceed the value in the influent.

The anaerobic or facultatively anaerobic bacteria may be classified into three physiological groups:

acid forming (acetogenic) bacteria
products of fermentation consist of lactic acid, formic acid, butyric and acetic acid. Formic acid is decomposed by gas-forming bacteria, for example *Aerobacter*, into hydrogen and CO_2;
alcohol-forming bacteria
products of fermentation consist of ethanol, glycerol, butanol, acetone, isopropanol and sometimes hydrogen as well.

The end products of the metabolism of acetogenic and alcohol-forming bacteria are primarily low-molecular weight compounds, such as acids and alcohols which are readily utilised by aerobic bacteria. Industrial effluents, which contain such substrates in larger amounts, generally give rise to bulking sludge formation. It may be inferred that the formation of bulking sludge in some municipal treatment plants is therefore encouraged or facilitated by the presence of similar products of anaerobic metabolic reactions.

Methane-forming (methanogenic) bacteria
Methane bacteria are very sensitive to the presence of free oxygen, so that their proliferation in activated sludge must be dismissed. Gas analyses performed on the exhausted air from aeration tanks, however, have demonstrated the presence of methane. However, human and animal excreta, the layers of sewer slime and also foul sewage, may contain methane bacteria which are subsequently washed into the aeration tank, where they exhibit ante-mortem activity or may be incorporated into the flocs in a viable form. The methane bacteria utilise as substrates for the formation of methane those compounds produced by the acetogenic bacteria, such as acetic acid, hydrogen and carbon dioxide. Their peak activity can only be developed in slightly alkaline media at pH values of 7.2–8.0, with strictly anaerobic conditions.

4.3.3 Metabolism of nitrogen compounds
Municipal sewage contains about 50 mg/l of nitrogen of which about 40% is organic nitrogen, chiefly in the form of proteins and urea. The rest is composed of ammonium ions and often a few milligrammes of nitrite and nitrate. In industrial effluents nitrogen may be a growth-limiting factor, with the result that it may need to be added to permit biological treatment to proceed. Alternatively it may sometimes occur in extremely high concentrations in different oxidation states, or as more or less poorly-degradable heterocyclic and polycyclic aromatic compounds. In biological treatment there are three processes to be distinguished:

— ammonification: the formation of ammonia by breakdown of organic nitrogen compounds;
— nitrification: the formation of nitrite and/or nitrate from ammonia;
— denitrification: formation of free niotrogen from nitrite and/or nitrate.

4.3.3.1 Ammonification

The principal fraction of organic nitrogen is contained in protein which may be composed of up to 20 different amino acids having the structural formula:

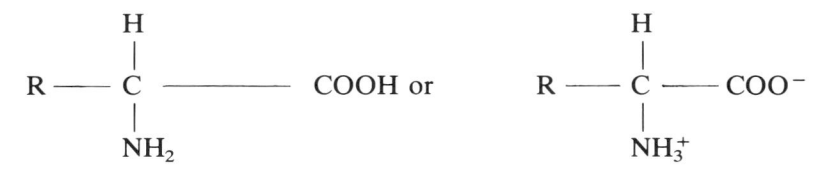

$$R-\underset{\underset{NH_2}{|}}{\overset{\overset{H}{|}}{C}}-COOH \quad \text{or} \quad R-\underset{\underset{NH_3^+}{|}}{\overset{\overset{H}{|}}{C}}-COO^-$$

The group R may have a ring or chain structure. In the proteins the individual amino acids are linked together by means of peptide bonds, —CO—NH—.

Proteins may act as enzymes, cell reserve materials, mechanical building blocks or even as pathogenic toxins, such as those of botulism, tetanus, and so on. The various proteins occur as a consequence of the different arrangements of amino acids in the peptide chain. During the biochemical decomposition of proteins, three successive reaction steps take place:

cleavage of the protein molecule into polypeptides due to the action of proteases with introduction of a water molucule into the —CO—NH— bond and rupture of the bond as a result;
cleavage of polypeptides to amino acids due to the action of peptidases;
decomposition of amino acids, with either ammonia or other organic nitrogen compounds being formed.

The metabolism of amino acids can take the form of oxidative deamination with formation of keto-acids, deamination with formation of unsaturated fatty acids, hydrolytic deamination of amides and imides, for example urea splitting, or reductive deamination and decarboxylation with formation of amines.

The amines formed as a product of many anaerobic processes are oxidatively metabolised by numerous species of anaerobic bacteria. The terminal products of their decomposition are NH_3, CO_2 and H_2O. The turnover of amines proceeds more rapidly the less substituted they are; the reaction rate thus decreases in the order methylamine, dimethylamine, trimethylamine.

Ammonification is out of the question following complete biological treatment. However, biologically treated sewage effluents often contain 3–6 mg/l of dissolved organic nitrogen which is not removed even in plants with controlled denitrification (Fig. 4.26). Besides unchanged biologically resistant nitrogen-containing substrates, which make up from 60% to 80% of the organic nitrogen in the effluent, there will also be, according to Parkin and McCarty (1981), the metabolic products of activated sludge, due to substrate oxidation, dead organisms and the so-called equilibrium nitrogen, by means of which an equilibrium is maintained between the nitrogen content of the internal and external call media. According to

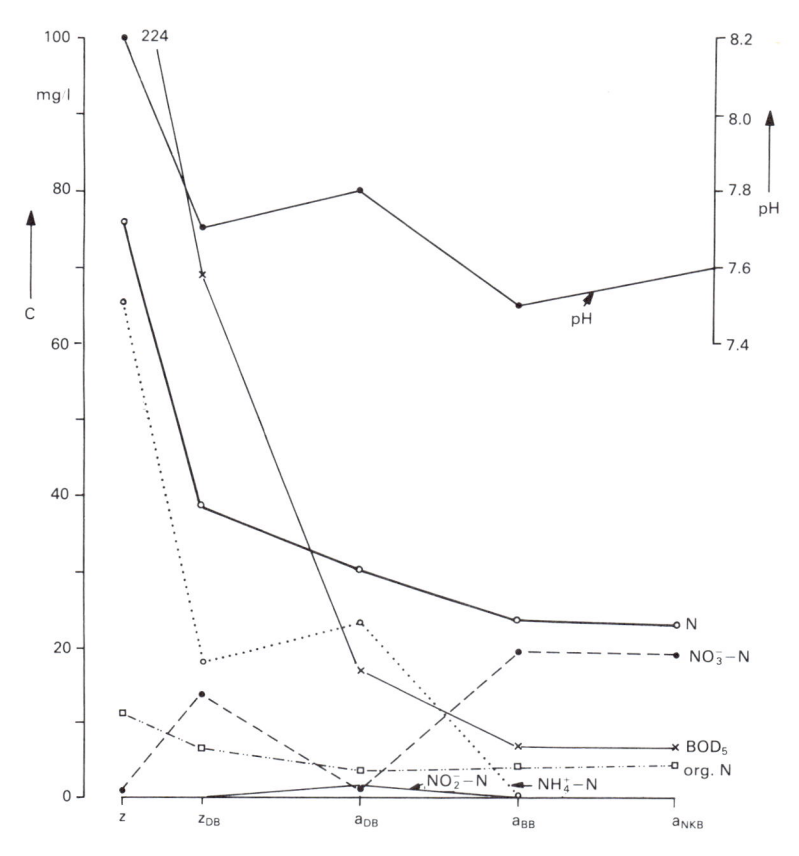

Fig. 4.26 — Changes in the nitrogenous components, BOD$_5$ and pH in response to a preliminary denitrification stage. Operating conditions: municipal sewage. B_{TS}=0.05 kg/kg d; T=9.5°C; V_{RS}=2.5 V_h; denitrification time 1.7 h; denitrification rate 1.41 g/kg h for 24 h combined sample.

these authors, the lowest concentration of dissolved organic nitrogen is obtained for an aeration time of about six hours and a sludge age of four to ten days (for sludge ages of six or ten days, organic nitrogen contents of 3.5–4 mg/l or 5.5–6 mg/l respectively may be expected).

4.3.3.2 *Nitrification*

While most of the organisms contained in the activated sludge biomass are heterotrophs, that is they feed on organic carbon compounds, the nitrifying bacteria are *autotrophic organisms*. For the formation of biomass they utilise carbon dioxide as their carbon source. The energy for the reduction of carbon dioxide is obtained from the *oxidation of ammonia* to nitrite or nitrate. Higher concentrations of organic substrate inhibit these organisms, this being the most frequent example of the diauxy phenomenon on activated sludge. For the activated sludge process this implies that the nitrification first commences when the organic substrates have been largely eliminated; this is the case, depending on the temperature, when the sludge loading has fallen to 0.15–0.3 kg/kg d. The *nitrification rate* at 15°C amounts to about 1.4 g/kg h and follows the Arrhenius rule. The reaction proceeds in two stages, often clearly distinct processes in activated sludge, according to the following overall equations:

— Nitrite formation by *Nitrosomonas*
$$NH_4^+ + 1\tfrac{1}{2} O_2 \rightarrow NO_2^- + H_2O + 2H^+ + 243 - 352 \text{ kJ/mol}.$$

The hydrogen formed in this reaction is immediately combined by the bicarbonate present in the medium; in this way the bicarbonate–CO_2 buffer system and hence the acid neutralising capacity of the sewage is reduced by an amount equivalent to 7.14 mg $CaCO_3$ per 1 mg of NH_4^+-N oxidised. Should the bicarbonate be used up, then the pH value sinks, in some cases to values as low as 4.0.

— Nitrate formation by *Nitrobacter*
$$NO_2^- + \tfrac{1}{2}O_2 \rightarrow NO_3^- + 63 - 99 \text{ kJ/mol}.$$

According to [16], *Nitrosospira*, *Nitrosococcus* and *Nitrosolobus* may also be responsible for nitrite formation, and *Nitrospira* and *Nitrococcus* may take part in nitrate formation.

For nitrification a considerable amount of oxygen is required; 3.43 mg O_2/mg N for nitrate production and 1.14 mg O_2/mg N for nitrate production. The oxygen content essential for nitrification in the mixed liquor is usually stated in the literature to be 2 mg/l, although it is possible for nitrification to proceed quite fast at only 0.5 mg/l of dissolved oxygen.

The nitrifiers are certainly among the organisms most sensitive to the presence of toxic substances, or to low temperatures in the activated sludge. Table 4.8 lists the LC_{75} values for some fairly common toxic constituents of sewage. Fig. 4.29 shows the inhibition of nitrification as a result of increasing concentrations of ions of the heavy metal cadmium. Both *Nitrosomonas* and *Nitrobacter* exhibit a pronounced substrate sensitivity; *Nitrosomonas* is inhibited by ammonia, and *Nitrobacter* by both nitrite and ammonia. Owing to the latter effect nitrification may come to a halt at the nitrite stage, until the ammonia which is toxic to *Nitrobacter* has been oxidised. In laboratory cultures both groups of organisms exhibit a very marked pH-dependence: at pH values below 7.5 *Nitrosomonas* ceased metabolic activity, and below pH 5.7 *Nitrobacter* also ceased their activity. In an activated sludge suspension the optimal pH-range lies between 7.0 and 8.8. With nitrogen-rich effluents from the brown coal industry, however, nitrification can cause the pH to fall to 4.0.

Should nitrate production be inhibited by toxins in contrast to nitrite production and also during the start-up of the nitrification reaction, it may happen that nitrite may be formed at concentrations of over 100 mg/l and as a result may give rise to interference in the COD and BOD determinations.

The temperature effect is most pronounced; in many activated sludges nitrification stops, depending on the sludge loading, at temperatures below 7°C, the temperature exerting both a direct and also an indirect effect, as a result of the curtailing of the sludge age at low temperatures. Many nitrifiers, such as *Nitro-*

Table 4.8 — Partial (about 75%) inhibition of nitrification by some constituents of raw effluents. From the data given by Tomlinson *et al.* (1966) and Frangipane and Urbini 1978)

Compound	Concentration (mg/l)
Acetone	2000
Ally alcohol	19.5
Allyl chloride	180
Allyl isothiocyanate	1.9
Aniline	7.5
Ethanol	2400
Chloroform	18
2,4-Dinitrophenol	460
Guanidine carbonate	16.5
O-Cresol	12.8
m-Cresol	11.4
p-Cresol	11.5
Hydrazine	58
Methylisothiocyanate	0.8
Sodium cyanide	1.2
Sodium methyldithiocarbamate	0.9
Phenol	5.6
Carbondisulphide	35
Skatole	7
Thioacetamide	0.53
Thiosemicarbazide	0.18
Thiourea	0.076
Trimethylamine	118

somonas, may however, survive periods of adverse conditions by means of cyst formation.

The nitrifiers are relatively sluggish organisms where their metabolism is concerned. Their cell growth is slow, while oxygen consumption is high. The growth of the bacteria and the oxidation of ammonia are a coupled reaction pair; for *Nitrosomonas* the following reactions are possible, from which the role of the hydrogen becomes more obvious than in the preceding overall equation

Nitrite formation

$$2NH_4^+ + 4H_2O \rightarrow 2NO_2^- + 4H^+ + 12H.$$

Oxidising hydrogen
(a) by oxygen

$$12H + 3O_2 \rightarrow 6H_2O;$$

(b) by carbon dioxide

$$12H + 3CO_2 \rightarrow 3(CH_2O) + 3H_2O.$$

Hofman and Lees (1952) obtained the following approximate equation for the formation of organic carbon:

Log C ≡ 0.73 log N−0.35
(C = organic carbon produced, mg/l
N = nitrate produced, mg/l).

This equation shows that during nitrification only very little nitrifying biomass is produced. Consequently the nitrifiers can develop fully only activated sludges with a high sludge age and hence a low level of waste activated sludge production.

The sludge age of nine days quoted in Table 10.1 applies to the yearly average value for municipal sewage. For complete nitrification of this type of sewage at 20°C a sludge age of about six days is necessary, but at 8°C this increases to 16 days; the corresponding sludge loadings are 0.24 and 0.08 kg/kg d respectively. In the boundary region the process may stop at the nitrite stage (see Fig. 2.3).

4.3.3.3 Denitrification

By denitrification we understand the enzymatic production of an oxidised nitrogen compound: nitrate→nitrite, nitrite→nitric oxide, nitric oxide→nitrous oxide, nitrous oxide→nitrogen. In sewage treatment technology it is usually confined to the reduction of oxidised nitrogen compounds to gaseous nitrogen which then escapes to the atmosphere. Denitrification offers to heterotrophic bacteria the opportunity under anoxic conditions, of obtaining the oxygen contained in the oxidised forms of nitrogen, as a means of respiration and the formation of new cell biomass. The denitrifiers can thus complete the oxidation of organic substrates using the enzymes of the respiration chain even in the absence of dissolved oxygen. Most denitrifiers immediately cease denitrification when an oxygen content of a few tenths of a milligram per litre is reached; as can be seen from Fig. 4.13 denitrifiers may carry out denitrification in the boundary zones of the activated sludge flocs even if dissolved oxygen is not present in the aeration tank. It proceeds according to the overall equation:

$$NO_3^- + 5H + 5e^- \rightarrow \tfrac{1}{2}N_2 + 2H_2O + OH^- + 360 \text{ kJ/mol},$$

the following successive stages being involved

$$HNO_3 + 2H \rightarrow HNO_2 + H_2O$$
$$HNO_2 + 2H \rightarrow \tfrac{1}{2}H_2N_2O_2 + H_2O$$
$$H_2N_2O_2 + 2H \rightarrow N_2 + 2H_2O.$$

Where denitrification is incomplete owing to a deficiency of hydrogen donors or shortage of time, then nitrite may be produced in appreciable amounts as an intermediate. (Fig. 4.26, Koné and Behrens 1981; for nitrate reduction with acetate an intermediate nitrite concentration of 700 mg/l NO_2^- occurred.)

$H_2N_2O_2$ is unstable and probably breaks down, particularly at elevated temperatures, to nitrous oxide as follows:

$$H_2N_2O_2 \rightarrow N_2O + H_2O.$$

Nitrous oxide is either released to the atmosphere along with nitrogen, or it remains in solution. *Pseudomonas stutzeri* and *P. aeruginosa* perform the denitrification of nitrous oxide to gaseous nitrogen.

Technical denitrification processes may be viewed as subject to interference from the metabolic reactions of organisms which reduce oxidised nitrogen compounds to ammonia. This is not uncommon (first equation in Table 4.9) and may for example be

Table 4.9 — Reaction equations and substrate requirements for denitrification of nitrate. From Pöppinghaus (1975) with slight alteration

Hydrogen source	Theoretical equation for denitrification	Substrate demand (g/g NO_3–N)
Raw sewage	—	1.5–6.0 g BOD_5
Sludge/biomass	$C_5H_7NO_2 + 4NO_3 \rightarrow$ $5CO_2 + NH_3 + 2N_2 + 4OH$	2.0–3.5 g BOD_5
Methanol	$5CH_3OH + 6NO- \rightarrow$ $5CO_2 + 3N_2 + 7H_2O + 6OH^-$	1.9 g BOD_5
Glucose	$5C_6H_{12}O_6 + 24NO- \rightarrow$ $30CO_2 + 12N_2 + 18H_2O + 24OH-$	2.58 g BOD_5
Methane	$5CH_4 + 8NO- \rightarrow$ $5CO_2 + 4N_2 + 6H_2O + 8OH^-$	0.71 g BOD_5

caused by *Azotobacter*, *Desulphovibrio*, *Denitrobacillus* and *Escherichia coli* according to the following equation:

$$HNO_3 + 4H_2 \rightarrow NH_3 + 3H_2O.$$

In the plants investigated so far this process has indeed been a relatively minor one, but is nevertheless of considerable importance where extended nitrogen removal is concerned (Fig. 4.26 — inlet and outlet of the denitrification tank).

In contrast to nitrification, many types of bacteria are capable of performing denitrification; practically no activated sludge exists without denitrifiers. Besides the capacity for denitrification already employed in the systematic names, for example *Denitrobacillus* or *Micrococcus denitrificans*, it is also possible for members of the taxa such as *Spirillum*, *Bacillus*, *Pseudomonas*, *Escherichia*, etc., to denitrify (see table 4.6). Hydrogen-supplying substrates can take the place of organic compounds; even methane and molecular hydrogen, together with the cell reserves of the bacteria, can be used as hydrogen sources (Table 4.2). According to Koné and Behrens (1981) however aromatic substrates are less suited for this purpose. Theoretical equations for the reactions involving different substrates are shown in Table 4.9.

The equations for denitrification demonstrate that OH^- ions are liberated.

These form bicarbonate by reaction with carbon dioxide, so that the previously diminished bicarbonate-CO_2 buffer capacity resulting from nitrification is restored to the extent of 3.6 mg/l of alkalinity per 1 mg of NO_3^--N reduced and consequently the pH value may increase somewhat.

The denitrification reaction is of course an enzymatically controlled process. The activation enzymes concerned are termed nitrate — or nitrite — reductases; for the former both molybdenum and iron are required, and for the latter iron and —SH groups are necessary.

Denitrification is a reaction of zero order $\frac{dN}{dt} = KTS_{DB}$, where $\frac{dN}{dt} =$ denitrification rate (g/h); K = rate constant (g/kg h); TS_{DB} = sludge solids content in the denitrification tank.

The rate constant K is dependent on the substrate supply, the temperature, the activity of the organisms and the quantity of denitrifiers present in the sludge biomass, and thus varies from one activated sludge to another. In the English literature the correlations listed in Table 4.10 between the sludge age (related to

Table 4.10 — Denitrification rate as a function of temperature and sludge age. From Jones and Sabra (1980) and Sutton et al. (1974)

Temp °C	Sludge age (d)	Denitrification rate (g/kg h)
7	3	3.68[†]
	3.4	1.14
	6	1.95[†]
	10.3	0.48
15		1.4
	3	5.22[†]
	4	2.39
	6	4.27[†]
25	1.8	5.39
	3	10.20[†]
	5	4.13
	6	8.44[†]

[†] Values quoted by Sotton et al.

sludge loading according to Table 10.1) and denitrification rate have been reported. The dependence on *sludge age* is explained by the higher concentration of active biomass in the more heavily-loaded sludges. For complete nitrogen removal in a nitrification-denitrification system, sludge ages as shown below are required:

at 25°C 3 days $\simeq$ B_{TS} 0.35 kg/kg d;
at 15°C 6 days $\simeq$ B_{TS} 0.24 kg/kg d;
at 7°C 20 days $\simeq$ B_{TS} 0.08 kg/kg d.

The dependence of denitrification rate for nitrate on the BOD_5-sludge loading under anoxic conditions at 20°C is indicated by the observations of Burdick (1982):

at B_{TS} = 0.4 kg/kg d → 1.7 mg/g h;
at B_{TS} = 1.2 kg/kg d → 2.5 mg/g h;
at B_{TS} = 1.7 kg/kg d → 3.3 mg/g h.

Sewages characterised by a low BOD_5-loading or a high sludge age have a low denitrification rate on account of the preponderant use of the BOD_5 for dissimilatory processes. Fig. 4.26 shows the change in nitrogenous constituents in such a lightly-loaded denitrification-nitrification system with a high influent N-content and a relatively short denitrification time.

The denitrification rate may be much increased if the growth conditions for the denitrifiers are optimised. If for example nitrate-rich biologically treated sewage is introduced into the feed to an anaerobic trickling filter receiving primary settled sewage, then the reaction rate is increased to 10–60 g/kg h, because the biofilm in the trickling filter is composed almost entirely of denitrifying organisms.

4.3.4 Metabolism of phosphorus compounds

Sewage and effluents contain phosphate, polyphosphates and organic compounds of phosphorus. During biological treatment, the polyphosphates and the organic phosphates are converted to orthophosphate to roughly the same extent as the BOD_5 is eliminated.

The phosphorus demand of the organisms arises chiefly from the special role of phosphrus in the energy budget of the organisms and from the formation of the cell-wall structures. If the P-content in the aeration tank falls below 0.2 mg/l P, then phosphorus becomes rate-limiting; the rates of nitrogen metabolism are also reduced and it becomes necessary to introduce the element artificially. Based on the empirical formula for sludge biomass, $C_{108}H_{180}O_{45}N_{16}P$, the quantity of phosphorus required for the formation of fresh biomass is stoichiometrically defined. It amounts normally to around 0.6–1.5 wt%. Plants with a high sludge wastage rate (see Figs 4.2 and 4.17) accordingly require, and also eliminate, more phosphorus than those with lower rates of sludge production. For separate aerobic sludge stabilisation due to the reduction of biomass in a fashion similar to that in sludge digestion plants, an amount of phosphorus equal to 1 mg P/g biomass is liberated daily and transferred to the liquid phase. The phosphorus requirement in sewage treatment terms is usually expressed as a proportion of the amount of COD or BOD_5 to be removed. From experience 0.5–1.0 mg P appears adequate for the elimination of 100 mg/l BOD_5 or 200 mg/l COD-Cr. Those quantities which are widely reported in the literature are, however, incorrect as the actual P-demand is not determined directly by the amount of substrate removed, but by the waste activated sludge production.

Among the organisms present in the sludge biomass, some (Osborn and Nicholls 1978, list 36 species or taxa) can store phosphorus as polyphosphate above the stoichiometric ratio implied in the formula cited above, up to about 7% of the solids

Sec. 4.3]   **Materials conversion processes**   149

content. This process, which was termed 'luxury uptake' by Levin and Shapiro (1965), takes place according to Nicholls and Oaborn (1979) when the organisms find themselves in a stressed condition owing to a deficiency of an essential element such as sulphur, nitrogen, oxygen, phosphorus or carbon, but when the available energy allows the transport of phosphate into the cells. Where the 'missing' element is phosphorus, then the storage of polyphosphate after a period of phosphate deficiency is termed 'over-compensation'. The phosphate deficiency may arise on account of the sewage characteristics or due to anaerobic conditions, or it may be produced artificially.

Fig. 4.27 shows the enzymatically controlled phosphate budget of bacteria, and

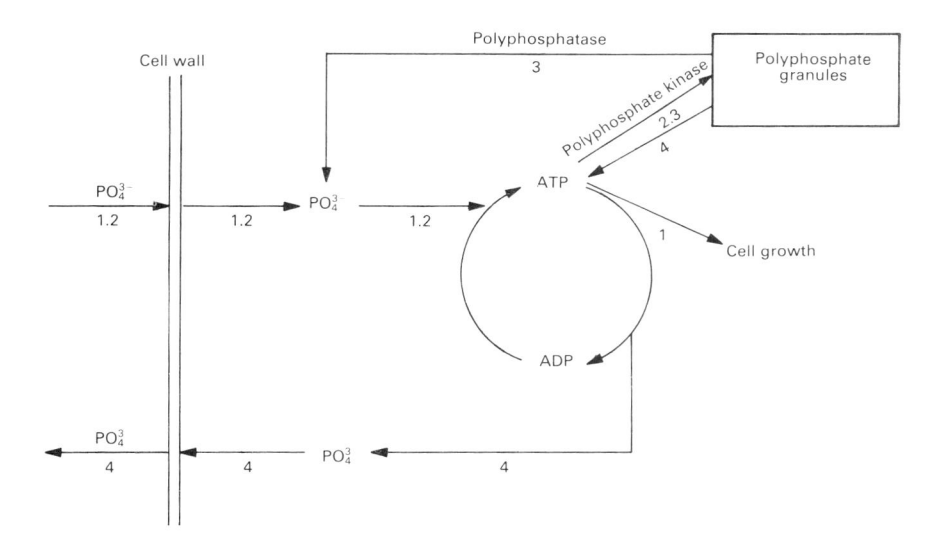

Fig. 4.27 — Simplified phosphate cycle for bacteria. Based on Osborn and Nicholls (1978). Pathway 1: Synthesis of biomass with the P-uptake in stoichiometric proportion to the P-content of the biomass and inhibition of simultaneous production of polyphosphates. Pathway 2: Where organic substrates are deficient, cell growth stagnates while with a simultaneous abundance of P, polyphosphate is produced by the action of the enzyme polyphosphate kinase, the so-called 'luxury P uptake', following a previous deficiency of P, and with elevated levels of kinase, a large quantity of polyphosphate is stored in a very short time, the so-called 'polyphosphate over-compensation'. Pathway 3: With the onset of anaerobic conditions or a drop in pH, the enzyme polyphosphatase liberates energy-rich phosphate from the stored polyphosphate while on regaining aerobic conditions, P-storage takes place again according to the 'over-compensation' mechanism of Route 2. Pathway 4: Under anaerobic conditions phosphate is taken out of storage and utilised in the form of ATP by means of an inverse kinase reaction; the phosphate released by ADP-formation is excreted.

Fig. 4.28 the phosphorus metabolism of activated sludge which is governed by the availability of oxygen. Phosphorus over-compensation, luxury uptake and subsequent liberation under anaerobic conditions are inducible as a result of long-term acclimatisation. In addition to the enzyme acclimatisation process, phosphorus

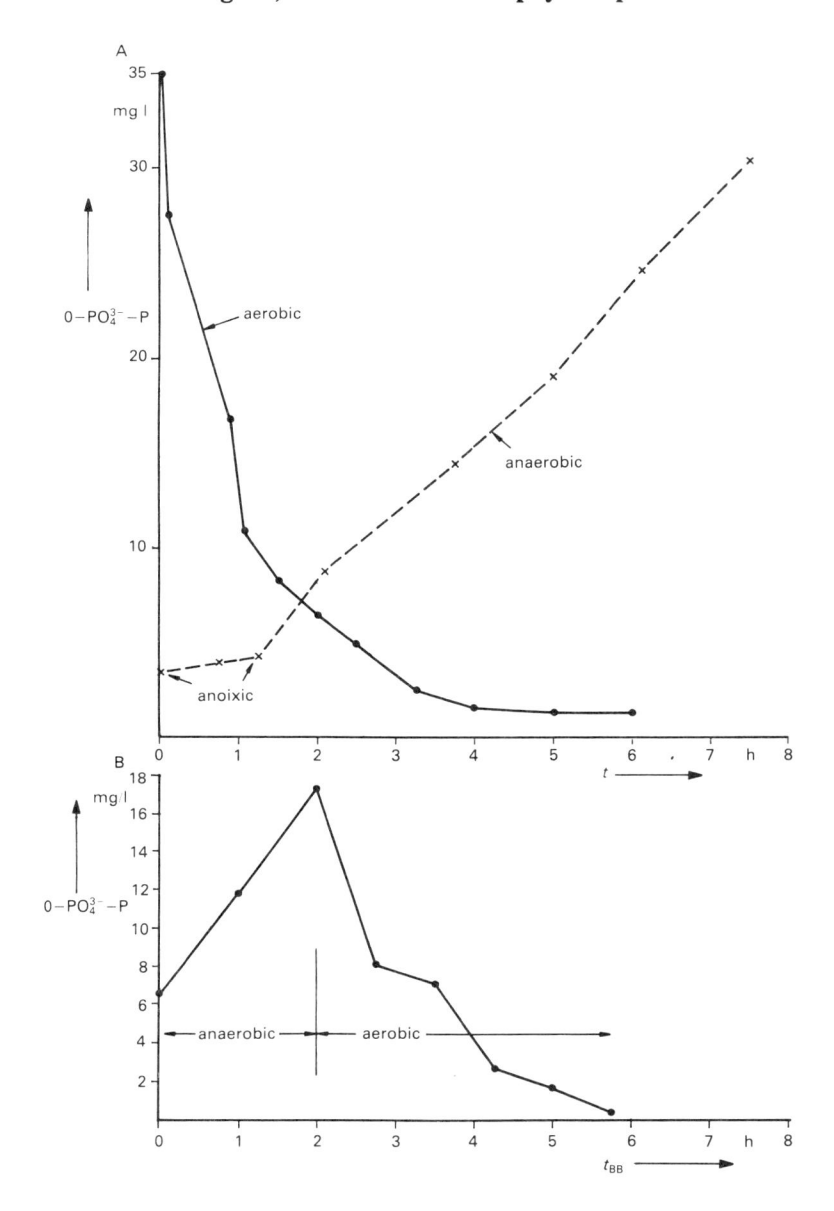

Fig. 4.28 — Metabolism of orthophosphate by sludge biomass from a plant performing biological phosphorus removal. A: phosphorus uptake under aerobic conditions and liberation under anaerobic conditions. Laboratory test $T=22$–$24°C$; $TS=7.2$ g/l. B: orthophosphate profile in two anaerobic and five aerobic compartments in series. $T=21.3°C$; $TS=2.3$ g/l.

metabolism is affected by the substrate supply, the pH value, the oxygen content and the temperature. According to Ludwig *et al.* (1982), phosphorus uptake under aerobic conditions roughly follows the Arrhenius Rule, while phosphate liberation under anaerobic conditions at 20°C amounted to 1.9 mg/g sludge solids, but only 0.3 mg/g sludge solids at 12°C; this implies a strong temperature dependence for biological processes for phosphorus removal.

The liberation of phosphorus is inhibited by nitrate concentrations greater thasn 2 mg/l NO_3-N. Denitrification is thus a prerequisite for the liberation of phosphorus under anaerobic conditions and hence also for a high rate of phosphorus removal. On the other hand the input to the final settling tank should contain more than 2 mg/l of NO_3-N, in order to prevent the otherwise customary release of 0.2–1.5 mg P/l in the final settling tank.

Acinetobacter, a coccoid to oval aerobic bacterium growing in glutinous colonies, can also exhibit luxury phosphorus uptake without an intermediate anaerobic phase (Fush and Chen 1975, Nechvátal and Sladká 1981) provided the necessary substrates, such as acetate, fatty acids, ethanol, etc., are available.

Acinetobacter lwoffi cultured in the laboratory and then introduced into activated sludge can achieve effluent-P concentrations of less than 1 mg/l P for several days; however, these bacteria are unable to hold their own in the activated sludge biomass [32].

Phosphorus elimination can also be performed by chemical means. Menar and Jenkins (1970) for example, for hard calcareous effluents containing 220–300 mg/l Ca CO_3, were able to achieve P-removal percentages of up to 80% as a consequence of the combination of phosphate as insoluble calcium phosphate. Chemical phosphate coagulation may also be effected by heavy metal ions which may present in the sewage, such as iron salts from metal working processes, iron or aluminium hydroxide sludges from water treatment [25], or may be induced by so-called simultaneous coagulation with artificial dosing of coagulants according to TGL 27886/01.

4.3.5 Metabolism of sulphur compounds

Putrescent municipal sewage and many industrial effluents contain sulphides, including hydrogen sulphide, in milligram quantities and in some instances other oxidisable sulphur compounds in addition. About 1% of the organic solids fraction consists of sulphur.

The sulphur-containing amino acids cystine, cysteine, methionine and lanthionine are formed during the breakdown of proteins and their sulphhydryl groups are released by desulphurases during further breakdown. This process of hydrogen sulphide formation usually commences inside the sewerage network. Under strongly anaerobic conditions, even sulphate may be reduced to hydrogen sulphide:

$$SO_4^{2-} + 8H^+ \rightarrow H_2S + 2H_2O + 2OH^-.$$

The above process resembles denitrification. The sulphate-reducing bacteria live in a medium with a very low redox potential so that they will only very rarely be able to find suitable conditions for survival in an aeration tank.

The resulting hydrogen sulphide may be oxidised under aerobic conditions either chemically or biologically. For biological oxidation, for which a few microscopically recognisable bacteria are adapted, for example *Beggiatoa* and *Thiothrix*, there are two reaction processes, either of which may proceed depending on the oxygen and sulphide supply:

$$2H_2S + O_2 \rightarrow S_2 + 2H_2O + 251 \text{ kJ};$$
$$S_2 + 3O_2 + 2H_2O \rightarrow 2H_2SO_4 + 1080 \text{ kJ}.$$

Where the oxidation stops at the first stage, then sulphur granules are deposited in the bacterial cells; if the second stage reaction occurs, the sulphur is oxidised to sulphuric acid.

Grünwald et al. (1983) demonstrated that in the treatment of sulphide-rich waters, a considerable proportion of the sulphide is oxidised *chemically* in the aeration tank. The resulting thiosulphate formed as an intermediate is also completely oxidised at sludge loadings (thiosulphate) of up to 0.35 kg/kg d. For sulphide sludge loadings of 0.3 kg/kg d the sulphides are also 100% oxidised, and for 1.42 kg/kg d the corresponding figure is 98%.

H_2S-containing discharges from the final clarifier are indicative of operating failures, for example oxygen deficiency in the aeration tank, septic zones in tanks and gullies, or surge inputs of digester liquor.

4.3.6 Waterborne toxic substances

Waterborne toxic substances are those which, even at low concentrations, exhibit:

— toxic effects on warm-blooded animals or aquatic organisms;
— interference with the self-purifying action (or are not susceptible to it);
— harmful effects on the intended use of water [7].

From the standpoint of sewage treatment, as a rule only those substances which interfere with the sewage treatment process are regarded as toxic. However, this is only a sub-process within the context of the hydrological cycle, which takes in the reuse of water either for industrial processes or domestic use. The treated water may exhibit undesirable properties of a subjective nature (colour, turbidity, temperature) undesirable chemical ingredients (e.g. salts, hardness) and substances harmful to health (e.g. pathogens, carcinogens, nitrate). With an annual growth of around 100 000 organic compounds worldwide, the hazard from individual substances, and their possible synergistic effect in combination with other substances, is hardly calculable. In addition, the traditional compounds of inorganic chemistry are still insufficiently understood with respect to their cumulative action and their effect under certain physiological states for individual human and animal subjects; out of the 107 known elements of the Periodic Table, over 80 are in current use (Paucke and Bauer 1979) and can thus gain entry to the sewer. Formerly only a few organic compounds were capable of being accurately determined and hence studies of their elimination could only be performed in the case of chemically measurable and physiologically suspect substances.

While some twenty or thirty years ago the major focus was on phenols and surfactants, nowadays the emphasis has shifted towards pesticides and carcinogenic substances. Heavy metals are also the subject of much greater interest since the fairly recent recognition of the toxicity of cadmium to humans. As a consequence of the availability of mass spectrometers it is now possible to determine many of the organic substrates present in sewage, and also to measure their elimination rates in activated sludge plants, together with the formation of stable intermediates.

From a medical viewpoint the waterborne toxic substances may be subdivided into the following categories:

— pathogenic organisms — disease producers;
— carcinogenic substances — cancer producers;
— teratogenic substances — leading to deformities;
— mutagenic substances — affecting the gene system;
— allergenic substances — giving rise to allergies;
— irritant substances — skin rash producers

Apart from the pathogenic organisms it is often difficult to assign the other pollutants to one or other of the above categories, especially since experimental evidence can only be obtained from animal experiments or studies *in vitro* and their results can only be applied with reservations to human beings.

Heavy metals and metals

Those heavy metals which are relevant to environmental protection include copper, cadmium, zinc, nickel, chromium, mercury and lead. These mostly very toxic elements should ideally be eliminated at the point of origin and not merely during the process of sewage treatment. Heavy metal concentrations of a a few hundred milligrammes per litre of municipal sewage have been reported in the literature many times. (McDermott *et al.* 1963).

Activated sludge biomass has a considerable affinity for all metal ions. According to literature data collected by Neufeld *et al.* (1977) the extent of accumulation of metal salts can amount to as much as 10–20% of total solids for Hg, Zn and Cd, 17% for uranium; 12.5–34% for Cu, Co, Fe, Ni (for *Zooglea ramigera*); 12.5% for Hg; 27% for Cd, and 18.4% for Cr^{6+} (Moore *et al.* 1961). In batch experiments the following order of affinity for metal binding to sludge biomass was obtained: $Pb > Cd > Hg > Cr^{3+} > Cr^{6+} > Zn > Ni$. Owing to the high binding power of activated sludge towards heavy metals the limiting values specified in TGL 26056/02 in respect of heavy metal contents of sludge applied to land may be exceeded.

The binding of heavy metals by sludge is mainly due to physicochemical forces and is only slightly influenced by biological mechanisms for transport and accumulation.

The equilibrium distribution of heavy metals between the flocs and the liquid phase can be described sufficiently accurately by means of *adsorption isotherms*. The adsorption constants derived from these isotherms vary from metal to metal and are also pH-dependent.

For the binding of two particular metals the following equations were derived by Neufeld *et al.* (1977):

For mercury $r = 0.247 C^{0.877} - 9.084 \left(\dfrac{Q}{120} - Q \right)^{1.38}$

For cadmium $r = 0.2287 C^{0.9936} - 31.04 \left(\dfrac{Q}{2.74} - Q \right)^{1.154}$

where the symbols have the following meaning:
r = extent of metal accumulation by sludge biomass (mg/g sludge solids);
C = concentration of metal in liquid phase;
Q = mass of metal already bound to the sludge biomass (mg/g sludge solids).

The detoxification and immobilisation of metals by the sludge biomass probably takes place preferentially in the bacterial slime and at the cell wall due to the following processes:

— physicochemical attachment to extracellular polymers;
— physical attachment to insoluble sulphides formed in the sewer system or in the primary settling stage;
— formation of insoluble salts, for example Pb^{2+} and Ba^{2+} with SO_4^{2-}, Ag^+ with Cl^-;
— hydroxide formation, for example for Fe^{3+}, Al^{3+}, Mn^{4+};
— inclusion in an insoluble complex, in which the heavy metal forms the central atom, or by pH-dependent chelate formation with proteins at the amino and carboxyl groups.

As a result of the tendency of heavy metal ions to form complexes with varying biochemical properties by combination with different ligands (Nähle 1980), monitoring of elimination rates is rendered more difficult. In non-treated sewage, polypeptides and polysaccharides may mask the presence of heavy metals due to complex formation, and this can also happen in fully-treated sewage owing to the presence of stable complexes with humic acids.

For physicochemical bonding the process is largely complete after about three to ten minutes. Owing to equilibrium changes between the liquid and activated sludge phases, the elimination of heavy metal ions does not reach a constant level; the metal ions bound by the activated sludge may be released into the aqueous phase in response to a decline in the heavy metal concentration in solution. Consequently, heavy metals which have been bound to the exocellular polymers, and hence detoxified, may be suddenly liberated as a consequence of ion-exchange reactions following an input of large amounts of common salt or caustic soda, and hence their toxic potential may reappear. High activated sludge levels for coping with shock loads and the wasting of activated sludge following the ingress of such materials will accordingly increase the elimination rate of heavy metal ions.

The effect of sludge solids content on the elimination of cadmium was demonstrated by the measurements of Port (1978):

MLSS g/l   1.4, 0.7, 0.35, 0.17;
Removal %   92 82   58   25.

The effect of pH is indicated below:

for Cu at: pH < 7, η = $<78\%$
pH > 7, η = 78–98%;

pH 7, η = 6%; pH 8, η = 17%;
pH 9, η = 70%.

As the metal ions are bound to the sludge biomass, the retention of the sludge flocs in the final settling stage is of particular importance. However, exactly the reverse effect is indicative of the toxic effect of metal salts, when there is an immediate increase in turbidity of the treated effluent, associated with the deflocculation of the activated sludge. The resulting microflocs, heavily contaminated by metal ions, which are washed out of the final tank, are very difficult to separate from the final effluent using the normal sedimentation techniques.

Table 4.11 presents a summary of some literature data concerning toxic effects of heavy metals on activated sludge plants, while Fig. 4.29 shows the effect of cadmium at different concentrations on nitrification. The toxic effects in activated sludge plants may, however, be very much greater as a result of the combined action of several toxins, heavy metals being often associated with mineral oils and cyanides. Nevertheless, activated sludges, due to their capability for elimination of metal salts, exhibit a resistance which is roughly 100 times greater than that of free-swimming bacteria (Reimann 1969). The recovery time for activated sludges following injury from heavy metal ions is dependent, among other things, on the nature, concentration and affinity of the metal ions for the sludge. It usually takes from 12 to 48 hours according to Neufeld (1976) but may take up to 14 days. Within certain concentration ranges a noticeably enhanced resistance to renewed toxic shocks may be displayed by a sludge which has recovered from the toxic effects of heavy metals (Offhaus 1968).

According to Bagby and Sherrard (1981) the toxicity of heavy metals, under conditions of equal metal concentration and sludge age, is greater in the case of sewage with low COD values than for more concentrated effluents, for which the increased coagulation and/or complex formation in the concentrated effluents can be held responsible. Moreover, plants with a low sludge age were found to be more susceptible, other things being equal, than those with a high sludge age.

Phenols

Phenols are very undesirable constituents of drinking water, owing to their adverse effect on taste, and also of surface waters because of the accumulation of taint in fish flesh at concentrations of 0.02 mg/l and upwards in the water. Following acclimatisation of sludges to phenol, it may be broken down by numerous bacteria at rates of 5 to 20 mg/g h (Fig. 4.6). For weakly-buffered sewage, a drop in pH, possibly accompanied by bulking sludge formation, may be observed. Higher concentrations of phenol are often linked with the presence of other toxic materials such as cyanide, thiocyanate, ammonia and hydrogen sulphide. The total phenol content of effluent from the brown coal industry, even after phenol extraction, may still be as high as 1000 mg/l, of which up to 90% may consist of polyphenols. The large-scale stirred reactors used to treat effluent at volumetric loading rates for phenol of up to 2.6 kg/m^3 d achieve an elimination performance of 80–95% for substrates exhibiting a BOD value. For complete treatment multi-stage systems are required. Metabolic

Table 4.11 — Effect of metal ions on activated sludges

Metal ion	Authors	Effect observed
Silver, Ag^+	Reimann (1969)	1 mg/l causes 2% inhibition, 10 mg/l causes 57% inhibition after two days (Ag as $AgNO_3$, in Sapromat)
Copper, Cu^{2+}	Barth et al. (1965)	Limiting concentration for continuous input: 1.2 mg/l
	McDermott et al. (1963)	Limiting concentration for continuous input: 1.0 mg/l Elevation of turbidity of treated effluent from 0.8 mg/l Limiting concentration for shock loads (4 h with $t_{BB} = 6$ h) 50 mg/l for $CuSO_4$, 10 mg/l for copper cyanide complex
	Reimann (1969)	10 mg/l causes 38% inhibition, 50 mg/l causes 85% inhibition of substrate respiration after five days (Cu^{2+} as $CuSO_4$, Sapromat)
Zinc, Zn^{2+}	Neufeld (1976)	Above 40 mg/l deflocculation occurs
	Brown and Andrew (1972)	Under continuous input, 5–10 mg/l no adverse effect, 20 mg/l causes inhibition
	Barth et al. (1965)	Limiting concentration for continuous input 5–10 mg/l
	Offhaus (1968)	91.9% inhibition of substrate respiration for 20 mg/l after 24–48 h, 81.6% inhibition after 48–72 h, falling to 66.8% after 72–96 h and 45.7% after 96–120 h. Corresponding values for acclimatised sludge 46.6%, 2.8%, 16.0%, 17.6%. Measurements made in Sapromat with Zn as $ZnSO_4$
Nickel Ni^{2+}	McDermott et al. (1963)	Limiting concentration for continuous input 1–2.5 mg/l; with continuous input of 2.5–10 mg/l, $\eta\ BOD_5$ is 5% lower
Mercury Hg^{2+}	Neufeld (1976)	Deflocculation occurs from 100 mg/l; see Fig. 2.10
Chromium Cr^{6+}	Bonomo (1974)	For continuous input at 10 mg/l, $\eta\ BOD_5 = 90\%$, at 40 mg/l, $\eta\ BOD_5 = 83\%$, at 60 mg/l, $\eta\ BOD_5$ very low
	Moore et al. (1961)	Limiting concentration for continuous input: 10 mg/l at 50 mg/l; $\eta\ BOD_5$ drops to about 3% For shock loading: at 10 mg/l no effect: at 100 mg/l deflocculation
Chromium Cr^{3+}		Non-toxic; in tannery effluents at 40 mg/l Cr^{3+} and $B_{TS} \leq 0.5$ kg/kg d, $\eta\ BOD_5$ exceeds 90% and $\eta\ cr^{3+}$ 90% [39]
Cadmium Cd^{2+}	Neufeld (1976)	Deflocculation occurs from 20 mg/l
	Weber and Sherrard (1980)	Limiting concentration for continuous input: 10 mg/l. For effect on denitrification see Fig. 4.29
Cobalt Co^{2+}	Reimann (1969)	Total inhibition of respiratory activity at 2000 mg/l At 1 mg/l substrate respiration after 5 days reduced by 5%, at 10 mg/l by 26%, at 20 mg/l by 43%, at 50 mg/l by 59% (Co^{2+} as $CoCl_2$ in Sapromat)

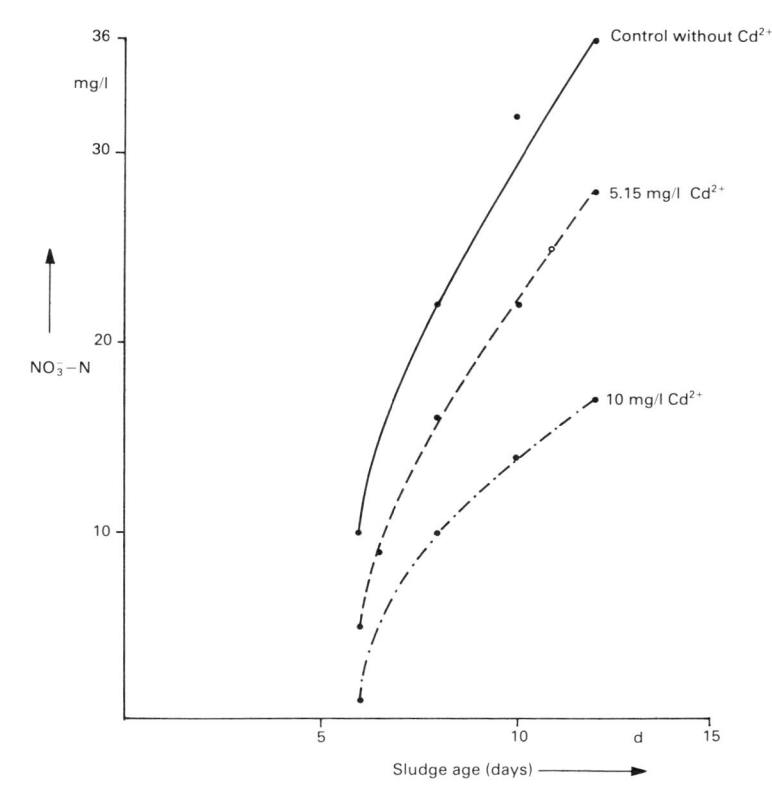

Fig. 4.29 — Inhibition of nitrification by cadmium-containing influents. From Weber and Scherrard (1980).

decomposition of the higher phenols may lead to the formation of mucins which can give rise to severe foaming problems in the aeration tank.

The process of ring-opening for benzene derivatives is more complicated than the decomposition of aliphatic substrates (see Fush 1961). For biocoenoses of low species diversity pronounced diauxy effects may occur as a result.

Nitro-substituted phenols (and sometimes also chlorinated phenols) are often severely toxic and metabolised only with difficulty. Jakóbczyk *et al.* (1984) observed pronounced toxic responses during the decomposition of nitro-substituted phenols, benzenes and naphthalene, with acclimatisation times of anything from five to eight months, followed, however, by removal efficiencies of up to 99.9% relative to the nitro compounds, with COD sludge loadings of 0.15–0.30 kg/kg d.

Acids and alkalis
Tolerable limits for the pH in activated sludge aeration tanks are from 6.0 to 9.0. Even influent pH values outside this range are of little or no practical significance, according to Figs 2.2 and 4.30, provided either a completely mixed reactor or a

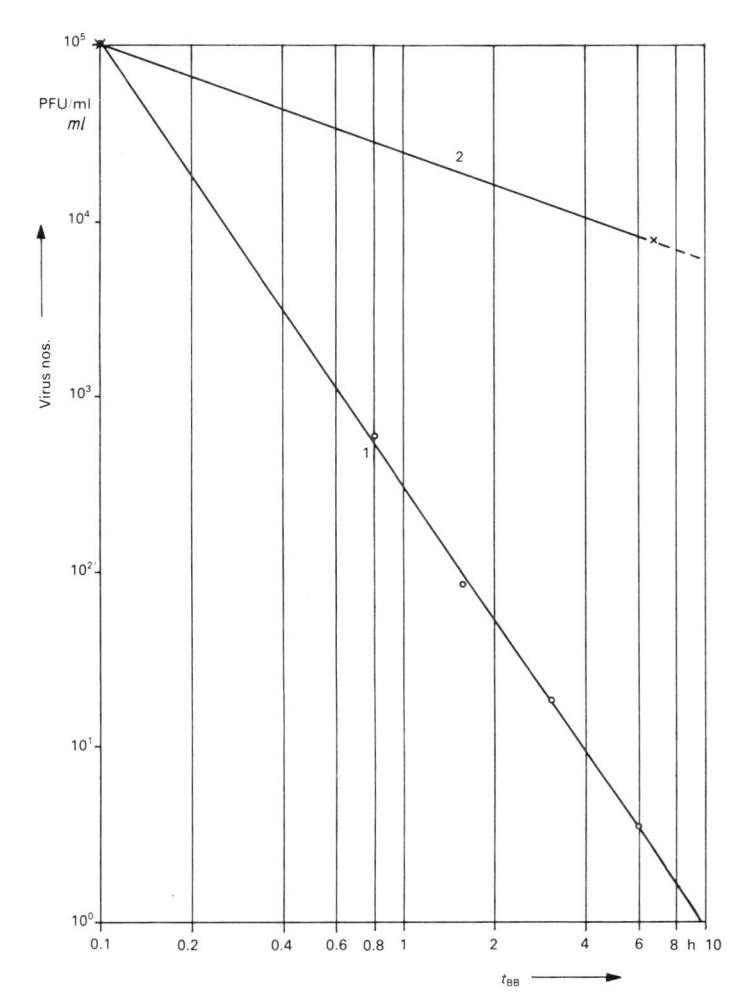

Fig. 4.30 — Reduction in virus forms in sludge biomass at 23–27°C in laboratory tests. Curve 1: Coxsackie virus Ag, $oTS = 0.4$–1.5 g/l; Curve 2: Poliovirus 1, $oTS = 1.2$ g/l. From Clarke et al. (1961).

sufficiently high buffer capacity is available. According to Zülke (1967), turnover rates for many organic substrates in industrial treatment plants are unaffected in response to pH variations from 4.0 to 9.5 in the aeration tank.

Organic halogen compounds
Some organochlorine compounds are carcinogenic, or exhibit toxic or chronic effects on organisms and may also, to some extent, accumulate within the organism. As a general rule the ease of metabolism declines with increasing chlorine substitution of the substrate. In addition, the position of the substituent chlorine atoms in the molecule may sometimes cause a marked change in the ease of decomposition. Thus

2,5-dichlorobenzoate exhibits a K_S value of 1.5 mg/l, while that of 3,5-dichlorobenzoate is 25.3 mg/l. The substantial elimination of some organochlorine compounds, for which discharge to sewer is prohibited by reason of their solvent properties, is shown in Table 4.12. At higher concentrations, however, these compounds, which are also employed as fixatives, are highly toxic.

Table 4.12 — Degree of elimination of low molecular-weight organohalogen compounds during biological sewage treatment. Recalculated from literature data of Sprenger (1982)

Compound	Formula	Influent concentration (mg/m³)		Effluent concentration (mg/m³)	η %
Trichloromethane (chloroform)	$CHCl_3$	Mean	6.8	0.9	86.8
		Max	31.3	2.9	90.7
Dichloromethane (methylene chloride)	CH_2Cl_2	Mean	(10.2)	(5.5)	(46.7)
		Max	27.2	5.5	79.8
Trichloroethane	$C_2H_3Cl_3$	Mean	0.43	0.03	93.0
		Max	1.7	0.05	97.0
Carbon tetrachloride	CCl_4	Mean	0.17	0.04	76.5
		Max	0.4	0.04	90.0
Trichloroethylene	C_2HCl_3	Mean	4.6	0.5	89.1
		Max	23.9	0.3	98.8
Tetrachloroethylene	C_2Cl_4	Mean	65.0	4.5	94.6
		Max	331	11.7	96.4
Organic chlorine (tot)	—	Mean	68.4	5.5	92.0
		Max	295	10.8	96.5

Pesticides

Pesticides also frequently consist of chlorinated organic compounds, such as chlorinated hydrocarbons (lindane, methoxychlor), halogenated phenoxy carboxylic acids 2,4-D, 2,4DP), polychlorinated biphenyls (PCBs), heterocyclic six-member rings with three heteroatoms (simazine, prometryn), organophosphorus compounds (methyl parathion), and so on. Extensive studies [3] have shown that pesticides are adsorptively bound to activated sludge and are largely metabolised both under aerobic and anaerobic conditions (Table 4.13). They only become toxic to the sludge biomass at higher concentrations providing they are not bactericidal compounds, nor characterised by a high heavy metal content such as copper in Cupral or zinc in Zineb, for which the toxicity is dependent on the heavy metal ions.

Oils

Fats and oils of animal or vegetable origin consist exclusively of glyceryl esters (glycerides) of the higher fatty acids, such as palmitic, stearic and oleic acids. The alkali salts of these fatty acids are soaps. The fatty acids and the glycerol liberated after hydrolysis are susceptible to metabolic breakdown, as already indicated in Fig. 4.18. For very fatty effluents from abattoirs the percentage removal rates for BOD_5 and fats are roughly of equal magnitude. However, where the fat is present in the form of large particles, it is hardly accessible for utilisation by the relevant organisms

Table 4.13 — Elimination of pesticide compounds in an activated sludge plant. Operating conditions: $t_{BB} = 2.8$ h; $B_R = 1.8$ kg/m^3 d; $B_{TS} = 0.9$ kg/kg d. Compiled from results reported in [30]

Dilution ratios for synthetic mixture of residual compounds in municipal sewage
(28 formulations and six artificially incorporated compounds)

Active ingredient	Inlet mg/l	1:1000 Outlet mg/l	η %	Inlet mg/l	1:500 Outlet mg/l	η %	Inlet mg/l	1:100 Outlet mg/l	η %
Lindane	0.347	0.01	97	0.72	0.008	98	4.67	0.05	99
2,4-D	1.034	0.102	90	1.25	n.d	100	3.74	0.05	99
2,4-DP	0.605	0.07	88	1.15	0.203	82	2.2	0.03	98
Simazine	0.134	0.019	86	0.3	0.007	98	2.5	0.24	90
Methyl parathion	0.21	n.d.	100	0.44	trace	100	5.07	0.06	98
Nitrofen	0.627	0.015	98	2.53	0.007	99	17.6	0.21	99
BOD$_5$	189	17	91	189	17	91	233	14	94
COD-Cr	400	68	83	322	87	73	429	120	72
COD-Mn	75	21	72	95	20	79	160	40	75

(see Fig. 4.3). If such particles enter the aeration tank, then owing to their lower density they accumulate at the surface and also on the surface of the clarifier and the intake structures.

Mineral oils are pollutants; one litre of oil can render as much as 25 m^3 of water undrinkable on account of the taint, and certain oil derivatives are toxic to animals and humans. Originally the opinion was held that mineral oils surround the flocs with an oil film and hence impede the liquid-sludge materials-transport process. The oxygen transfer rate is reduced by 36% even for only 0.1 or 0.2 mg/l of oil, as a result of an oil film at the air-water interface (Hashimoto 1975). However, microscopic examination reveals that the oil is deposited on the flocs in the form of droplets. A large amount of information concerning the ease of metabolism of petroleum and petroleum products has been obtained from studies of the activated sludge treatment of petroleum refinery effluents, and more especially from the continuous fermentation of petroleum products. According to Grünwald (1980) the rate of biological turnover declines in the order: alkanes > isoalkanes > cycloalkanes > aromatics. Cyclohexane is more rapidly decomposed than ethylcyclohexane, but ethyl benzene is more rapidly metabolised than benzene. For the aromatic hydrocarbons the order is: benzene > toluene > m-xylene.

For biological decomposition of mineral oils an acclimatisation period lasting from several weeks to months is needed. According to Lautenbach (1966) tests with 100 mg/l of petrol, crude oil or diesel oil achieved 9–100% removals with BOD$_5$ sludge loading of 0.3 kg/kg h, and an aeration time of three hours, without impeding nitrification or harming the biocoenosis. According to Table 4.14 removals of 75–85% were achieved, and from Table 4.15 about 85% of the lipophilic materials were eliminated [27].

A substantial part of the oil fraction accumulates in the form of microscopically small droplets in the sludge biomass. According to Huber (1968) an oil content of 10% of the sludge solids should not be exceeded, otherwise the density of the activated sludge is reduced to such an extent that it was washed out of the final settling tank. A tolerable limit for the oil content in the influent is 5 mg/l, while 100 mg/l may be considered just permissible.

In the event of accidental oil spills, the sludge biomass is seldom suffficiently acclimatised for the oil to be biologically decomposed. In such cases it is advisable to shut off the aerators and to displace the oil from the aeration tank by the entry of sewage. It then floats on the surface and passes to the clarifier from which it may be skimmed off, possibly following the addition of binding agents as specified in TGL 22213/06. Following such occurrences there will be a certain degree of sludge carryover into the settling tank and also a prolonged discharge of oil traces. In the mobile test rig (Plate 6) a 1:60 hydraulic oil:sludge mixture was once obtained with no apparent harm to the biocoenosis, although a rise in the effluent BOD$_5$ concentration from 15 to 30–40 mg/l resulted. For accidental inputs of heavy oil it is advisable immediately to waste the oil-contaminated sludge until the suspended solids content is reduced to 0.5 g/l, followed by about one week's operation in the logarithmic growth phase (Fig. 4.16).

For final treatment of oil-containing plant effluents following the normal biological treatment stage, Huber (1968) advocated the use of lagoons in which an oil content of 1–2 mg/l is further reduced by 40–55%.

Table 4.14 — Effect of increasing salt content on the elimination of organic substrates and on nitrification in activated sludges from the petroleum industry, using water from the Caspian Sea. Summary of data from [12]

Salt content in inflow g/l	B_{TS} rel BOD_{20} kg/kg d	BOD_{20} mg/l in	BOD_{20} mg/l out	COD mg/l in	COD mg/l out	Petroleum products mg/l in	Petroleum products mg/l out	NO_2^- % of inorg N in	NO_2^- % of inorg N out	NO_3^- % of inorg N in	NO_3^- % of inorg N out
0.2	0.42	625[†]	10	1230	75	58	9	0	8.6	0	71.8
6.3	0.43	645[†]	14	1280	86	58	9	0	5.8	0	57.2
23.0	0.42	625[†]	14	1260	90	58	13	0	12.4	0	56.6
34.3	0.30	460[†]	13	975	97	58	9	0	13.6	0	50.4
54.3	0.21	218	19	848	288	54	14	0	21.9	0	31.5
54.4	0.14	233	18	976	342	50	12	0	19.5	0	36.4
62.1	0.17	209	63	790	577	57	31	0	4.2	0	0

[†] Denotes inlet values raised by addition of 350 or 500 mg/l phenol.

Table 4.15 — Amounts of lipophilic substances in the inlet and outlet of municipal activated sludge plants (from [27]) (mg/l)

Plant ref		A	B	C	D	E	F
Lipophilic	in	1.28	1.37	1.08	1.42	0.98	1.03
substances	out	0.18	0.22	0.17	0.24	0.15	0.18

Surfactants (detergents)
In modern times household washing products have become notorious as ingredients of sewage due to the resulting foam generated on the water surface — sometimes banks of foam 1 m high occurred on the aeration tanks due to the action of the 'hard' non-biodegradable detergents (now banned) such as tetrapropylenebenzene-sulphonate — and also on account of their toxicity to fish (SC 1–9 mg/l). More recently the dangers of toxicity to humans and the risk of accumulation in the food chain have received enhanced recognition. Winter and Erbert (1981), besides giving a comprehensive summary of detergents and their ease of metabolic decomposition, mention their harmful effects on the activated sludge process, reduction of the interfacial and surface tensions, foam formation, dispersant properties, reduction of oxygen diffusion rates, etc. Toxic heavy metals are retained in suspension; emulsifying and dispersing agents retard the sedimentation process. Those substances produced with the intention of lowering the surface activation energy of water are also capable of destroying the structure of the activated sludge flocs.

Depending on their chemical composition it is possible to distinguish between anionic, cationic and non-ionic detergents. Anionic detergents are usually readily metabolised; those with linear alkyl chains more easily than those with linear polyether, branched alkyl and alkyleneoxide chains or aromatic rings. Diphenyl derivatives and alkylated polyalcohols are toxic. Cationic detergents (pyridine derivatives, quaternary ammonium compounds, etc., used as bactericides, fungicides and disinfectants) may inhibit nitrification at concentrations of 10 mg/l and over.

In municipal sewage the concentration of detergents is usually 5–25 mg/l, but in effluents from laundries they may reach 440 mg/l (Halle and Trautsch 1978). Even the most concentrated detergent solutions (Busse *et al.* 1970 quote values of up to 9000 mg/l) can be decomposed with an efficiency of over 90% in activated sludge processes employing stirred reactors. Non-ionic detergents based on polyethers are often eliminated only to about 20%, this being most probably entirely due to adsorption processes. For many surfactants such as NTA (Fig.4.8) the activated sludge requires acclimatisation periods of over 14 days. During the running-in phase for an activated sludge plant treating sewage rich in detergents or containing non-biodegradable surfactants, operation may be disrupted by froth-flotation and wash-out of the bacteria. In this case iron salts may be dosed into the aeration tank for adsorptive binding of the detergents and also to serve as inert supports for the bacterial growths.

Undiluted effluents from large-scale laundries which use the Fedapon washing process, with a content of 35 mg/l of surface-active material, may give rise to a surface foam of 30% in an aeration tank using fine-bubble aeration [28]. Although the anionic detergents, after acclimatisation of the activated sludge, were eliminated to a residual level of less than 1 mg/l (see Table 4.16) the treated effluent contained

Table 4.16 — Decomposition of anionic surfactants. Results of laboratory experiments by Fuka and Pitter (1981)

Type of effluent or surfactant	B_{TS} kg/kg d	Anionic surfactant inlet conc mg/l surface-active material	Decomp rate mg/g h
Municipal sewage	8.1	7.9	1.2
(15–20°C)	1.0	5.9	1.0
	0.15	8.2	0.8
Laundry effluents			
(Fedapon process)	0.4	76	2.6
(25–34°C)	1.8	82	7.2

Results from acclimatised sludges in laboratory tests

Product (brand name)	kg/kg d rel to COD	COD inlet	mg/g h rel to COD
Zenit	1	100	2.2
Alfa	1	100	2.8
Gong	1	100	2.5
Biomat	1	100	2.2
Peromat	1	100	1.8
Batul	1	100	1.4
Pollena	1	100	0.6
Sanso	1	100	1.3
Rakon	1	100	0.7
Tim	1	100	0.6
Dash	1	100	0.8

above-average quantities of suspended solids, so that a simultaneous coagulation or post-treatment separation stage may become necessary.

Salts
Injury to the organisms of activated sludge due to changes in osmotic pressure in the cells occurs in a stepwise manner, according to the data quoted in Table 4.14: up to 6 g/l, no effect; from 6 to 34 g/l incipient inhibition of nitrate formation and reduced

COD elimination; from 34 to 54 g/l definite inhibition of nitrate formation, marked reduction of COD removal, and slight inhibition of BOD_5 removal; and from 54 to 62 g/l severe inhibition of nitrite formation and COD removal with a distinct decline in the BOD_5 removal.

The salt content of municipal sewage, at 1–3 g/l is far below the toxic region. However, in certain industrial effluents such as those from the processing of salted fish (Cl^- of 11.9 g/l), or where small activated sludge units are employed on board ship and from time to time sea water is used for toilet flushing, extremely high salt contents may occcur, to which the organisms are incapable of adapting at once. In the latter case of shipboard packaged systems, acclimatisation periods of 10 to 14 days were necessary. Ventz (1971) confines the surge inputs from fish processing plants to 4.15 g/l of chloride. Table 4.17 presents various limiting values for chloride and sulphate as well as the non-negligible effect of the relevant cations.

Table 4.17 — Tolerance of activated sludge biomass toward chloride and sulphate ions. From [10]

Ion	Salt	Max tolerable concentration (g/l) Under steady conditions	Under shock loading
Cl^-	NaCl	18	9
	$CaCl_2$	13	6
SO_4^{2-}	Na_2SO_4	40	34
	$CaSO_4$	33	27

Only a negligible degree of elimination of soluble salts takes place during biological treatment as a result of incorporation in the biomass.

Cyanide and thiocyanate (CN^- and CNS_- compounds)
Whereas thiocyanates, according to studies by Schönborn and Lautenbach (1966), can be removed with an efficiency of 76–90% following acclimatisation, and only give rise to harmful effects on the biocoenosis at concentrations of 270 mg/l, the decomposition of cyanide is a serious problem. Although Lautenbach (1966) succeeded in acclimatising activated sludge to 100 mg/l KCN, nevertheless all the protozoa were killed at a concentration of 20 mg/l. A considerable portion of the cyanide is removed by gaseous stripping. Actinomycetes, for example *Streptomycete* spp, are certainly capable of degrading cyanide, but the process is very variable and unstable in character (0–25% removal rates) in the sludge suspension. Complex iron cyanides are hardly removed at all. Sudden inputs of over 0.3–2 mg/l of simple cyanides are toxic, the first indication being provided by deflocculation of the activated sludge. The toxic effect of cyanides is often still further enhanced by that of heavy metal ions.

Dyes

In contrast to most of the other constituents of sewage, dyes and colouring agents are actually visible in the treated effluent from the final settling tank and are accordingly most undesirable even though their toxic effect may be slight. Dyes are characterised by a great variety of chemical structures; their broad classification into sulphur, azo and triphenylmethane dyes, porphyrins and similar types does not exhaust the multiplicity of possible structures. A factor common to most dyeing establishments which militates against the effective removal of dyestuffs from the effluent is the frequency of changes in the type of dye used, so that a stable acclimatisation of the biocoenosis is not feasible. It is therefore often advisable to select organic sludge loading values which guarantee an extended sludge age (Table 10.1), which in addition to long reaction times also ensures a very diverse community with a multiplicity of inherited characteristics favouring the metabolism of selected dyestuffs. On the other hand many dyes can be adsorptively bound, and are then disposed of along with the waste activated sludge, so that, particularly where poorly biodegradable substrates are concerned, a two-stage activated sludge process with a high sludge wastage rate in the first stage can be beneficial.

A major role in the structure of many dyestuffs is exerted by aniline, $C_6H_5NH_2$, which according to Fig. 4.6 is biologically degradable, but is nevertheless toxic to nitrifiers at concentrations greater than 7.5 mg/l (Table 4.8). A wide range of dyestuffs is thus biologically non-degradable or degradable to only a limited extent. The author, for example, was unable to achieve any significant elimination of the dye from a fluorescein-containing effluent during a week-long spell of treating this effluent in a bench-scale activated sludge system. In such cases it is advisable to prohibit the discharge of such substances in the sewage, or at least to insist on major reductions, since post-coagulation treatments may achieve only modest results, with reductions in colour intensity of only 40–60%.

As many textile chemicals are toxic to the anaerobic organisms in particular, the use of anaerobic treatments is usually out of the question; for this reason several authors recommend the use of aerobic stabilisation as a means of treating the waste activated sludge from textile effluent treatment systems.

Polycyclic aromatic hydrocarbons

To this group belong the suspected carcinogens such as 3,4-benzpyrene, 3,4-benzfluoranthene, 10,11-benzfluoranthene, 1,2-benzanthracene and indeno-(1,2,3,c,d)-pyrene. They are contained in petroleum and in coal and arise during the combustion and carbonisation of organic matter. In sewage they occur chiefly in particulates and to only a minor extent in the dissolved form. According to Borneff and Kunte (1967) a maximum of 300 mg/m^3 of polycyclic aromatic hydrocarbons, including 134 mg/m^3 of carcinogens, was detected in municipal sewage, with an average of 50 mg/ml^3 or a specific level of 0.8 mg per capita day. Although these compounds belong to the class of naturally occurring substances they are only metabolised with difficulty on account of their ring structures. The elimination of polycyclic aromatics during activated sludge treatment is principally due to adsorption processes although the biological degradability of 3,4-benzpyrene for example has been reported. Activated sludges remove from 40 to 90% of the polycyclic compounds, or 85% on average (Table 4.18). Effluents with a high content of

Table 4.18 — Elimination of polycyclic aromatic hydrocarbons and accumulation of aromatics in activated sludges from five different plants. From Borneff and Kunte (1967) also Reichert et al. (1971)

Plant ref and date	Inlet concentration mg/m^3	Outlet concentration mg/m^3	Proportion of carcinogens in outlet (%)	Aromatic concentration of sludge mg/kg
A (1967)	2.62	1.47	20	28.9
B (1967)	67.1	0.81	41	15.15
C (1967)	9.28	0.883	20	6.65
D (1971)	6.1	0.91	43	—
E (1971)	10.9	0.967	26	—

polyciclics following biological treatment should be subjected to further treatment by physicochemical methods, including the action of flocculating agents, activated carbon treatment or chlorination. In contrast to chlorinated aliphatic compounds, the chlorination of polycyclic carcinogens eliminates their carcinogenic potential.

Disinfectants

Disinfectants are employed in hospitals, in the cleansing of food processing establishments, in stables and in animal by-product plants. The disinfectants are usually applied and flushed away intermittently, which may be the occasion of particular danger. However, long transport distances and primary settling times, coupled with the dilution effect, usually exhaust their toxic potential.

The following disinfectants are among those most frequently employed.

Alkaline disinfectants (caustic solutions)
   as a rule these are harmless on account of the buffer capacity of bicarbonate and the continuous formation of carbon dioxide in the aeration tank (see Fig. 2.2).

Chlorine-containing disinfectants
   (Cl_2, chloramine, etc.). As a rule these are harmless to the sludge biomass because of the chlorine uptake of the raw sewage of 30–50 mg/l. Tolerable chlorine contents in the influent are between 0.2 and 1.0 mg/l. Chloramine, a chlorine preparation with long-term action, if introduced to the aeration tank in surges of 3–10 mg/l may give rise to stressed respiration with an increase in respiration rate of 17%. At levels of 50 and 100 mg/l respiration may be partially inhibited by as much as 7% to 17%. The severest inhibition was recorded during batch experiments following a contact time of 60 min. It may also be possible for the tolerable concentration limits for the sludge biomass to be exceeded following the discharge of chlorine bleach solutions from the textile industry.

Peroxide containing disinfectants
   (O_3, H_2O_2, peracetic acid). These are usually not harmful owing to their rapid breakdown in the sewerage network and in the primary treatment stage. For peroxide-containing effluents such as those from the polymerisation of plastics, Streichsbier and Washüttl (1980) advocated the use of COD sludge loadings of less than 0.33 kg/kg d.

Formalin-containing disinfectants
(e.g. Fesiasol, Wofasept). Fig. 2.8 illustrates the toxic effect and the utilisation of formaldehyde as substrate by a non-acclimatised municipal activated sludge biomass. According to Skidrov *et al.* (1980) the conversion rate for an acclimatised sludge may be 10.1 mg/g h. With formaldehyde as the only toxic substance, levels of up to 500 mg/l in the influent could be dealt with, there being no apparent toxic effect. In the presence of benzaldehyde as a second toxic ingredient, however, the threshold level fell to 200 mg/l.

Phenolic disinfectants
(cresol, guaiacol, carbolic acid): as a rule these are harmless to activated sludge and are degradable.

Radioactive substances

Radioactive isotopes are incorporated by adsorption in or on the activated sludge flocs to the same extent as the corresponding stable isotopes. For example, ^{32}P was almost 100% removed by a phosphorus-limited activated sludge, whereas ^{24}Na and ^{32}P were scarcely eliminated at all by an activated sludge treating laundry effluents. In addition ^{106}Ru and ^{134}Cs were reportedly not bound to activated sludge (Ruf 1958).

4.3.7 Roles of different groups of organisms
4.3.7.1 Bacteria

Some of the metabolic tasks of bacteria are discussed in sections 4.3.1 to 4.3.6. Table 4.6 indicates by extension certain selected taxa and their nutrient substances, together with specific metabolic reactions.

As a rule several hundred species or strains of bacteria may be detected in the sludge biomass. A high species diversity coupled with a broad physiological subsistence spectrum are essential provisos for the elimination of large numbers of organic substrates.

Depending on the type of subsistence, the bacteria present in activated sludge are classified as follows:

chemoorganotrophic bacteria: organic substrates are used as carbon and energy sources;

chemolithotrophic bacteria: carbon dioxide is the carbon source and oxidisable inorganic compounds are the energy sources;

mixotrophic bacteria: different combinations of organic and inorganic substrates function as simultaneous sources of carbon and energy.

Besides the process of bacterial decomposition and transformation, the elimination of non-biodegradable or poorly biodegradable compounds by absorption and adsorption processes (section 4.5) must be taken into account.

The communities of viable organisms in the activated sludge biomass are 'wild' biocoenoses. Many of the organic substrates present in the sewage are utilised only slowly or not at all. Although there are indeed quite often specialist or chance mutations with the capability of metabolising such substrates, these organisms are suppressed on account of the population pressure from organisms which use easily metabolised substrates and hence multiply more rapidly. Given the present state of microbiology however, it is also possible, by mutation and selection, to cultivate certain bacterial strains which are able to utilise poorly degradable substrates. These artificially cultivated bacteria are preserved while maintaining their specific properties and rendered capable of lasting for long periods. The bacteria are reactivated prior to use by suspension in water at a slightly elevated temperature and then introduced into the activated sludge. Jähnig (1980) reported the existence of a species bank with over 150 strains of microorganisms, while Heyden (1982) described hundreds of bacterial species with the ability to metabolise compounds such as cyanide, aromatics, tannins, lignins, acrylonitriles and other biologically refractory substances. The use of the bacterial mutants results in the more rapid running-in of industrial activated sludge plants during start-up and following the effect of toxic shock loadings. An inoculum of about 17 g of bacteria/m^3 of sewage was advocated, followed by 0.125 g/m^3 as a maintenance dose. It is also possible to use cultures of *Acinetobacter lwoffi* at intervals to enhance the elimination of phosphorus [32].

4.3.7.2 Fungi

Fungi are heterotrophic organisms: their simplified chemical formula, $C_{11}H_{17}O_6N$ shows that their nitrogen demand is less than that of bacteria. Although the fungal cells are larger than those of bacteria, thy have a large surface area in relation to their volume. One gramme of fungal mycelium has a surface area of 4.2 m^2 and a ball of mycelium occupying only 1 cm^3 has a total fibre length of 40 km.

Fungi are organisms of an aerobic nature, only the yeasts possessing the capability for anaerobic metabolism. In the floc interiors where there is an oxygen deficiency the yeast cells may accordingly carry out anaerobic fermentation. Under aerobic conditions they are of course in competition with bacteria for the acquisition of organic substrates from municipal sewage; the large numbers of individuals frequently observed in the sludge biomass are occasioned by the discharge of yeast-laden effluents (Sladká and Hänel 1965).

Fungi can penetrate into solid organic matter such as wood, horn, plant and animal tissue, since they open up these materials by the action of extracellular enzymes, beak them down into simpler constituents, and then incorporate them into the cell by osmotic pathways.

In activated sludges two types of physiologically adapted fungi exert an appreciable role.

The saprophytes
   live on dead organic matter like cellulose, hemicelluloses, sugars, hydrocarbons, wood, horn, proteins and their breakdown products. Their metabolic products under aerobic conditions are ammonia, carbon dioxide and water. From the decomposition of lignin, hemicelluloses and cellulose, the substances produced include fulvic acid, polyuronic acids, and humic acid-like materials which impart to the sewage a reddish tint only removable by coagulants. From anaerobic metabolism, organic acids and alcohols are obtained. The nature of the substrates used by the saprophytes is such that they occur preferentially in the activated sludge suspensions for certain industrial processes, such as those from the pulp and paper industries, rather than in municipal sewage treatment plants, where

they are often present but in smallish numbers. In sewage or effluents with a pH of less than 5 they are, on the contrary, ahead of the more sensitive bacteria, so that they may even form almost pure cultures.

The predators
are predatory fungi which capture organisms such as living protozoa, for example rhizopods, and also metazoa such as nematodes and rotifers, by means of specially adapted projections. They are then ingested and subsequently digested.

The powerful fungal threads may, just like thread forming bacteria, hinder the settling of the sludge flocs and consequently give rise to *bulking sludge*. Massive growths of fungi can usually be impeded or inhibited by adjustment of the pH in the aeration tank to a value between 7.0 and 8.0.

4.3.7.3 Protozoa and metazoa

Based on the governing factors detailed in section 3.7.2.2, healthy activated sludges are heavily colonised by protozoa, and often by metazoa as well. The role of the protozoa in the sewage treatment process was for a long time in doubt. Many authors demanded their elimination as protozoa feed on bacteria which are the most important organisms for the treatment of sewage. From a comparison of the cumulative volumes of the various organisms with that of the sludge, it appears that from 0.8 to 3.4% of the sludge volume is accounted for by these organisms. When allowing for water occluded by the sludge, the proportion by volume of protozoa and metazoa increases by eight to ten-fold. As discussed in section 4.2.2, only 1–3% of the bacteria present in activated sludge are actively involved in the aerobic sewage treatment process. The volume fraction of the protozoa and metazoa is thus equal to or greater than that of the active bacterial fraction.

Apart from massive growths of polysaprobic protozoa in activated sludges, which are occasioned by periodic oxygen deficits, there are often direct correlations between effluent turbidity and the numbers of protozoa, especially ciliates. This leads to the inference thast the potozoan consumers of bacteria and detritus contribute to a reduction in the turbidity of the final effluent, supported by the fact that simultaneous coagulation and oxygenated aeration systems with relatively few protozoa often exhibit a high turbidity in the final effluent. By using heat-sterilised sewage and a hermetically enclosed experimental treatment plant, Curds *et al.* (1968) were able to obtain activated sludges either with or without ciliates, which provided positive evidence for their beneficial effect on the treatment process (Table 4.0).

The positive effect of protozoa involves:

Elimination of particulate sewage constituents
- Besides the generally applicable rate-determining factors for the metabolic processes, which can also be applied to protozoa in the form of the Michaelis–Menten equation, the particle size and protozoan dimensions are also important. As Fenchel (1980) showed, an optimum size range exists for ciliate food particles, which is determined by the particular mouth structure. The holotrichal ciliates which occur regularly in activated sludges preferentially ingest particles in the size range 0.3–1 μm, the spirotrichal forms those of 1.0–1.5 μm, and the peritrichal forms those of 5–10 μm. Colourless flagellates ordinarily feed on particles of 0.2–1

μm in size. Those protozoa which were not attached to the flocs exert a specific positive influence on the treatment process.
- Viruses — for this aspect see section 4.4.1.
- Bacteria — the data assembled by Hänel (1979) for colourless flagellates indicated feeding rates of 20–70 bacteria per organism per hour. The feeding rate for the ciliate *Paramecium caudatum* of 30 000–42 000 bacteria per hour quoted in the older literature was also confirmed by recent measurements (Schönborn 1981, pers. comm.). Studies by Fenchel (1980) show that small ciliates may consume hourly over 100% of their own volume, although for most species the figure is 50–100%. From the multiplication rate of the most active colourless flagellates it may be deduced that they consume more than 200% of their own volume per hour. As the bacteria-laden municipal sewage contains from 10^8 to 10^{10} bacteria/ml, their elimination from a healthy sludge would be complete in a few hours, were there not a pronounced multiplication of solitary bacteria in the activated sludge (see Fig. 4.14).

As the pathogenic bacteria are able to multiply only in exceptional cases under the conditions inhospitable to them in the sludge, their elimination of 90–98% is relatively modest despite the high feeding activity of the protozoa. Investigations by Bahr (1944) concerning a range of bacteria confirmed what several other authors had suspected, namely that *B. subtilis* (a non-pathogenic hay bacillus) was taken in preference to *Staphylococcus aureus* (a cause of wound infection) and this was again preferred to *B. coli* (rod-like organism from the digestive tract).

Elimination of dissolved and colloidally dispersed constituents
   A direct uptake of dissolved and colloidal substrates has been reported for many protozoa. Colloidal substrates, such as viruses, are first of all adsorptively bound to the protozoan surface and then actively transported to the cell mouth. According to McKinney (1962) direct uptake of dissolved substrate only occurs at substrate concentrations above 5 g/l.

Production of glutinous substances
   which assist the formation of the sludge flocs. The production of glutinous matter has been reported for the ciliates *Balantiophorus minutus* and *Paramecium caudatum* together with accelerated floc formation by *Monus termo* (Hardin, 1943) and also for *Vorticella microtsomata* (Sudo and Shuichi 1973).

The numbers of metazoa are mostly very much smaller than those of protozoa, nevertheless owing to their large volume they occupy a considerable proportion of the activated sludge volume (Table 4.19). According to Erman (1963) *Branchionus calyciflorus* for example takes in daily 180% of its wet weight in the form of feed and hence only about 75% of that of the small ciliates. The metazoa are capable of ingesting the smaller activated sludge flocs, together with their associated pathogenic organisms, larger detritus particles, protozoa and also, depending on their size, some of the smaller metazoa.

Fig. 3.11 shows the food chain which is established in a sludge with a high sludge age. During passage from one level to the next one above, such as from bacteria to flagellates, from flagellates to ciliates, from ciliates to insect larvae, rotifers, fungi,

Table 4.19 — Individual numbers and volumes of protozoa and metazoa found in five activated sludge plants treating different types of wastewater. V_J = volume of an individual; J = No. of individuals; V_g = Total vol. of all individuals of a species

Activated sludge plant		Municipal sewage plant						Textile dyehouse effluents		Laundry wastewater	
		1		2		3					
B_{TS} [Kg/kg·d]		1.0		0.7		0.3		0.4		0.2	
SV [ml/l]		280		210		190		310		122	
Organisms	V_J 10^3 μm^3	J 10^3/ml	V_g $10^3 \mu m^3$/ml	J 10^3/ml	V_g $10^3 \mu m^3$/ml	J 10^3/ml	V_g $10^3 \mu m^3$/ml	J 10^3/ml	V_g $10^3 \mu m^3$/ml	J 10^3/ml	V_g $10^3 \mu m^3$/ml
Flagellates:											
Bodo angustus	0.16										
B. caudatus	0.18	216	38.9	125	22.5	67	12.1			9	1.4
B. saltans	0.16	510	83.2	5340	854	36	5.8				
Helkesimastix faecicola	0.08			18	1.4						
Monas guttala	0.52	16	8.3								
M. termo	0.11			18	2.0	18	2.0	18	1.1	0.1	0.01
Petalomonas pusilla	0.06	16	1.0	107	6.4	9	0.5				
Rhizopods:											
Amoeba limax	25			0.1	2.5						
Diplophrys archeri	0.5										
Hartmaniella cantbrigiensis	0.7	8	5.6				9	4.5		18	9
Thecamoeba verrucosa	60	8	480								
Trichamoeba cloaca	15								135		
Vahlkampfia spp	0.3			285	85.5	107	32.1	36	12		
Vanella mira	4.4			18	79.2						
V. platypodia	2.8							196	549		
Ciliates:											
Aspidisca costata	5.5			3.6	19.8	4	22	2.7	14.8	2.2	30.8
Euplotes affinis	14										
Opercularia coarctata	60	0.25	15	0.9	54	0.9	54	2.4	16.8	0.4	2.8
Trachelophyllum pusillem	7	0.25	1.8			1	7		0.1	0.7	
Uronema marinum	7										
Vorticells convallaria	64	1	64			3	192	3.2	205		
V. microstoma	54	8	432								
V. putrina	48	9	96	10	480	0.1	4.8	0.3	16		
Metazoa:											
Nematodes	11000	0.1	1100	0.06	660	0.1	1100	0.4	680	0.94	3264
Philodina roseola	1700									1.98	990
Cephalodella gracilis	500										
Volume of Protozoa and Metazoa											
(ml/l)			2.33		2.26		1.43		1.63		4.2
(% SV)			0.83		1.08		0.75		0.53		3.4

etc., from 50% to 90% of the energy content of the lower stage is consumed, that is 10–18 kJ/g organic solids.

Protozoa and metazoa thus in general have a positive role in activated sludge. The more species there are, the greater is the chance that even the more refractory substances will be utilised by the sludge and decomposed, or else deposited in the form of non-biodegradable matter within the organisms, for example in the fatty tissue of the metazoa. In this light we should also consider those organisms which according to our present knowledge do not exert any positive influence on the treatment process, such as protozoa-consuming fungi, floc-consuming Chironomid larvae, etc.

4.4 PATHOGENIC ORGANISMS

Pathogenic organisms may be present in municipal sewage which includes effluents from a variety of sources such as hospitals for infectious diseases, sanatoria, abattoirs, stables, tanneries, milk factories, etc. They are the infective agents for a wide range of human and animal disease. Increasing tourism, foreign travel and international transport of livestock all lead to an enhanced incidence of exotic disease organisms. The numbers and variety of pathogens present in sewage provide a good indication of the level of infection in the catchment area, as the causative agents of all infectious illnesses will be present in the sewage. Activated sludge installations possess a considerable anti-pathogenic potential by reason of the actions of protozoa and matazoa and of saprophytic bacteria, coupled with the use of correctly-dimensioned final settling tanks. The possible risk of infection from the pathogens present in sewage and activated sludge is dependent on the number of organisms and the size of the infective dose; both factors are specific to a given disease-producer, and the infective dose is also dependent on the state of health of the human or animal affected. Infection may take place by a number of routes, for example via mucous membranes, oral ingestion, inhalation and also percutaneously (e.g. by wound infection).

Questioning of several factory doctors employed by the State Concern for Water Supply and Sewage Treatment showed that sewage plant operatives did not exhibit increased sickness indices on account of infectious illness. It must, however, be borne in mind that the *immune status* of sewage works' personnel is very high and the occasional visitor is at risk from sewage-related infectious disease. According to the requirements laid down in TGL 26730/01, strict attention must also be given to personal hygiene when dealing with sewage.

4.4.1 Viruses

Our knowledge of the risk of infection associated with sewage-borne viruses is still defective in many respects. Thus at the present time we do not know for certain the minimum infective dose for any of the viruses; for this reason we cannot state with certainty to what extent the viruses present in activated sludge may actually produce illness. However, it is well known that viruses give rise to life-threatening forms of illness in both man and animals, and hence the greatest care is called for.

Human faecal excreta may contain from 10^6–10^9 viruses/g ($\sim$100 g/day) and even, in the case of rotaviruses, as many as 10^{11}/g (Müller 1981); Table 4.21 shows the most

Table 4.20 — Effect of ciliates on treated effluent quality for laboratory scale plants. From Curds et al. (1968)

Parameter	With ciliates	Without ciliates
TS (mg/l)	26–34	86–118
BOD$_5$ (mg/l)	7–24	53–70
COD (mg/l)	124–142	198–250
Bacterial count ($\times 10^6$/ml)	1–9	106–160

Table 4.21 — Viral human pathogens which enter sewage with urine and faeces. From Müller (1981). Reoviruses from Schäfer (1970)

Virus groups	No. of types	Nature of illness due to viral infection
Enterovirus		
Poliovirus	3	Paralysis, meningitis, fever
Echovirus	34	Meningitis, diarrhoea, respiratory infections, fever
Coxsackie virus A	24	Meningitis, herpetic angina, respiratory infections, fever
Coxsackie virus B	6	Myocarditis, heart disorders in neonates, meningitis, fever, respiratory illness, pleurisy
New enteroviruses	4	Meningitis, encephalitis, acute haemorrhagic conjunctivitis, respiratory illness, fever
Hepatitis A	1	Infectious hepatitis
Gastroenteritis virus (Norwalk type)	2	Epidemic vomiting and diarrhoea; fever
Rotavirus	?	Epidemic vomiting and diarrhoea, especially in children
Reovirus	3	Respiratory andf gastrointestinal illnesses
Adenovirus	>30	Respiratory illness, eye infections
Parvovirus	3	Not confirmed; perhaps respiratory illness in children

important sewage-borne viruses which are pathogenic to man. Animal viral pathogens are for example and causative agents of foot-and-mouth disease, cattle-, swine-, and fowl-pest, infectious anaemia of horses, cowpox and related diseases in sheep and pigs. Owing to the very large numbers present in sewage, the infective sewage volume is possibly in the millilitre or even the microlitre range. According to Dobberkau and Waller (1979) the enterovirus content of sewage is equivalent to five times the infective dose per millilitre.

Viruses are bound to particulate matter such as detritus and bacteria, so that for their elimination, coagulation and flocculation processes together with final settling are of overriding importance. Berg (1965) states 'The activated sludge process appears to be the one most suited to the removal of viruses from sewage'. Especially appropriate in this connection is the fact that no enrichment of the viruses in the sludge takes place, but only a purging effect (Fig. 4.30). Nevertheless the remaining viruses — in contrast to their behaviour during the sludge digestion stage — do not lose their pathogenicity, for example the foot and mouth virus and swine pest virus (Hussel 1963).

Adsorption onto the activated sludge flocs as a means of viral elimination proceeds very rapidly, together with the grazing activity of the protozoa and also the bacterial metabolism of ribonucleic and deoxynucleic acids forming the viruses.

Möse et al. (1970) give a review of the literature concerning the inactivation of viruses by protozoa. Experiments performed with *Tetrahymena pyriformis* showed that influenza and vaccinia viruses were inactivated, while poliovirus — probably on account of its small size — was unaffected (see Fig. 4.30).

The viruses are first adsorptively attached to the cell surface and then transported into the nutrient vacuoles. Numerous studies concerning the inactivation of viruses by amoebae have shown that amobae are also capable of inactivating certain other viruses, such as polioviruses.

In contrast to some pathogenic bacteria, neither human nor animal viral pathogens can multiply in the activated sludge. The multiplication of bacteriophages, which kill pathogenic bacteria, such as Salmonellae, is of course much to be desired. Whether the bacteriophages also sometimes attack those bacteria responsible for the treatment performance to such an extent that the performance is impaired is conceivable but not yet substantiated. Little is also known about the possible occurrence of protozoan viruses. According to studies by Kovacs et al. (1969) some viruses can multiply within yeast cells and in the ciliate *Tetrahymena pyriformis*, without actually being released into the surrounding medium.

4.4.2 Bacteria

Sewage contains both obligate and facultative pathogenic bacteria. To the latter category belong some of the forms frequently encountered in activated sludge from the taxa *Pseudomonas Proteus*, *Aerobacter*, *Streptococcus*, etc. which belong to the normal resident flora of humans and animals and only occasionally give rise to illness, the so-called banal organisms. In the case of illness a causal relationship with sewage can hardly be demonstrated.

Table 4.22 shows the most important human pathogenic bacteria in sewage, and Table 4.23 shows some frequently-occurring animal pathogens. About 150 species can cause zoonosis, that is they are pathogenic to both humans and animals. From a comparison of the numbers of pathogens present in sewage and the number of organisms required to initiate an illness, according to Table 4.22 it emerges that the minimum volume of sewage capable of causing infection following oral ingestion is generally greater than 0.1–1.0 litre. During epidemics sewage or effluents from hospitals and scientific establishments will naturally contain substantially greater numbers of these organisms.

During treatment of sewage by activated sludges which incorporate protozoa the

Table 4.22 — Bacterial human pathogens which enter sewage with urine and faeces. Data extracted from Müller (1981)

Bacterial pathogen	Illness	Nos./litre of raw municipal sewage	Nos. to initiate infection in man	Infective volume of sewage [1]
Salmonella typhimurium	Enteritis	$<1.8 \times 10^8$	10^5–10^8	>1
S. typhi	Typhoid	$<10^3$	10^3	>1
S. paratyphi B	Paratyphoid	$<10^3$	$>10^3$	≥0.1
Shigella spec.div.	Dysentery	$<10^2$	10–10^2	>0.1 (>1)
Escherichia coli	Enteritis	$<10^2$	10–10^2	>0.1
Campylobacter fetus	Gastroenteritis	?	?	>0.1
Vibrio spec. div.	Cholera Diarrhoea	$<10^3$	10^4–10^6	>1
Yersinia enterocolitica	Gastroenteritis	$<10^6$	$>10^6$	>1
Y. psueudotuberculosis	Pseudotuberculosis	$<10^6$	$>10^6$	>1
Mycobacterium tuberculosis	Tuberculosis	$<2.5 \times 10^2$	$\sim 10^2$	>0.1

Table 4.23 — Bacterial animal diseases and zoonoses capable of transmission by sewage. From Amlacher (197), Cena (1975), Hussel (1963) and (16)

Bacterial pathogen	Nature of illness	Transmission to:
Salmonella spp. (>600 types)	Enteritis, typhoid, etc.	Cattle, horses, ducks, gulls, rodents, humans
Mycobacterium spp. (various)	Tuberculosis	Cattle, horses, pigs, dogs, cats, humans
Bacillus anthrax	Anthrax	Sheep, pigs, cattle, horses, camels, humans
Erysipelothrix insidiosa	Swine fever	Pigs, rabbits, humans
Brucella spp. (various)	Brucellosis	Ruminants, pigs, horses, poultry, humans
Listeria spp. (various)	Listeriosis	Pigs, rats, humans
Leptospira interrogans	Leptospirosis	Cattle, horses, rodents, humans
Francisella tulariensis	Tularaemia	Hares, rodents, humans
Pseudomonas putida	Red murrain	Carp, eels
P. fluorescens	Ulceration	Many fish species
Aeromonas punctata	Infectious disease of the abdominal fluid	Carp
Mycobaterium piscium	Fish tuberculosis	Many fish species

colony counts of pathogenic bacteria are generally reduced by 90–98%. Colony counts of the inherently resistant and notorious agents of diarrhoeal disease of the *Salmonella* group, are reduced for example by 0.7–2.2 powers of ten. Fleissig (1960) using batch tests was able to show that in a mixed cultue of *Escherichia coli* and *Salmonella enteriditis* the latter had completely disappeared within 4 hours *E. coli* is capable of inhibiting or even killing various cocci, tuberculosis and anthrax producers. The inhibitory substances produced by the bacteria and acting outside the confines of the cell are termed *colicins*. Besides the grazing activity of the protozoa and metazoa, as well as the action of bacteriophages, there are also certain bacteria and fungi which are antagonists of the pathogenic bacteria and can eradicate them. The more fastidious pathogens are also at a disadvantage relative to the saprophytic bacteria with respect to competition for substrates as well as the effect of environmental factors in the sludge, so that they are unable to multiply at the same rates. They are therefore overwhelmed by the floc-forming bacteria, so that they become occluded within the sludge biomass and hence are ultimately drawn off with the wasted sludge. On the other hand their inclusion within the flocs provides a certain degree of protection for pathogenic bacteria against the effect of the unfavourable environment. The same protective effect can also be obtained from protein and coarse solid particles. Hence particles present as carryover in the effluent from the final settling tank pose a constant danger of infection.

The above-mentioned growth inhibition of pathogenic bacteria by saprophytic bacteria is, however, not of general validity. Thus bacteria from the taxa Staphylococcus and Bacillus may stimulate the growth of the tuberculosis-producing mycobacterium. In addition there are numerous reports in the literature of Salmonellae being able to multiply in activated sludges at elevated temperatures and with a good supply of protein.

A comprehensive literature survey concerning the survival time for various pathogenic organisms was presented by Bargel (1965). The periods of 50 years for anthrax bacilli and those of two years for a number of Salmonellae indicate that caution is needed in the disposal and re-use of sewage effluent. Despite this however, the infective volumes apparent from Table 4.22 are so large that any serious danger or risk of the spread of bacterial infections due to treated sewage effluents is unlikely.

In the control of water quality as a rule, groups of bacteria rather than individual species are enumerated. The guidelines proposed by Fiedler (1981) for bacterial counts for biologically treated sewage (Table 10.2) of municipal origin are attainable by activated sludge plants only in exceptional cases. For irrigation purposes in agriculture governed by TGL 6466/01, municipal sewage which has been treated by the activated sludge process must accordingly be rated as belonging to the category E_b5.

4.4.3 Protozoa

The possible disease-producers comprise *Balantidium coli* (a ciliate) as the causative agent of dysentry, *Entamoeba hystolytica* (a rhizopod) as the agent of amoebiasis, and *Giardia lamblia* (a colourless flagellate) as the agent of giardiasis. Numbers of *Giardia* cysts in municipal sewage, according to data reported by Müller (1981) amount to 90–520/litre, while the minimum infective dose varies widely from 10–25 cysts up to 1 million. For the other protozoa referred to reliable counts are not available, but as they are less common than *Giardia* it is probable that the risk of infection is smaller. For *Entamoeba histolytica* in Central Europe, about 12–15% of the population must be regarded as carriers, so that there is a certain risk of infection in this case. Concerning the elimination of protozoa it must be assumed that a high percentage of the total is bound to the sludge biomass and hence is removed during the treatment of sewage.

4.4.4 Helminth ova

The number of parasitic ova in sewage is primarily determined by the indirect discharges from abattoirs (max. 23 000/litre of slaughterhouse effluent according to Hussel 1963) and occasionally by the runoff from livestock rearing establishments. Around 90% of the ova in sewage are of animal origin (Table 4.24). For eleven

Table 4. 24 — Animal parasites transmissable by water in temperate climatic regions. From Amlacher (1972) and Hussel (1963)

Animals affected	Helminth ova present in sewage
Horses	Palisade worm, various Strongylides, *Parascaris equorium*
Cattle	*Ascaris vitolorum, Haemonchus-Trichostrongylus* spp and *Nematodirus* spp.
Sheep	Large and small liver parasites, stomach worms *Protostrongylus* and *Dictycaulus* ssp.
Pigs	*Ascaris lumbricoides*, several *Metastrongylus* spp. tapeworms (*Trichuris trichura*)
Fish	*Dibothriocephalus latus* (fish tapeworm)

different sources of municipal sewage from smaller communities, Knaack and Ritschel (1975) obtained the following results.

Total no. of ova/litre=27.3, of which *Ascaris*=6.7, *Enterobius*=9.9, *Trichuris*=3.1, *Taenia*=4.0, Others=3.6; Max. no. recorded=123.

The density of ova is greater than 1 g/cm^3, hence they are about 90% eliminated at the primary settling stage, assuming a retention time of 1.5 h. Tapeworm ova, however, owing to their small size settle at a rate of only about 10–20 cm/h, so that they quite often pass through the primary settling tank. The ova which are thus washed out of the primary settling tank (according to section 12.2.2 this may sometimes be omitted) then accumulate in the activated sludge. However, a sufficiently large and properly designed final settling tank will achieve practically 100% removal of the remainder. According to Knaack and Ritschel (1975) the number of ova in the treated effluent from the final clarifier is correlated with the amount of settleable solids. With SVI⩾0.1 ml/l the retention effect diminishes markedly. There are also particular dangers associated with Dortmund tanks under denitrifying conditions and also with small activated sludge units incorporating separate sludge compartments.

Parasitic ova retain their viability and infectivity so that ova washed out of the final settling stage constitute a continual risk of infection. For sewage works' staff there is only a minor infectious hazard, as nearly all the ova require a special intermediate host for their further development (Table 3.2). Strictly speaking only the relatively harmless and easily counteracted intestinal parasite *Enterobius vermicularis*, frequently found in children, can be directly transmitted between humans; at

Sec. 4.5]	**Physicochemical consequences**	179

temperatures below 23°C the ova die within about two days. The otherwise directly transmissible dwarf tapeworm *Hymenolepsis nana* is rare and hence of no practical significance. In lightly-loaded activated sludge plants with a good oxygen supply and high liquid temperatures the conditions may favour incubation of the eggs of *Ascaris* with possible infectious consequences (at 15°C~30 days, or at 35°C~12 days).

4.5 PHYSICOCHEMICAL CONSEQUENCES

As a result of the metabolic activity of the organisms and the presence of dissolved salts and gases in the liquid, a wide variety of physicochemical processes can take place in the activated sludge tanks. As a rule these processes are hidden so that they are not obvious in the course of operating checks, but may nevertheless be the cause of operational disturbances.

pH

The following four examples may serve to illustrate the main features.

During studies of the treatment of sewage from the city of Mülhausen in 1973 there were pH fluctuations between 8.5 and 13 in the raw sewage between 7.00 a.m. on Monday and 11 p.m. on Saturday, while at the weekend the pH was about 7; the pH in the aeration tank, however, was invariably between 7.8 and 8.5 (Fig. 4.31(A)).

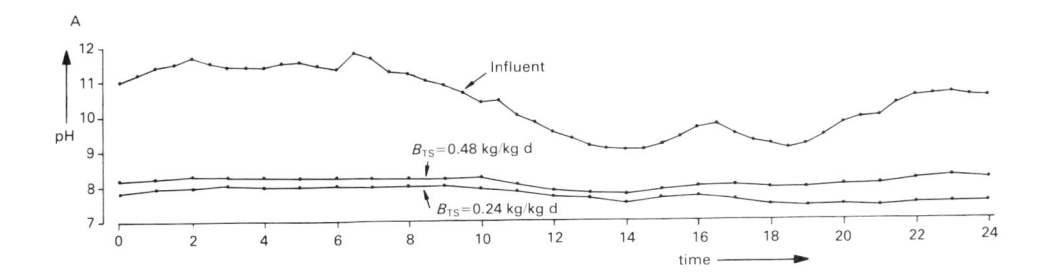

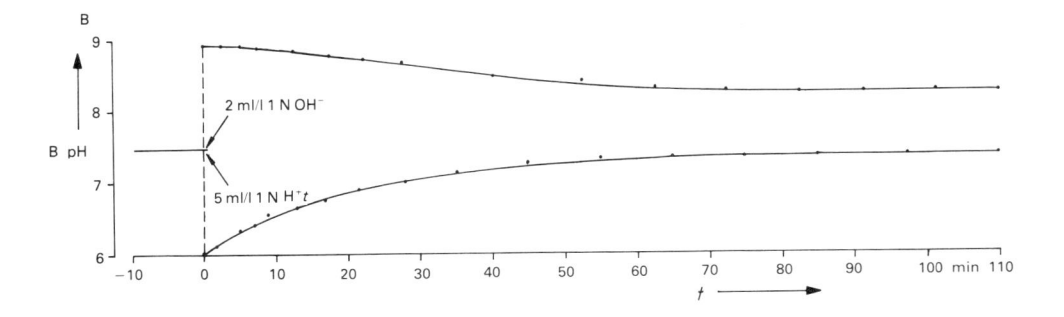

Fig. 4.31 — pH regulation as a result of the metabolic activity of the sludge biomass. A: Extensive neutralisation of an alkaline effluent from the textile industry in two continuously operated stirred reactors with different loading rates. B: Buffering of acids and alkalis during the endogenous respiration phase under laboratory conditions.

Similar studies for Karl-Marx-Stadt during 1964–1965 disclosed extremes of pH in the region of 2–3 on almost every weekday, which were, however, invariably buffered to values of 6.5–7.5. Sládeček (1955) also reported for the Hostivar sewage works, Prague, that the pH of a sewage of purely municipal origin, with an alkalinity of 10–20 meq/l, fell to 5.0 in the aeration tank. For ammoniacal coke-oven effluents, containing 4.5 g/l of fatty acids, Zülke (1967) reported a pH rise from 5.9 to 8.1.

The dissolved salts contribute to relatively stable and reversible equilibria (Fig. 4.31(B)) which help to counteract any change in pH. Among the buffer systems present in activated sludge the carbonate/carbonic acid system covering the range pH 4.5–8.3 is normally paramount. Carbon dioxide is formed in considerable quantities in the course of the citric acid cycle (Fig. 4.18) of biochemical changes, and also as a result of anaerobic reactions, 367 mg/l of CO_2 being formed from the decarboxylation of 100 mg/l of organic carbon. Its physical solubility varies with temperature:

Temperature (°C)	0	10	20
Sol. of CO_2 (mg/l)	110	84	68

A substantial part of the carbon dioxide produced dissolves in the water according to the equation:

$$CO_2 + H_2O \rightleftharpoons H_2CO_3.$$

The carbonic acid undergoes dissociation as follows:

$$H_2CO_3 \rightleftharpoons H^+ + HCO_3^- \rightleftharpoons 2H^+ + CO_3^{2-}.$$

Those forms of carbonic acid and their salts which occur in water and their dependence on pH are shown in Table 4.25.

Table 4.25 — Proportions (%) of free and combined forms of CO_2 as a function of pH. From [2]

pH	Free CO_2	HCO_3^-	CO_3^{2-}
4	99.5	0.5	—
5	95.4	4.6	—
6	67.7	32.3	—
7	17.3	82.7	—
8	2.0	97.4	0.6
8.3	1.0	97.8	1.2
9	0.2	94.1	5.7
10	—	62.5	37.5
11	—	14.3	85.7

The buffering capacity evident in the first two examples cited above depends primarily on the activity of the bicarbonate ion, while above pH 8.3 a further decisive factor is the solubility equilibrium of $CaCO_3$:

$$HCO_3^- + OH^- \rightleftharpoons H_2O + CO_3^{2-};$$
$$HCO_3^- + H^+ \rightleftharpoons H_2O + CO_2.$$

From the data given in Table 4.25 the bicarbonate buffer is effective over the pH range from 4.5 to 11. Where the buffer capacity is decreased as a result of the decomposition of HCO_3^- by alkalis, then by the laws of mass action, more and more carbon dioxide goes into solution as bicarbonate, and the buffer capacity is thereby restored. From 300 to 350 mg/l of bicarbonate may already be present in the incoming sewage.

A further source of bicarbonate ions may arise in the aeration tank as a consequence of the hydrolytic cleavage of urea according to the following equations.

$$CO(NH_2)_2 + 2H_2O \rightarrow (NH_4)_2CO_3;$$
$$(NH_4)_2CO_3 + H_2O \rightarrow NH_4HCO_3 + NH_4OH.$$

Besides the bicarbonate buffer system, the phosphate buffer system and the amphoteric behaviour of proteins are also important contributors of buffering activity.

$$HPO_4^{2-} + OH^- \rightarrow H_2O + PO_4^{3-}$$
$$HPO_4^{2-} + H^+ \rightarrow H_2PO_4^-.$$

The buffering action of the first two plants cited above, however, amounted to only about 2–4% of that of the bicarbonate buffer system.

In the third of the above-mentioned examples the pH buffering acitivity was disrupted by the nitrates and nitric acid produced as a result of nitrification, until finally the free acids reduced the pH, even though according to Table 4.5, bicarbonate can still be supplied at very low pH values.

In the fourth example, the pH which reflected the interaction between the fatty acids and the ammonia finally rose following the decomposition of the fatty acids. Organic acids may also be present at high concentrations in effluents from the food industry and in digester liquor, and are the cause of very low pH values. When these acids are metabolised by the activated sludge organisms, the pH may rise by 1 or 2 units.

The drop in pH as a consequence of nitrification in weakly-buffered media may give rise to bulking sludge problems. For a preliminary estimate of the loss of alkalinity, Kapp (1983) took into consideration the average carbonate hardness of the mains water, the carbonate combined with ammonium ions, and the phosphate buffer, according to the following equations:

— alkalinity in the influent

$A_i = 17.86\ H_T + 3.57\ NH_4-N_i + 1.4\ P_i$ (mg/l $CaCO_3$);

— alkalinity after biological treatment
$= A_i - 7.14\ (NO_x)_e - 3.57\ [NH_4-N_i - NH_4-N_e] - 1.4\ (P_i - P_e)$ mg/l $CaCO_3$;

where the symbols have the meanings below

Subscripts i and e = influent and effluent, respectively;
A = alkalinity;
H_T = total carbonate hardness of mains water (°dH);
P = total phosphorus content (mg/l);
NH_4^+–N = content of ammonium ions (mg/l);
(NO_x) = content of NO^- + NO_3^- (mg N/l).

When the alkalinity falls below 70 mg/l $CaCO_3$ in the course of biological treatment, in order to avoid operational problems from bulking sludge formation it is advisable to arrange for prior denitrification. With such weakly-buffered systems hardly any buffer capacity is available to counteract acidic effluents.

Precipitation reactions
Some metal ions can form insoluble salts or hydroxides by combination with a range of anions, such as sulphate, phosphate etc., and these are deposited on the surface of the flocs or occluded by them and finally wasted with the surplus sludge. Some of the salts help to weight the sludge flocs and hence give rise to improved settling behaviour.

From the water quality aspect the coagulation of phosphates with calcium, iron or aluminium salts is the most important example. The coagulation products may be of variable composition in chemical terms, owing to the presence of other ions, the pH value, the oxygen regime and the ratio of phosphate to coagulant concentrations.

Chemical phosphate removal with iron and aluminium salts depends on the fomation of poorly-soluble metal phosphates and hydroxides as well as on the sorption of monomeric, condensed and organically-combined phosphates onto the metal hydroxide flocs.

Oxyhydrates of trivalent iron and aluminium tend to form condensation products with elimination of water, the number of OH-groups being reduced and a lattice structure composed of —O—Me—O—Me— chains produced, the non-combined third 'reactive arm' of the metal ions having the ability for binding phosphate, for example:

Only a small fraction of the nominal reactive centres is used to bind phosphate, which is the reason why for simultaneous coagulation often 2–3 moles of iron are required for each mole of phosphate. A progressive ageing of the hydroxide precipitate is associated with a loss of water and cross-linking of the chains into strips, so that a release of phosphate may occur:

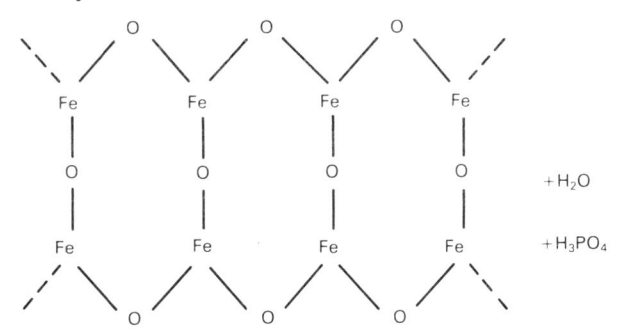

$$\begin{array}{ccccc}
\diagdown / \mathrm{O} \diagdown & / \mathrm{O} \diagdown & / \mathrm{O} \diagdown & / \mathrm{O} \diagdown & \\
\mathrm{Fe} & \mathrm{Fe} & \mathrm{Fe} & \mathrm{Fe} & \\
| & | & | & | & \\
\mathrm{O} & \mathrm{O} & \mathrm{O} & \mathrm{O} & +\mathrm{H_2O} \\
| & | & | & | & \\
\mathrm{Fe} & \mathrm{Fe} & \mathrm{Fe} & \mathrm{Fe} & +\mathrm{H_3PO_4} \\
\diagup \diagdown \mathrm{O} & \diagup \diagdown \mathrm{O} & \diagup \diagdown \mathrm{O} & \diagup \diagdown & \\
\end{array}$$

Absorption and adsorption, biosorption
Sorption processes play an important part in the activated sludge system (Figs 4.7 and 4.10) in which it is not possible to distinguish exactly between the physical adsorption mechanism and the biochemical atttachment of the substrate to the enzymes. With municipal sewage and lightly-loaded plug-flow reactors, about 60% of the degradable BOD_5 is removed in a matter of two to three minutes by sorption processes and in lightly-loaded stirred reactors the proportion may be as much as 80–90%. According to Kroiss (1981) the adsorption follows a saturation process, in which full saturation is obtained with 200–300 mg of COD per gram of organic solids. The organic sewage constituents taken up by sorption processes during the first few minutes are then degraded by enzymatic reactions of much greater duration, so that finally the sorptive capacity of the flocs is liberated again.

By introducing activated carbon into the aeration tank the adsorption processes for removal of substances metabolised only slowly may be greatly enhanced. It is supposed that the substances attached to the activated carbon particles can act as substances for specialised strains of bacteria which become attached to the carbon.

Ion exchange
By reason of the surface structure of the cell membrane, different ions may be adsorptively bound to the organisms. With a sudden excess of other ions, ion-exchange reactions may be initiated, which may — as indicated under section 4.3.6, on heavy metals — lead to toxic effects.

Stripping of volatile sewage constituents
The gases given off from the aeration contain a variety of volatile constituents such as the odourless compounds CO_2, CH_4, H_2O, N_2O, N_2, H_2, O_2 as well as compounds which are detectable by smell like H_2S, organic acids and ammonia. The concentation of these compounds in the off-gas is dependent on the nature of the sewage, and the conditions obtaining in the aeration tank, in particular the pH, the redox potential, the temperature, the energy input, the tank geometry, and also on the various operating principles and types of aeration system. According to Table 5.5,

effluents from oil-refining operations may contain n-pentane, toluene and benzene, while Engelhardt (1978), for effluent from a chemical factory, cited aniline derivatives, alcohols, nitrobenzene, esters, ortho-cresol, etc., as being present in the off-gas. For the latter establishment a total of 60 kg organic carbon was released to atmosphere per hour. These emissions could be reduced to about one-sixth by the fitting of hoods of various kinds to the mechanical aerators, without creating any significant reduction in the oxygen input or treatment performance.

Table 5.5 shows that for effluents from oil refinery operations some substrates may be more effectively stripped than metabolised. Petroleum hydrocarbon fractions are stripped in the following order of intensity: petrol > aviation spirit > petroleum > engine oil > crude oil > gearbox oil. For reasons of environmental protection, where high concentrations of such volatile substances are present, it may be necessary to use enclosed reactors with exhaust gas scrubbing, or else to aerate with technically pure oxygen.

5
Oxygen

The presence of oxygen is the crucial factor which determines whether the biochemical processes taking place in the activated sludge compartment proceed according to the scheme outlined in Fig. 4.18 or that in Fig. 4.25, that is whether the treated effluent quality will correspond to the composition indicated in Table 10.1 or whether the actual values will be several times greater, As, moreover, the aeration system accounts for up to 80% of the total energy requirement of a sewage treatment plant [14], oxygen supply is of overriding importance for satisfactory plant operation both from the economic and the water quality aspects.

5.1 OXYGEN INPUT

The oxygen input, that is the rate of oxygen transfer from the atmosphere or from air bubbles into the activated sludge suspension, is governed by Fick's Law of Diffusion, according to which the rate at which a given amount (m) of oxygen diffuses per unit of time across a surface of area A is proportional to the concentration gradient (dc/dx) and the diffusion constant D:

$$\frac{dm}{dt} = D \times A \frac{dc}{dx}.$$

The diffusion constant D is system-dependent, such that the bubble size, the temperature and the constituents of the liquid phase are controlling factors. Both the bubble size and the surface area of the phase boundary can be controlled by technical means.

The general form of Fick's Law was adapted by Pasveer (1958) for application to the oxygen transfer into pure water as follows:

$$OC_{10} = 25.9 \frac{1}{t_t - t_0} \log \frac{C_s - C_0}{C_s - C_t} \sqrt{\frac{K_{10}}{K_t}} (g/m^3 \, h)$$

where
- OC_{10} = oxygen input at 10°C (g/m³ h);
- $t_t - t_0$ = period of measurement (h);
- C_s = oxygen saturation concentration for the test conditions (g/m³ h);
- C_0 = oxygen content at time t_0 (g/m³ h);
- C_t = oxygen content at time t_t (g/m³ h);

$\sqrt{\dfrac{K_{10}}{K_t}}$ = conversion factor for the diffusion velocity at the temperature of the test to 10°C (Table 5.1).

Table 5.1 — Saturation values for dissolved oxygen in distilled water at a total pressure for the saturated atmosphere of 1013 hPa. From Elmore and Hayes (1960) and American Public Health Association (from 35°C) (Nösel 1978), together with temperature correction factors for oxygen input measurements (Pasveer 1958)

Temp °C	C_s (mg/l)	$\sqrt{K_{10}/K_t}$	Temp °C	C_s (mg/l)	$\sqrt{K_{10}/K_t}$
0	14.65	—	20	9.02	0.830
2	13.86	1.160	22	8.67	0.799
4	13.13	1.118	24	8.33	0.770
6	12.46	1.077	26	8.02	0.74
8	11.84	1.037	28	7.72	0.71
10	11.27	1.000	30	7.44	0.69
12	10.75	0.964	35	7.10	0.63
14	10.26	0.928	40	6.60	0.57
16	9.82	0.895	45	6.10	—
18	9.40	0.861	50	5.60	—

In the sewage/sludge suspension the oxygen saturation value (β-value = 0.9–0.98) and the diffusion constant D both differ from those in pure water and hence the oxygen input rate alters accordingly. The change in the diffusion constant is expressed by means of the factor α, where $\alpha (= D'/D$ or $OC'/OC)$ may take values from 0.3 up to 1.6. The value of α falls in response to inhibition of the oxygen transfer rate by surfactants (Scherb 1965, especially above 5 mg/l), oil (Hashimoto 1975), or salts (Pöpel 1971, Zlokarnik 1975) and also by activated sludge biomass (Fig. 5.1). On the other hand it may sometimes be increased, as in the case of hindered coalescence especially with strongly foaming media. The α-value is also dependent on the type of aeration system and is liable to very pronounced variation. Thus the literature review by Billmeier (1983) cited the following values:

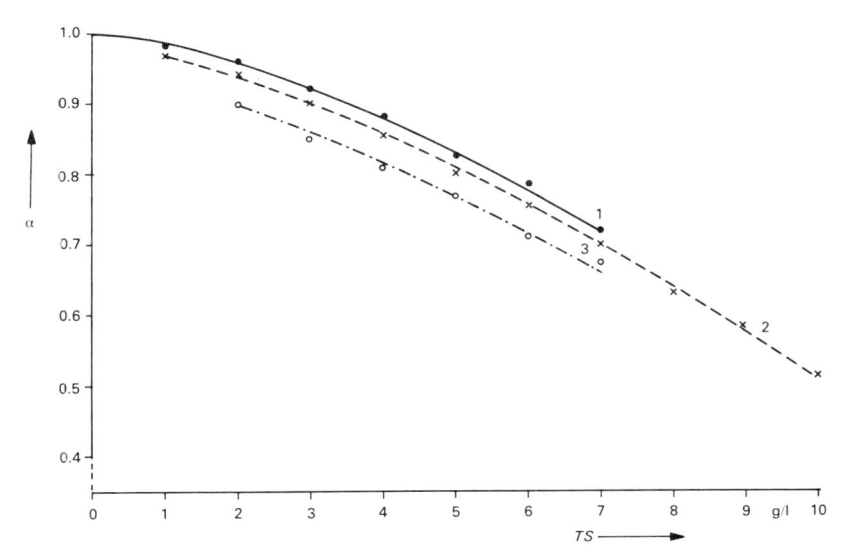

Fig. 5.1 — Effect of the solids content of the activated sludge suspension in the aeration tank on the α-value. Curve 1: surface aerators; Curve 2: compressed-air injection, from Kayser (1967), and Curve 3: compressed-air injection from Wolf and Nordmann (1978).

— primary treated municipal sewage, 0.2–0.6;
— fine-bubble aeration 0.29–0.77;
— medium-bubble aeration 0.4–0.94;
— horizontal rotors 0.38–0.93;
— centrifugal aerators 0.85–0.90 (see Plate 8).

The pronounced influence of temperature on the oxygen saturation value C_s can be seen in Table 5.1. Temperature has a somewhat smaller but still noticeable effect on the oxygen input which varies with the intensity of aeration.

5.2 MEASUREMENT OF OXYGEN INPUT

Measurements in pure water are performed as described by Pasveer (1958) with the aid of the formula given in section 5.1. The principle of measurement is based on the fact that the water is first of all deoxygenated by the addition of sulphite in the presence of a cobalt catalyst, or by purging with nitrogen gas; then following the start-up of the aerator the rate of increase of oxygen concentration with time is measured up to saturation level. A plot of $(C_s - C_t)$ against time on semi-logarithmic co-ordinates gives a straight line, the slope of which is correlated with the rate of oxygen input (Fig. 5.2). If the measurement values of the oxygen concentration C_t are entered on a specially programmed calculator, then the calculation of the OC values can be greatly simplified. For very exact calculations, further factors must be taken into account, compared with Pasveer's formula, such as the atmospheric pressure, height above sea level, etc. A factor of overriding importance for

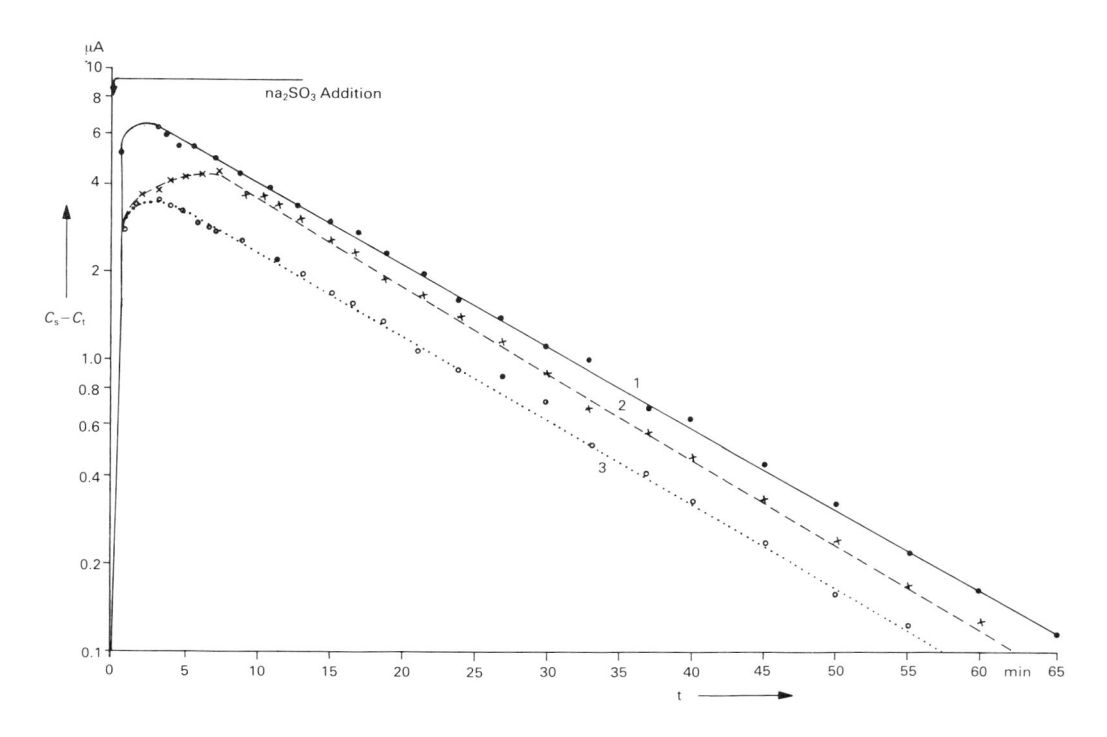

Fig. 5.2 — Measurement of oxygen input into clean water and tests of the degree of mixing in an aeration tank using non-temperature-compensated O_2 electrodes. Measurement pos. 1: $OC_{10} = 49.8$ g/m³ h; Measurement pos. 2: $OC_{10} = 52.7$ g/m³ h; Measurement pos. 3: $OC_{10} = 56.7$ g/m³ h. The rate of change at pos. 2 between 4 and 7 min is indicative of incomplete mixing of the contents of the tank at this point.

minimising errors is the exact determination of the C_s value. Measurement of the oxygen input under pure water conditions is particularly applicable to performance comparisons for different systems of aeration under comparable conditions. Contrary to several literature reports that this method is only suitable for measurements of OC values up to 150 g/m³ h, by the use of sensitive oxygen electrodes in conjunction with continuous recorders, OC values of up to 1650 g/m³ h can be reliably determined. It is thus adequate for all rates of oxygen input likely to be encountered in sewage treatment operations (see Fig. 4.19).

For measurements of oxygen input under *operating conditions* which may become necessary in the case of upsets in the oxygen budget, or as a prelude to planned increases in capacity, a sufficient degree of accuracy is only achieved, according to Kayser (1967), if the sludge biomass is in the endogenous respiration phase. For this it is necessary, prior to taking measurements, to disconnect the sewage input and sludge recycle lines and then to aerate the sludge suspension for about two hours. After achieving a constant rate of respiration the aeration system is shut off or turned right down, until the sludge biomass has used up all the available oxygen. After switching on the aerator again, the increase in the oxygen content is then monitored as for the case of pure water.

The oxygen input is then calculated from the equation:

$$OC' = K'\left(C_s^x + \frac{OV}{K'}\right) \text{ g/m}^3 \text{ h}$$

where
- OC' = oxygen input under operating conditions at the test temperature;
- C_s^x = terminal value of the oxygen content attained under constant rates of respiration and oxygen input (g/m³);
- OV = respiration rate of the activated sludge biomass (g/m³ h);
- K' = aeration constant (h⁻¹);

$$K' = \frac{60'}{t_{90\%}} \times 2.303$$

where $t_{90\%}$ is the time (minutes) for 90% of the saturation deficit to be eliminated.

Should the disconnection of the raw sewage and sludge recycle inputs not be practicable, then the oxygen input can be determined roughly from the oxygen mass balance equation:

$$OC' = OV \frac{0.9\, C_s}{0.9\, (C_s - C)} \text{ g/m}^3 \text{ h}$$

where
- C_s = theoretical oxygen saturation value (mg/l);
- C = actual oxygen content (mg/l).

The calculation of OC' becomes more exact the lower the actual oxygen concentration is (see Billmeier 1983).

Both methods for determination of the oxygen input under operating conditions are only valid at present for OC' values of up to 200 g/m³ h owing to the necessity for measuring the respiration rate.

5.3 AERATION SYSTEMS

The aeration systems perform the function of introducing *oxygen* into the aeration tank so that aerobic treatment of sewage can proceed; at the same time a sufficient degree of *turbulence* must be created so that the settling out of the sludge suspension is prevented, an intensive materials transfer between the sludge flocs and the liquid being treated is encouraged and concentration surges are balanced out by the general dilution effect for incoming sewage.

There are a multitude of different types of aeration system and often several variants of any particular system. In addition. the aeration system influences the performance of the organisms by the degree of macro- and micro-turbulence generated. The former affects the nature of the reactor and the circulation pattern prevailing, while the latter governs the floc size and hence the area available for mass transfer between the sludge flocs and the liquid. As the level of macro-and micro-turbulence must be matched to the oxygen demand of the activated sludge and to the desired level of treatment, and because of the variable sizes of different installations and limited availability of space which result in different tank configurations, there is no single aeration system which can be regarded as ideal. The choice of a particular aeration system will be based on a number of different factors, such as economy of oxygen supply, equipment costs, the demands of tank geometry and hence the type of construction, the cost of service and maintenance and not least the effects on the environment due to factors such as noise, aerosol formation or foam build-up.

According to Grosse et al. (1978) and also Hänel and Schmidt (1980) the following methods of aeration have proved effective in the municipal and industrial sectors on process-engineering and economic grounds.

Compressed air (Plates 7 and 10)
Forms of widespread application include fine-bubble aeration with apertures of less than 0.1 mm diameter for large-scale treatment plants, and medium-bubble aeration systems with apertures between 3 and 6 mm diameter for small-scale activated sludge systems. Fine-bubble aeration makes extensive use of porous ceramic diffusers, and more recently of sintered plastic, while the medium-bubble systems employ perforated plastic or steel sparge pipes. The attainable oxygen inputs and yields can be seen from Table 5.2. For the recently developed sintered plastic diffusers now in common

Table 5.2 — Representative values for specific oxygen input and aeration efficiency under operating conditions. From [14]

	Favourable conditions		Average conditions	
	OC' (g/m^3 m)	OC'/P kg/kWh	OC' (g/m^3 m)	OC'/P kg/kWh
Fine-bubble	10	1.8	8	1.3
Medium bubble (low level)	5.5	1.1	4.5	0.8
Medium bubble (high level)	7.5	1.5	6.5	1.2
Coarse bubble	4.5	0.9	4.0	0.7

use, Fiedler (1981) quotes an optimum throughput of 16 Nm^3/h and an oxygen input of 0.23 kg/h per diffuser. For countercurrent aeration, in which fine bubbles of compressed air are introduced against the direction of liquid circulation, oxygen yield coefficients of 3.75 kg/kWh have been recorded [11].

Many compressed-air installations suffer from an inadequate oxygen input, the causes for which often include a non-uniform and restricted escape of air from the diffuser, with α-values of perhaps only 0.6, and the very high level of oxygen demand close to the sewage inlet, since aeration tanks for very large plants are usually designed as plug-flow reactors.

Horizontal rotors (Plate 11)
These may be designed as spiked rotors or cage rotors and are installed in trench-like aeration compartments, such as oxidation ditches or aerated lagoons. According to [17] the rate of oxygen input for a standard design of cage rotor amounts to 6.75 kg/h, the OC per unit power consumed is from 1.5 to 2.2 kg/kWh.

Centrifugal aerators (Plates 8 and 12)
The type B centrifugal aerator specified in TGL 36763 has become standard equipment for activated sludge plants; the oxygen input and energetic performance are indicated in Table 5.3. The particular benefits of the centrifugal impeller without

Table 5.3 — Performance indices for centrifugal aerators of Type B according to TGL 36763/02

Model No.	n (min^{-1})	OC_{10} kg/h	P kW	OC_{10}/P kg/kWh
B 900-89/60	89	6.9	3.4	2.0
	60	2.5	2.2	1.1
B 1500-57/38	57	26	12.3	2.1
	38	9	42	2.1
B 2400-46/31	46	95	57	1.7
	31	27	23	1.2

a draught tube lie in the high level of oxygen input and in the ease of control of the rate of input, that is, varying the rate of electric power consumed for oxygenation in response to variations in operating conditions, and not merely for achieving a completely mixed reactor system. The alternative variant Type A [18], previously in common use, is now no longer recomended — except for very deep tanks — on account of its turbulence-limiting draught tube; with very wide tanks, failure of one aerator, as well as disconnection for reasons for oxygenation control, may lead to sludge deposition on the floor of the tank.

For treatment of industrial effluents with an oxygen demand of appreciably more than 150 g/m^3 h several specialised aeration systems have been developed [8]. In contrast to the performance indicated in Fig. 4.19, such effluents may exhibit very high oxygen requirements:

— at high temperatures, in line with the Arrhenius Rule;
— in the case of special toxic effects, such as decoupling of the respiration chain;

— in cases of deflocculated biomass growth owing to the increased specific surface of the organisms;
— for sludge solids concentrations of over 4 g/l;
— for aerobic mesophilic and thermophilic sludge stabilisation.

In such cases the following types of aerator have been employed.

Submerged centrifugal aerators of the Franz pattern (1968)
For activated sludge suspensions with strongly foaming tendencies in relatively shallow tanks (≤2.5-m deep) this type of aerator, sometimes in conjunction with compressed air, which helps to generate the necessary turbulence on the floor of the tank, has proved satisfactory. Typical applications are in treating effluents from low-temperature carbonization, feed yeast production and the manufacture of fattening diets for pigs. For such conditions the energy requirements are equivalent to 1.5 kg O_2/kWh, but for municipal and similar types of sewage, their performance falls to less than 1 kg O_2/kWh.

Submerged jet aerators (Plate 9)
The special advantages of submerged jet aerators, including their modification to inclined impulse jet systems, lie in their very low noise emission and in the lower level of aerosol formation and release of osmogens as well as the possibility of exhaust gas treatment. At low energy inputs the specific power consumption is of the order of 1.1 kg O_2/kWh [37]; however, for power inputs ≥200 W/m^3 and for effluents from the detergent manufacturing industry with impaired coalescence tendencies, values as high as 2.8 kg O_2/kWh may be achieved.

Disadvantages of the submerged jet design include incomplete mixing of the contents of the reactor on account of the localised input of energy, the difficulty of achieving control of the oxygen rate, and a distinct tendency to promote foam formation; in order to counteract foaming it is often necessary to employ foam breakers, foam bleed-off devices or anti-foaming agents.

5.4 REGULATION AND AUTOMATIC CONTROL OF THE OXYGEN INPUT

Fig. 5.3 presents curves showing the dissolved oxygen concentration in the aeration tank as a function of time in response to a constant rate of oxygen input. The rise and fall in the oxygen content is attributable to repeated changes in the respiratory activity of the biomass, which tends to fluctuate between endogenous respiration and substrate-induced respiration in response to diurnal changes in the BOD_5 loading (see Fig. 1.1). By comparing Fig. 5.3 and Table 5.4 it becomes clear that with a constant rate of oxygen supply, large amounts of energy must be squandered. As a result of the regulation or control of the rate of oxygen supply, however, it is possible to achieve major reductions in the overall energy consumption for most plants. The control of oxygen input for smaller plants (up to 500 000 PE) usually takes the form of a pre-programmed cycle. This is particularly appropriate where the fluctuations in oxygen demand occur in a regular periodic fashion, with the quantitative changes being a function of time. For different seasons of the year the programme should be varied to correspond to the changes in ambient temperature as at higher tempera-

Regulation and automatic control of the oxygen input

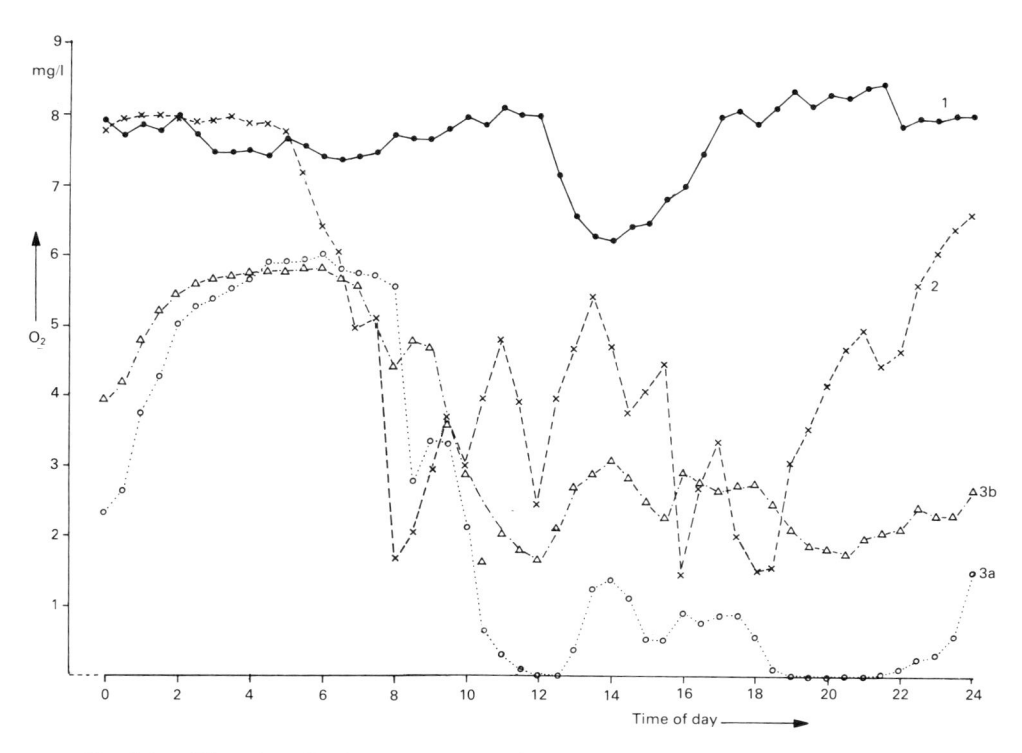

Fig. 5.3 — Diurnal profiles for oxygen content in the aeration tank in the absence of control of the O_2-injection rate. Curve 1: Stirred reactor, $B_R = 0.2$ kg/m³ d, at an industrial plant with 3-shift operation; balanced BOD_5 input and too high a rate of oxygen input. Curve 2: Stirred reactor, $B_R = 0.6$ kg/m³ d. Centrifugal aerator to TGL 22767, BOD_5 profile roughly as in Fig. 1.1A. Curve 3: Nearly complete mixing, $B_R = 2.3$ kg/m³ d tank fitted with 2 × BK B200 aerators. (a) O_2-content for rotor near inlet $OC' = 60$ g/m³ h; (b) O_2-content for rotor near outlet $OC' = 45$ g/m³ h, BOD_5 profile roughly as in Fig. 1.1(B).

Table 5.4 — Oxygen uptake efficiency as a function of oxygen content

O_2 (mg/l)	0	1	2	3	4	5	6	7	8
OC/P (kg/kWh)	2.0	1.8	1.6	1.4	1.2	1.0	0.8	0.6	0.4

tures the biomass, according to the Arrhenius Rule, will have a higher oxygen demand at a constant BOD_5 loading. A disadvantage of this method of control is that unexpected shock loadings will not be catered for, and that in the case of combined sewage flows, an unnecessarily large quantity of oxygen may be introduced in wet weather.

For large installations and irregular fluctuations in oxygen demand a feedback control system for regulating the oxygen input as a function of the instantaneous dissolved oxygen level is desirable. The positioning of the control electrode must be ascertained from routine measurements. The control parameters in the case of surface aerators comprise the speed of rotation, the water level or depth of immersion of the aerator, and the number of aerators in use; for compressed air systems the number of fans on stream may be the appropriate choice for control purposes. For upper and lower limits values of 0.5 and 2 mg/l are usually adopted. From the various details presented in Table 5.3 it is apparent that for a reduction in the energy consumption, not only the absolute dissolved oxygen level, but also the corresponding energy requirement must be taken into account. For example, using BKB900 aerators at 60 rev/min setting for a dissolved oxygen level of 0.5 mg/l is less economic than the use of 1 BKB900 aerator at 89 rev/min to maintain 2 mg O_2/l in the aeration tank.

For nitrifying systems the energy requirement during peak loading times may be appreciably reduced with the aid of an aeration control system if during such periods only the energy required to maintain the necessary degree of turbulence is supplied; the biomass must then have recourse to the oxygen contained in the nitrites or nitrates present in solution, which are thereby reduced to nitrogen as the end-product of denitrification.

5.5 USE OF TECHNICALLY PURE OXYGEN

On a worldwide scale more than 200 activated sludge plants, some of them serving large cities, use technically pure oxygen in place of air for 'aeration'. This process was advocated by Okun as early as 1948. Advantages cited in the literature for the use of oxygen in place of air are:

— extent of O_2 utilisation increases to 90%, O_2 contents raised to 5–10 mg/l with the saturation value raised to 4.7 times that for air;
— reduction of exhaust gas volume and hence of the emission of volatile sewage constituents to only about 1% of the amount for conventional aeration;
— higher BOD_5 removal at equal BOD_5 sludge loadings due to higher activity of the biomass, or alternatively higher performance for the same treated effluent quality, or aeration tanks of smaller dimensions;
— more compact, flocculating sludge with a low sludge volume index, as a result of which the MLSS content may be increased to 6–8 g/l and both the aeration and final settling tanks can be made smaller;
— a lower rate of surplus sludge production and hence a less costly sludge treatment operation;
— improved breakdown of refractory sewage constitutents;
— energetic performance increased to values of 1.7–3.5 kg O_2/kWh;
— lower phosphorus requirement, although this is only of benefit in the case of phosphorus-deficient industrial effluents;
— possibility of using enclosed reactors.

Owing to the absence of nitrogen, the partial pressure of oxygen for a pure oxygen system is increased from 217.7 to 1013.2 kPa so that both the solubility and the degree of penetration into the sludge flocs are increased. As a result a larger proportion of the bacteria resident in the floc interiors can be supplied with oxygen so that they are capable of a metabolic performance roughly 12 times that for anaerobic conditions. In addition the growth of thread-forming organisms is inhibited as far as concerns those organisms which utilise the products of anaerobic metabolism in the centre of the flocs. The higher aerobic proportion of metabolism for the biomass as a whole leads to a shift in the respiration and synthesis rates (Fig. 4.2) so that the rate of surplus sludge production falls at the expense of an increase in the oxidation reactions. The same effect has been reported in the case of activated sludge installations operating at elevated pressures.

As the stripping of low-boiling compunds is impeded on account of the smaller gas throughput, in the case of municipal sewage there is also a drop in pH to around 6.5 as a consequence of CO_2 retention in the liquid (Gassen 1982). The consequent adverse effect on nitrification may require some form of pH correction; the use of oxygen for nitrifying activated sludge systems thus shows no advantage over air both on account of the pH shift and also because of the necessity for a high sludge age. According to Kalinske (1978) biologically treated effluent from pure oxygen systems is more turbid, and there may also be interference with nitrification owing to the high levels of dissolved oxygen in the aeration tank; where these exceed 15–20 mg/l of O_2 there may also be interference with the elimination of C- and H- compounds. Further drawbacks associated with the use of technically pure oxygen are the corrosion and occupational safety problems which arise.

Under the conditions prevailing in the GDR the use of oxygen as a means of 'aeration' is uneconomic [34] and hence is only applicable to the treatment of odour-intensive or other types of effluent giving off polluting substances.

5.6 ENVIRONMENTAL POLLUTION AS A RESULT OF AERATION

An important concern of occupational hygiene and environmental protection measures for activated sludge plants is to protect both the sewage plant workers and the neighbouring population, together with the plant and animal life in the vicinity, from the adverse effects of aerosols, pathogens and other harmful constituents of sewage, as well as from the effect of excessive noise.

Noise
Surface aerators may give rise to noise levels of 70 to 92 dB in their immediate vicinity, while blowers for the larger activated sludge systems may give rise to 110 dB. According to TGL 10687/02, a maximum noise level for the hours of darkness of 40 dB is prescribed, while for short periods a total level of 115 dB may be permitted. Soundproofing or noise-reducing equipment is available from numerous firms. Recommendations concerning noise reduction are contained in [29]. The hoods developed by Koska (1984) for the BK 2400 aerators lead to a reduction in the local noise level of 10 dB. Regardless of the technical solutions, an activated sludge plant should in principle be provided with a perimeter screen of trees and shrubs with a high proportion of brushwood, which in addition to its noise abatement function

also serves to intercept aerosols and unpleasant odors given off by the treatment plant (Haupt 1976). A hedge with a width of 3 m reduces the noise level by about 1 dB.

Odour

The release of unpleasant odours from the mixture of sewage and activated sludge arises from the stripping of volatile constituents, the so-called primary osmogens (see Table 5.5) and of the intermediate products of metabolism, the so-called secondary

Table 5.5 — Disappearance of aliphatic and carbocyclic hydrocarbons in open aeration tanks (From Grünwald (1980))

Hydrocarbon	Elimination after 5 h aeration time (%)		K_s (mg/l)
	Exhaust air	Biol. decomp.	
n-Pentane	54.5	44.3	0.28
n-Hexane	9.8	31.7	0.32
n-Heptane	10.4	70.5	0.53
n-Decane	3.7	61.2	0.77
Cyclohexane	5.5	40.0	0.33
Ethyl-cyclohexane	13.1	35.5	0.27
Benzene	43.3	37.0	0.71
Toluene	43.5	51.9	0.40
m-Xylene	19.0	58.4	0.13
Ethylbenzene	15.9	47.0	0.14

osmogens. These osmogens are low-boiling organic substances with molecular weights in the range 20–350. The activated sludge tank is the strongest emitter of the entire sewage treatment plant, as the high degree of turbulence and frequently elevated temperatures greatly encourage the release of such compounds. The nose is capable of detecting odoriferous substances at much lower concentrations than either a gas chromatograph or a mass spectrometer, for example, 1.1×10^{-6} g acetone or only 3.5×10^{-12} g of skatole in a litre of air.

Every activated sludge plant has its own smell. Municipal sewage treatment plants give off unpleasant smells of a putrid, urinary or faecal sickly nature in cases of oxygen deficiency. With an adequate supply of oxygen, however, the smell of raw sewage is lost immediately on contacting the activated sludge suspension as the primary osmogens become bound to the sludge biomass and are metabolised. For some odour-intensive industrial effluents such as those from fermentation and chemical synthesis plants, and from animal by-products plants, the primary osmogens may be of such a penetrating nature that the use of enclosed reactors may be advisable, often coupled with oxygen for aeration together with scrubbing of the exhaust gases.

Secondary osmogen are often chemical compounds of an ill-defined nature; they may be classed as hay-like, earthy, soapy, etc. They are chiefly formed from the decomposition of non-specific substrates by organisms such as Streptomycetes, *Pseudomonas aerugmosa*, *Proteus* spp and similar organisms.

Pathogenic compounds
Aerosols given off by activated sludge plants may function as carriers for pathogenic organisms. Especially during cool, humid weather, visible aerosol-containing mists may be formed above the aeration tank (Plate 8). Aerosols are composed of water droplets suspended in the atmosphere with droplet sizes of up to about 500 μm; they may be carried for distances of up to 200 m, or 400 m maximum, depending on the strength of the wind. Müller and Bartocha (1978) indicate transmission distances under unfavourable conditions for a number of organisms, such as coliform bacteria up to 300 m, Clostridia up to 50 m, Salmonellae up to 10 m from the edge of the aeration tank. These figures are probably dependent on the levels of these organisms present in the sludge suspension.

The aerosols are exposed to UV irradiation from sunlight and also to other bactericidal factors in the atmosphere. The propagation and evaporation of aerosols were described in detail by Jarnych (1976) as function of their size and the atmospheric conditions, as well as the time-dependent decrease in the infectivity of the organisms concerned.

Table 5.6 — Emission of bacterial organisms as a function of the type of aeration system. (From Wanner (1975) standared deviation = 50–80% of x)

Aeration device		Total oarganism count in air (colony forming units/m^3)	
		at 1 m above aeration tank	at 2 m from aeration tank
Compressed air	fine-bubble	1 900	600
	medium-bubble	15 000	4 000
Turbines	upflow	56 000	1 700
	downflow	184 000	1 500
Horizontal rotors		47 000	19 000

Table 5.6 shows the microbial emissions of different aeration systems, and Fig. 5.4 the decline in bacterial numbers in the atmosphere as a function of climatic conditions and the distance from an aeration tank employing a medium-bubble compressed air system of aeration. The values measured at a height of 1 m above the tank include the larger aerosols which are not transmitted by winds of normal

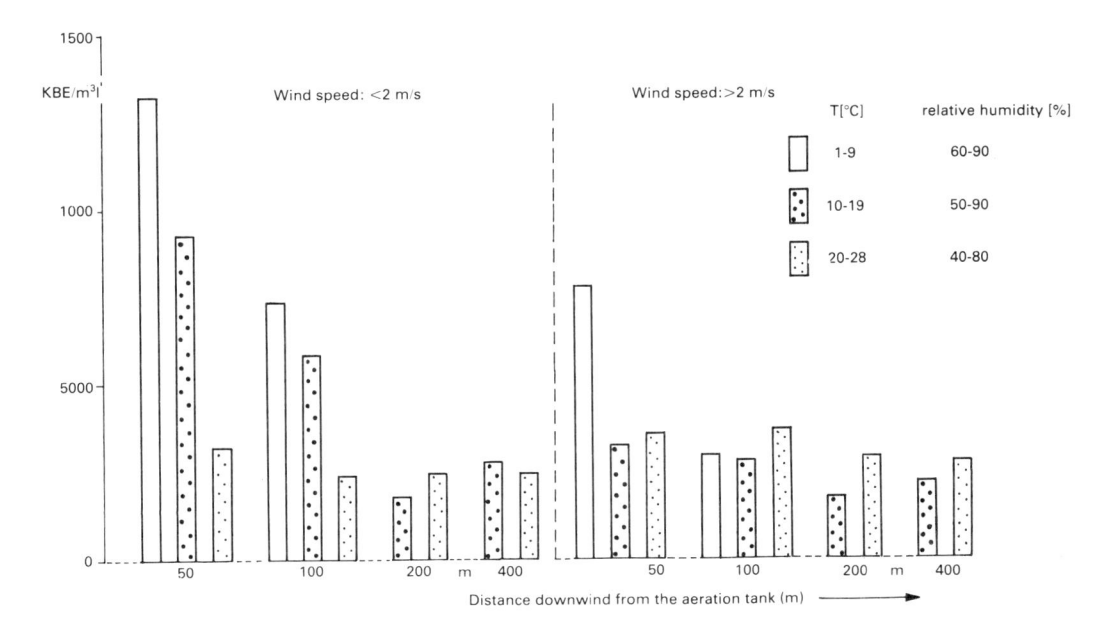

Fig. 5.4 — Microbial dispersion from an activated sludge system using medium-bubble aeration equipment under various meteorological conditions. KBE colony forming units at 22°C after five days. Average values for 13×2 hour measuring periods and 6–8 samplings with a slit sampler. Organisms present close to the tank at levels stated in Table 5.6. From Wanner (1975).

strength, but which may nevertheless be inhaled by the maintenance and operating personnel at the sewage works. At a distance of 2 m from the tank the larger aerosol droplets have precipitated, and the smaller ones are no longer detectable at a distance of 200 m, as long as the background level of 100–500 organisms/m^3 is allowed for. Fine-bubble aeration gives the lowest level of airborne contamination in the gas escaping from activated sludge tanks; however, irrespective of the type of aeration system, other factors such as the energy input into the tank, the air throughput, speed of rotation and geometry of the rotors of surface aerators, may affect the degree of aerosol formation so that no generalised conclusions are possible.

The largest *risk of infection* arises from the small aerosol particles with a diameter of ≤5 μm as these can be carried with the inhaled air directly into the alveoli of the lungs without being intercepted by the epithelium of the oral or nasal tracts. The numbers of organisms required to initiate infection in sewage works personnel or in neighbouring individuals are, however, not attained: for a Salmonella infection, for example, Müller and Bartocha (1978) estimated that a human exposure time of 10 000 hours at the point of strongest Salmonella emission would be required. More dangerous, however, are the roughly ten times greater numbers of organisms in covered aeration tanks. In such installations, when epidemics occur, or where infectious effluents are treated, the wearing of a protective mask over the nose and mouth is advisable in the vicinity of the aeration tanks.

Virus-containing aerosols have so far not led to any increase in the sickness rate among employees of sewage treatment plants. According to a literature review by Wullenweber and Joret (1983) only one virus per 10–10000 m^3 of air could be detected 2–10 m downwind from the aeration tanks.

6
Reactor types and process variants

6.1 CONTROL METHODS

Among the variants of the activated sludge process one includes differences in the pattern of residence time distribution, various positions for the input of recycled sludge and untreated sewage, and for multiple stage or cascade systems, the flow pathway through the successive tanks. By means of such factors it is possible to exert an appreciable degree of control over the substrate supply to the biomass, and also on the effect of toxic substances, oxygen distribution and supply, bulking sludge formation and treated effluent quality; even plant upsets, maintenance costs and energy consumption may be influenced in the same way.

Theoretical hypotheses concerning the effect of contact time between the activated sludge flocs and the liquid phase on the extent of substrate elimination, as well as the results of degradation rate experiments, lead to the assumption that the greatest possible degree of elimination of substrates is achieved only when the maximum retention time is fully utilised. This is realised in practice by means of a plug-flow reactor of the types indicated in Fig. 6.2(A) and (B). In an ideal reactor of this kind, which is of course not realised in practice because of the inevitable longitudinal mixing, the theoretical residence time can be expressed by means of the following formula

$$t_B = \frac{V_B}{V_h} \quad (h) ,$$

where: V_B = volume of aeration tank (m^3);
V_h = volumetric flowrate (m^3/h).

The theoretical residence time calculated on the basis of the sewage flowrate is, however, invariably reduced by a half or a third depending on the recycle sludge flow which may be 100% or 50% of the sewage flowrate. As can be seen from the schematic drawing in Fig. 1.4 the overall retention time of the sewage in the aeration tank and sludge settling compartment, as well as the settling time in the final settling tank, remain constant, independent of the recycled sludge flowrate. The incoming sewage is, however, more and more diluted with treated sewage as the sludge recycle rate is increased, concentration surges are evened out and the stirred tank reactor system is approached more closely by a linear reactor with a high sludge recycle rate.

Fig. 6.1 shows the residence time distribution in a linear reactor in comparison with one, two or three ideal stirred tank reactors arranged in series. At the end of the theoretical retention time for a linear reactor, 100% of the incoming sewage and recycled sludge introduced at $t = 0$ will exist from the reactor, but for an ideal stirred tank reactor the proportion is only 63% and for two and three-stage cascade systems with stirred tank reactors in series, the proportion falls to 37% and 18% respectively. In the stirred tank reactor, therefore, an appreciable percentage of the incoming liquid is retained beyond the end of the theoretical retention time. From the residence time distribution patterns and the results of biodegradation tests, it is incorrect, although commonplace, to ascribe any unique advantage to the plug-flow reactor (see Braha 1983). Fig. 4.7 proves that within the loading conditions specified in Table 10.1 for a stirred tank reactor with an MLSS content of 3–4 g/l practically the entire quantity of incoming BOD_5 is adsorptively bound and hence eliminated within a few minutes. According to section 4.2.1 the rate-limiting reaction for biological sewage treatment is the final disintegration of the enzyme-product complex. All the other reactions involved proceed considerably more rapidly than the limiting conditions implied by the retention time in the aeration tank. For a given ratio of sewage input rate to tank capacity therefore the varying residence times implicit in the different reactor types are of practically no consequence for the elimination performance with respect to the BOD_5-generating sewage constituents. In heavily loaded systems, however, where the elimination of organic substrates is in the transition region between zero and higher orders of reaction, the reactor type may have a considerable influence on the quality of the treated effluent. For certain biochemical reactions, such as nitrification in carrousel systems or the luxury uptake of phosphorus, (Fig. 4.28) plug-flow reactors are nevertheless distinctly superior.

The most frequently used process variants are indicated in Fig. 6.2.

Plug-flow reactor
Front-end feed of sewage and recycled sludge according to Figs 6.2(A) and (B), with slug-like or spiral transport paths through the tank without appreciable back-mixing, leads to strongly variable oxygen demand and concentration along the length of the tank (Figs 6.3(A) and (B)); there is frequently a deficiency of oxygen at the inlet; control of oxygen input rendered difficult on account of diurnal fluctuations of the O_2-demand; very often bulking sludge develops due to the action of *Sphaerotilus* organisms, which are favoured in the inlet region relative to other organisms on account of the O_2-deficit. For buffering of toxic shocks a higher rate of sludge recycle must be employed. This is the oldest tank design and is most common in connection with aeration systems based on compressed air or horizontally mounted rolls. To counteract the variable O_2-demand along the length of a plug-flow reactor, *stepwise aeration* according to Fig. 6.2(C) with graduation of the O_2-input is sometimes adopted.

Stepped feed
Based on the knowledge that the elimination processes have diminished to negligible levels near the outlet from a plug-flow reactor, many possible ways of ensuring a

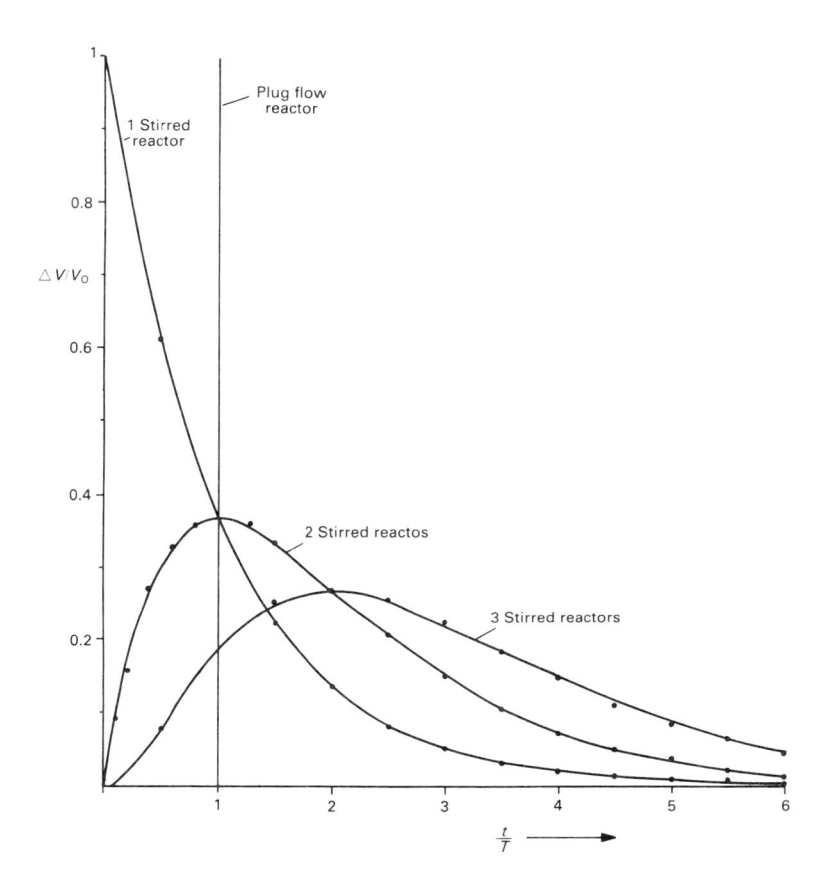

Fig. 6.1—Theoretical residence times for sewage and recycled sludge in a plug-flow reactor and in a succession of from one to three stirred reactors connected in series. Basis of calculations for stirred reactors as follows:

$$\frac{\Delta V}{V_0} = \frac{1}{m!}\left(\frac{t}{T}\right)^m e^{-\left(\frac{t}{T}\right)}$$

For $n = 1$ the reactor still contains at time t(h),

$$\frac{\Delta V}{V_0} = e^{-\left(\frac{t}{T}\right)}$$

and the volume already discharged is given by

$$\frac{V_0 - \Delta V}{V_0} = 1 - e^{-\left(\frac{t}{T}\right)}$$

where the symbols have the following meanings: V_0 = input plus recyle sludge flow rate (M³/h); ΔV = residual volume flowrate in the reactor at time t (m³/h); T = theoretical retention time $= \frac{V_{BB}}{V_0}$ (h); t = time (h); m = n − 1; n = no: of reactors; m = n − 1; n = reactor number.

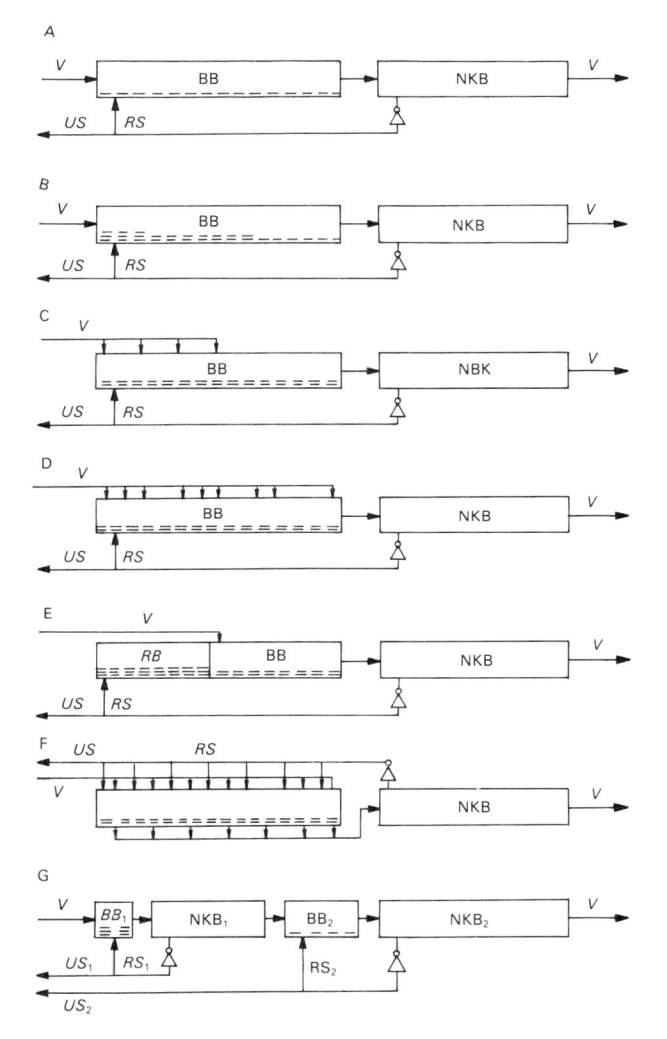

Fig. 6.2 — Sewage and activated sludge pathways together with air injection for compressed air operation. A: plug-flow reactor; B: stepwise aeration; C: stepwise feed; D: reactor with uniform sludge loading; E: separate sludge reaeration tank; F: completely mixed system; G: two-stage activated sludge plant. Key: BB = aeration tank; NKB = final settling tank; RB = sludge reaeration tank; V = sewage feed; RS = recycled sludge; US = surplus sludge.

more intensive utilisation of the tank have been devised, based on a series of different feed positions along the length of the reactor, Fig. 6.2(D). The stepped feed reactor design represents a transition in the direction of the stirred tank reactor.

Stirred tank reactor

In a similar way to the arrangement of Fig. 6.2(F) for compressed air and horizontal rotors, tanks with centrifugal aeration systems function as practically ideal stirred-

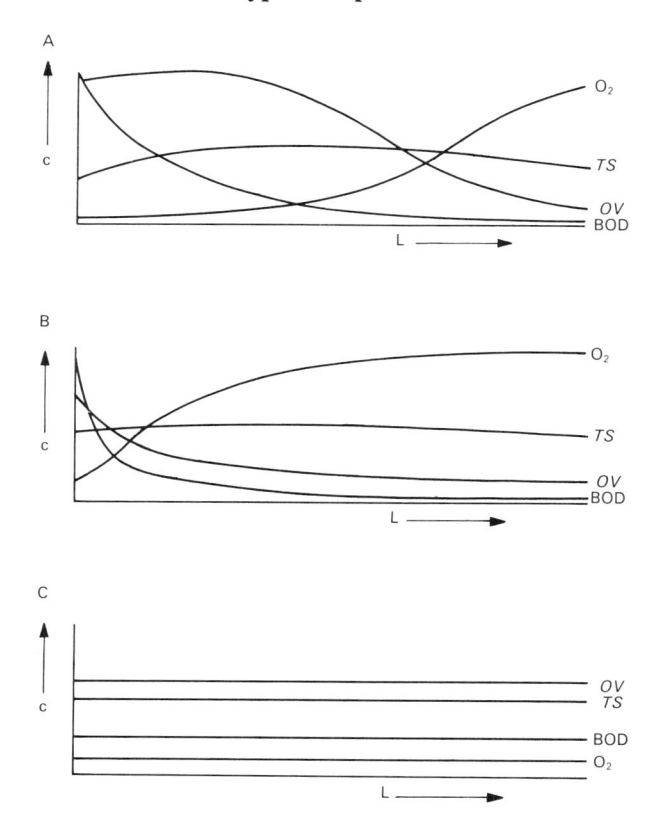

Fig. 6.3 — Schematic profiles for oxygen (O_2), respiration rate (OV) dissolved BOD_5 (BOD) and biomass content (TS) along the length of the aeration tank for ideal plug flow and completely mixed reactor systems. A: plug-flow reactor for daytime loading; B: plug-flow reactor for night-time loading; C: completely mixed reactor.

tank reactors. In process engineering technology, such reactors are considered as completely mixed when the mixing time is 10% of the theoretical residence time.

In completely mixed reactors the organisms are continuously presented with a very low level of dissolved substrates, of the same order as that in the effluent from the aeration tank and in the final settling tank. Toxic substances and concentration surges are immediately diluted with the entire liquid volume of the aeration tank, a decisive advantage in the case of those plants where the risk of such shocks is present. This *dilution effect* means that the completely mixed stirred-tank reactor is becoming the method of choice for practically all kinds of sewage and effluents (but see section 3.7.2.3). As a result of the immediate mixing of the incoming sewage with the entire tank contents it would be logical to suppose that poorly or slowly metabolised substrates not capable of being adsorbed onto the sludge flocs would leave the reactor unchanged. In particular for nitrifying systems and also for denitrifying plants this should be very definitely apparent. Obviously, however, the enzymes in a stirred-tank reactor are properly utilised, so that no deterioration in the effluent quality takes place; what is more the reduction of any inhibitory effect due to toxins and shock loading, as well as diauxy and polyauxy effects in general enable the sludge biomass to become more active.

For extreme shock loadings, such as were simulated in [19] there is, as may be inferred from Fig. 6.1, an earlier rise in the effluent concentration values for a stirred-tank reactor than for a plug-flow reactor or cascade system.

For several series-connected stirred-tank reactors the buffer capacity decreases as the number of tanks increases.

In many cases of industrial effluents where the metabolism of one substrate is hindered by the presence of another, that is, for diauxy or polyauxy, a cascade system may be indispensable.

Carrousel plants

These represent a transition stage in between plug-flow and completely mixed reactors and in their simplest form are known as oxidation ditches. As far as the water circulation is concerned they appear closely to resemble plug-flow systems; however, owing to the widely variable velocity of flow across a transverse section, a high degree of longitudinal mixing takes place, especially in the curved sections, which induces a much closer resemblance to a completely mixed reactor, on account of the prolonged retention times.

Separate aeration of recycled sludge (Fig. 6.2(G)

Owing to the high level of BOD_5 elimination during the first few minutes of contact between the incoming sewage and the sludge biomass (Fig. 4.7) and the slow progress of substrate metabolism, by 1955 the first plants to incorporate *sludge regeneration* had made their appearance. With this method, after a contact time of 15 to 60 minutes or occasionally longer, the actual process of elimination of organic substrate was interrupted, and the sludge/sewage mixture introduced into the settling tank. The recycled sludge was then pumped to a regeneration tank where it was aerated for a period of two to five hours without further sewage input. During this aeration period, the substrates adsorbed onto the flocs or bound in the form of enzyme-substrate complexes, and any remaining unbound substrate were subjected to metabolic transformation and accordingly decomposed. The advantages of this method were:

— a 20–70% higher content of sludge biomass, that is a lower sludge loading in the overall system, as in the regeneration tank the MLSS is that of the recycled sludge;
— a higher buffer capacity towards toxins and concentration surges, as there is a continuous supply of 'regenerated' sludge available;
— a reduced tendency to bulking sludge formation.

The disadvantages include an enhanced oxygen demand on account of the endogenous respiration of the larger amounts of sludge biomass (the oxygen input to the regenerator during peak periods may sometimes cause problems, especially during periods of high sewage temperatures, for example for a sludge solids content of 10 g/l, OV_{spec} = 15 mg/g h increasing to OV = 150 g/m^3 h) accompanied by severe foaming in the reaeration tank. This mode of operation was widely adopted abroad, particularly in the USSR and the USA. In the GDR it has been successfully employed in a few cases for the prevention of bulking sludge (Paulermann 1964). Owing to the strong foaming tendencies in the reaeration tank and the increased energy consumption it has, however, not become widely accepted.

Two-stage processes (Fig. 6.2(G))
The use of different biocoenoses leads to variously optimised physiological functions, so that a broad spectrum of sewage constituents can be eliminated. In the first stage, at a BOD_5 volumetric loading of 5–30 kg/m^3 d a highly active, chiefly polysaprobic, sludge biomass is produced, which converts the major portion of the organic sewage constituents into cell biomass, and effects only a small degree of oxidation, as a result of which 1 kg BOD_5 can be eliminated for an energy consumption of only 0.1 to 0.2 kWh. The increased quantity of sludge biomass has an enhanced adsorption capacity for sewage constituents, including those which are harmful to the treated water quality. The sludge exhibits a low sludge volume index and hence settles very well. The constituents remaining after the first stage are then reduced in a second, usually lightly-loaded stage so that the desired effluent quality is attained. The sludge biomass in this second stage is ordinarily much lighter than in the first stage, as the coarse solids with a weighting effect on the sludge flocs are missing, and because of the tendency to encourage thread-forming bacteria. The advantages of the two-stage process are:

— better elimination of poorly degradable substrates, because two sludge types are available and the second-stage sludge biomass is characterised by a high sludge age so that numerous species of microorganisms are available with a wide variety of degradation programmes;
— lower energy requirement;
— up to organic sludge loadings of 0.35 kg/kg d, the total tank volume is reduced, despite the need for two final settling tanks [35].

The two-stage process has so far chiefly been used for the more highly concentrated effluents from industry and agriculture, but is also of benefit for municipal sewage [26,35] Böhnke 1983).

For dimensional design of two-stage plants it must be borne in mind that the BOD_5 of the effluent from the second stage should be greater than 50 mg/l (preferably over 100 mg/l) so that sludge flocs are still capable of being formed in the second stage. While for single-stage plants an inlet BOD_5 of 35–56 mg/l ensures a reliable sludge balance (Kurnilovič 1972), a far-reaching elimination of growth-promoting substrates in the first stage impairs floc-formation in the second stage [26]. The aeration tank for the first stage usually takes the form of a stirred reactor, with the second designed either as a plug-flow or completely mixed system. The sludge production of two-stage plants is as a rule about 5 g/PE day greater than that of single-stage systems, on account of the high level of surplus sludge formation in the first stage. Contrary to the recommendations in the older literature the surplus sludge from the second stage should not be recycled to the first stage, because the very sensitive organisms it contains will be killed and will thus increase the organic loading of the first stage.

Single tank systems, without a clarifier
Owing to the difficult task of constructing final settling tanks (clarifiers) a few single-tank installations have been constructed in which the sewage/sludge suspension is aerated continuously together with the incoming sewage for two to five hours; the aerator is then switched off according to a signal from a time-controller or level switch, the activated sludge allowed to settle, and after a further one to two hours the outlet valve opened. When the treated liquid has run out, the valve is closed again, the aerator switched on and the cycle recommenced.

Alternate operation of two aeration tanks
As in the case of single-tank systems, the final settling tank is omitted. By using two or three aeration tanks connected in parallel, one tank can serve as the treatment tank and the next as the settling tank. As, however, aeration tanks do not possess the same ideal hydraulic settling characteristics as clarifiers designed for the purpose, for such systems, just as for single-tank plants, the effluent must be expected to be of poorer quality than that indicated by the figures cited in Table 10.1.

Installations of this kind are: mostly difficult to design and manufacture to exact criteria: more prone to breakdown owing to the added control requirements; operationally unpredictable; they are thus not recommended.

Besides the several general variants described above there are also a large number of other versions often of a proprietary pattern developed by industrial concerns or patent holders, which cannot be described in detail here.

For successful results from the activated sludge biomass it is advisable to introduce the recycled sludge into the aeration tank rather than into the inlet duct, as otherwise the biomass would be subjected to a 'shock' from the concentrated raw sewage. However, it is possible for such a shock to suppress the specially exposed thread-forming bacteria, and bulking sludge formation may thus be inhibited. The inlet point for the recycled sludge should also be situated above the water level as this makes for simpler observation and sampling as a means of process control.

6.2 SPECIAL METHODS
Some industrial effluents and organic wastes can cause considerable operating disturbances in activated sludge plants, for example, deterioration of the treatment performance, formation of bulking sludge, accumulation of toxic or inhibitory substances, etc. For certain particular combinations of surplus sludge, operationally troublesome effluents and contact times, the interferences may be counteracted by one or other of the following methods.

Kraus process
Referring to the diagrammatic layout plan of a complete sewage treatment plant (Fig. 1.2), the sludge liquor from the anaerobic sludge digestion plant is returned to the primary settling tank and then proceeds to the aeration tank. Sludge liquors

contain from 1000–2000 mg/l of BOD_5, N-concentrations often in excess of 800 mg/l, and the products of putrefaction, such as organic acids and hydrogen sulphide, which can damage the aerobic organisms and lead to the formation of bulking sludge. Discontinuous input of sludge liquor is especially liable to cause disruption of the biomass, which may result in elevated effluent concentration values lasting for a day or so. For this situation, Kraus (1945) proposed to aerate the digester liquor together with the surplus sludge, and then to introduce them into the primary settling tank for solids separation (Fig. 6.4). Any possible impairment of the activity of the surplus

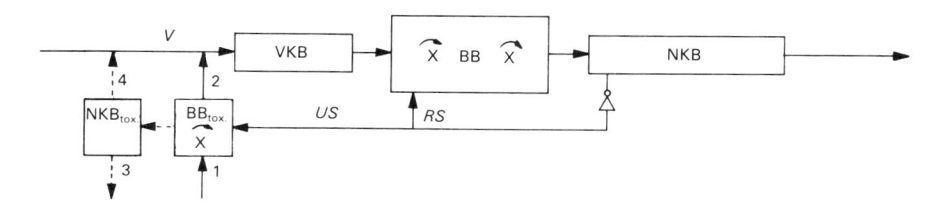

Fig. 6.4 — Process for detoxification of toxic effluents, pretreatment of highly concentrated industrial effluents or for preliminary treatment of effluents with tendencies to form bulking sludge. 1. Feed of toxic, high-strength or other effluent. 2. Partially treated effluent with biomass in suspension. 3. Sludge containing toxins or substances interfering with sludge digestion or utilisation stages. 4. Detoxified or partially treated effluent freed from suspended biomass. BB_{tox} = aerobic or anaerobic tank for detoxification of pretreatment; NKB_{tox} = final settling tank for separating detoxified effluent from toxin-containing sludge suspension.

sludge is of no consequence, as the sludge is being discharged in a spent condition from the aerobic section of the treatment plant. The necessary contact time can easily be determined by means of discontinuous batch experiments.

Separate adsorption and assimilation
Activated sludge is the cheapest means of coagulating and flocculating and one available in largish amounts, which, in contrast to synthetic coagulants, is capable of self-regeneration; with sludge treatment neither elevated sludge formation nor special more expensive constituents are required. There are no negative effects on the utilisation of the sludge, and activated sludge biomass causes no increase in the salinity of the receiving water.

Activated sludge may be employed according to Fig. 6.4 for *detoxification* of industrial effluents, including the adsoption of heavy metal ions (Port 1978), biocides [30] and petroleum products or the metabolism of toxic substrates such as cyanide, formaldehyde, thiocyanate and biocides. The surplus sludge should be introduced into the toxic effluent is such a ratio that the sludge retains its activity and the toxins are metabolised. The sludge may also be expended and subsequently discarded by tipping or other means, or possibly subjected to an anaerobic treatment as a means of detoxification. The process of detoxifying effluents in this way may be carried out as a single or two-stage process. Highly concentrated effluents or those tending to bulking sludge formation, for example, effluents from dairies, breweries, distilleries, etc., can be brought into contact with the waste activated sludge in a highly-loaded treatment stage to such good effect that they can be given primary treatment for removal of settleable solids and then be introduced into the aeration tank with no further trouble. The aeration time will be dependent on the concentration of organic substrate and the available activated sludge; any value within the range from 10 minutes to 24 hours is possible. The effluent to be treated must naturally be transported to the works via special pipelines or in tanker vehicles.

Elimination of oxygen-containing anions
Some oxygen-containing anions can be reduced to an anion consisting of the negatively charged primary atom by the action of microorganisms under anaerobic conditions. This process, which is employed on a large scale for the purpose of denitrification, may also be used for detoxification of chromates (Moore *et al.* 1961) as well as chlorates, perchlorates and iodates. The reduction times which govern the size of the aeration tank must be determined from bench-scale tests. For those effluents which contain these anions in the raw state, the process configurations indicated in Fig. 9.1 or for special types of industrial effluent, the arrangement shown in Fig. 6.4 may be adopted.

Although rarely employed on a commercial scale, sulphates may also be reduced under strongly anaerobic conditions. This process should, however, not be carried out as a sub-process forming part of an activated sludge system, as the very low redox potentials would cause excessive damage to the biocoenosis. For the anaerobic sludges according to section 6.3 or the process arranagement shown in Fig. 6.4, sulphate reduction becomes possible. The resulting hydrogen sulphide must be chemically or physically eliminated from the liquid in order to prevent it from undergoing re-oxidation.

Deflocculant growth
The elimination of organic substrate is performed by solitary microorganisms, that is not aggregated into flocs. Deflocculated growth occurs in the case of effluents with constitutents which inhibit floc formation, and also in the so-called intensive biological system according to WP 36748. It may also occur in a modified form in the case of non-toxic municipal wastewaters, in situations where the final settling tank is overloaded or there are deficiencies in the hydraulic design such that the flocs beginning to form are flushed away. As solitary bacteria cannot be separated from the treated effluent by sedimentation or flotation, the recycling of acclimatised microorganisms into the aeration tank can no longer occur. Only the microorganisms entrained in the incoming liquid, whose generation times are shorter than or equal to the aeration time, are capable of multiplying; the biomass concentrations for such systems are usually only about 0.1–0.2 g/l. As the organisms exist in isolation they are in closer contact with the liquid than if they were incorporated into flocs. The multiplication rate may in fact be much greater than that indicated in Fig. 4.15 for sludge flocs; respiration rate measurements may give values for 100 mg/l of dissolved organic solids of up to 1300 mg/g h. As the microorganisms are only poorly removed by sedimentation, the effluent concentrations may be anything up to 50% of the input BOD and COD values. Removal of microorganisms may be effected by the addition of coagulants, or possibly by flotation. For the intensive biological systems

referred to above, highly concentrated and hence toxic accumulations of substrate are converted into biomass, which may then be separated from the treated liquid in a consecutive, lightly-loaded aeration stage (Zülke 1967).

Aerated lagoons
An activated sludge process without sludge recycle but with the aid of activated sludge flocs also takes place in the completely mixed aerated lagoons of the type indicated in Fig. 1.3.

6.3 ANAEROBIC ACTIVATED SLUDGE PROCESSES

For highly organic effluents with BOD_5 values of over 2000 mg/l the formerly widespread anaerobic method of sewage treatment once again comes into its own owing to its much lower oxygen requirement. A correctly designed anaerobic activated sludge plant, with a consecutive terminal aeration stage in conjunction with biogas utilisation in the form of a regenerative waste heat and electric power plant, may in some cases enable industrial effluents to be treated without recourse to external power supplies.

Anaerobic treatment of sewage sludge with digestion times of 15 to 100 days under continuous fermentation conditions, that is without biomass recycle, is one of the most widely used sludge treatment methods in the field of sewage technology. The high solids content of the sludges to be digested, amounting to 40–60 g/l, enables the recycling of biomass to be omitted. For the anaerobic treatment of sewage, however, the sludge content must be enhanced by recycling of the thickened sludge from a settling tank in just the same way as for the aerobic activated sludge process, so that the sludge loading is reduced and the sludge age increased. Despite the high volumetric loadings, anaerobic sludges exhibit good settling and thickening properties so that sludge solids contents can be achieved which are greater than those of aerobic treatment systems.

The object of technically oriented anaerobic sewage treatments lies in the conversion of organic substrates into methane, water and carbon dioxide. As Fig. 4.25 shows, numerous reaction steps are involved in the process of anaerobic decomposition to the desired end-products, which to some extent call for different groups of bacteria.

The substrates needed by the methane bacteria are produced by acid-forming, facultatively anaerobic bacteria, for which the optimal conditions for growth are in the weakly acid region, and hence one in which the methane bacteria themselves are unable to exist. In order to optimise the growth conditions for the two physiologically diverse groups of bacteria, it has been frequently proposed that the process of anaerobic fermentation should be carried out in two physically separate digestion stages. However, owing to the product inhibition exerted by the hydrogen on the acid formers, an equilibrium between the two groups can be maintained in a single-stage system. This equilibrium is labile and can be disrupted by a sudden fall in the pH value. For monitoring the stability of the digestion porocess, a routine determination of the alkalinity, that is the buffering capacity towards acids, is advisable; a decline in alkalinity even while the pH is still satisfactory points to its imminent fall into the acid region.

Anaerobic activated sludge processes

For sulphate-rich effluents, the formation of hydrogen sulphide (almost the entire quantity of sulphate is reduced to H_2S or sulphides) will not only give rise to problems connected with stench, corrosion and toxicity to humans, but will also exert a toxic effect due to H_2S on the biocoenosis. According to Kroiss (1981) toxic effects on the digestion process are apparent at concentrations of H_2S in the digester gas of 8%, corresponding to 4 g/l of SO_4^{2-} in the liquid, but not at 1–2% of H_2S in the biogas. By recirculating the contents of the digester and H_2S-scrubbing of the digester gas, for example with $CuSO_4$, the toxic action of H_2S on the methane bacteria can be averted.

For the dimensional design of anaerobic activated sludge plants, Kroiss (1981) advocates the use of the COD-Cr value in place of the BOD_5. As the COD of CO_2 is equal to zero, the surplus production accounts for only 5–12% of the COD eliminated and the extent of H_2S formation is relatively unimportant at 1–2%, the COD elimination can be equated with the methane formation. For the elimination of 1 kg COD, 0.35 Nm^3 of methane gas is produced, with a calific value of 10 MJ/m^3; for the reduction of 1 kg SO_4^{2-}, 0.32 Nm^3 of hydrogen sulphide gas is formed and the gasification of 1 kg organic carbon produces 1.86 Nm^3 of gas, of which 50–70% consists of methane and the remaining 30–50% is carbon dioxide.

As Table 6.1 shows the anaerobic process may be preferred in the case of

Table 6.1 — Treatment performance of anaerobic activated sludge for industrial effluents. From Sixt (1979)

Industry	Digestion time (d)	B_R (kg/m³ d)	n (%)
Sugar refining	0.5	12	80–98% COD
	0.4	8	95% COD
Potato processing	0.4–0.8	4–45	94–96% BOD_5
			93–96% COD
Dairy products	7	0.54	85% BOD_5
Jam and fruit processing	4	0.5	90% BOD_5
Brewing	10	5.55	72% COD
Distilling	10	10	80% BOD_5
		9.9	78% COD
Yeast factories	6	1.8	80% BOD_5
Slaughterhouses	—	1.77	90% BOD_5
Wool scouring	10	0.8	90% BOD_5

effluents for which aerobic treatment gives rise to problems on account of bulking sludge formation. A further advantage of the anaerobic process for such effluents lies in the fact that the requirement for nutrients, owing to the much lower quantity of sludge produced, is only about a fifth of that for aerobic systems.

On the large scale three forms of construction have been devised:

— a digester with a final settling tank connected in series;
— a digestion compartment with an overhead settling chamber (principle of the Imhoff tank, with sewage entering the digestion compartment);
— a digester in series with a parallel-plate separator.

For obtaining adequate contact between the sludge and the sewage to be treated a gentle circulation with low shearing action is required. Digestion takes place at > 20°C, the optimum lying between 30°C and 36°C; owing to the warm nature of many industrial discharges, heating of the effluent is usually unnecessary. Owing to the slow rates of growth of the anaerobic bacteria, running-in times of two to three months are usually needed and the process is more susceptible to toxic effects than aerobic systems. Prolonged interruptions in the feed, however, do not apparently cause much deterioration in the activity of the anaerobic biomass.

The treated effluent is strong-smelling, has high COD and BOD_5 contents which will fluctuate widely in response to shock loadings, and also exhibits settleable solids contents despite good settling facilities of around 500 mg/l. This means that a further treatment based on the conventional activated sludge process is needed. Operational disturbances, for example due to bulking sludge on account of the large proportion of low-molecular weight substrates and H_2S, are not encountered in this aerobic post-treatment; obviously the digestion residues have a toxic effect not only on the nitrifiers but also on the thread-forming bacteria.

6.4 ARRANGEMENT OF THE ACTIVATED SLUDGE TANKS

Activated sludge tanks are as a rule of rectangular configuration, fabricated either from concrete *in situ* or else from prefabricated concrete sections. The aeration system determines the tank geometry. Systems employing compressed air are usually from 3 to 6 m deep; the depth width ratio is usually between 1:1 and 1:2. Centrifugal aeration systems have tanks between 1.5 m and 4 m deep according to TGL 36763/02, with side lengths of up to 25 m for each individual aerator. Tanks with horizontal roll aeration devices are relatively shallow with depths of only 2–2.5 m. The length is usually limited to around 60–100 m. For certain types of aerator, such as centrifugal rotors, helix devices, etc., embanked aeration chambers with sloping earth walls lined with polymeric or bituminous membranes may be employed.

Although many types of effluent exhibit corrosive tendencies, the material of construction for activated sludge tanks is of only minor importance, for the corrosive behaviour is normally rapidly counteracted by the physical and chemical characteristics of the sewage/sludge suspension present in the tank; an exception is presented by the aggressive behaviour of sulphate with respect to concrete, for which the stipulations of TGL 33408/01 should be observed. Those parts of the tank situated above the water level are, however, exposed to the action of the escaping gases and the constant re-wetting of the surface.

In addition to the large 'standard aeration tanks' there are the following widely prevalent alternative forms.

Small activated sludge plants (Fig. 1.4 and Plate 10)

Small (or compact) activated sludge plants are intended for sewer catchments with total populations between three and 2000 PE. They are ordinaraily factory produced out of steel. They were originally devised as compact plants without primary settling and with simultaneous aerobic stabilisation of the sludge biomass, whereby the settling and aeration compartments were linked in such a way by a slot at the base so that the settled sludge from the settling chamber was able to slide continuously along the base back into the aeration compartment. Owing to repeated hydraulic shocks, pronounced denitrification in the settling compartment, upwelling of 'floaters' carried in with the influent, and also mainly on account of inadequate inspection and expert attention, the treatment performance generally falls below expectations. Also the self-induced return of the activated sludge laden with 'sinkers' is liable to interruption, so that putrefaction processes can be initiated in the settling compartment (Hänel et al. 1981).

For eliminating or counteracting the operational disturbances as a consequence of discontinuous surplus sludge removal, deposits of coarse solids, upwelling of floaters and carryover of pathogenic organisms, two-storey settling tanks are nowadays frequently employed, or else modified Imhoff tanks are connected in advance of the small activated sludge plant, such as in the equipment produced by VEB Sewage Treatment Plant Manufacturers, Merseburg, according to TGL 22767, so that the resulting sludge deposits are stabilised anaerobically, and hence largely freed from pathogens. The connection of mechanical screening facilities in advance of compact activated sludge systems is the subject of legal controls in many countries (Scherb 1979).

Multi-comparmtent spetic tanks according to TGL 7762 may be converted into compact activated sludge plants, according to a suggestion by Klimpel (1976) by means of horizontal aerators as long as the sewage flow warrants the provision of a sufficiently large final settling compartment (see TGL 22767).

Oxidation ditches (Plate 11)

The ditches developed by Pasveer (1958) at the beginning of the 1950s originally lacked final settling tanks, although these were subsequently added, and even sludge receptacles were included in the course of reconstruction (Ostermann 1971). These consist of recirculating trenches with BOD_5 volumetric loadings of around 0.2 kg/m^3 d (TGL 24350), so that for municipal sewage without primary settling the retention time is about two days; the aerobically stabilised sludge has to be dewatered on sludge drying beds or trays. Populations served range from 500 to 6000 PE; ditches with BOD_5 volumetric loading rates of 0.5–0.7 kg/m^3 d are referred to as oxidation ditches (TGL 24350/06).

The *bed velocity* must of necessity be between 25 and 30 cm/s as otherwise sludge deposits, such as septic pockets with a detrimental effect on the biocoenosis, will occur; in extreme cases no activated sludge may form at all. The necessary horizontal velocity may be achieved by adjusting the depth of immersion of the rotors, by regulating the water level, by eliminating turbulence-limiting features such as sand deposits, weed growth on the sloping sides and so on. Owing to the prolonged retention time for the liquid, there is a risk of freezing during the winter months.

In the surface growth on sloping sides there are usually autotrophic organisms and others not usually encountered in activated sludge systems which can also become detached. With inadequate flowrates in the ditch or during periods of underloading, blooms of algae and metazoa may occur on the water surface (Hänel 1963) which are atypical for activated sludge systems.

Carrousel plants
In this case we are also concerned with trench-like oxidation systems which by means of a parallel arangement of several counter-positioned channels can provide for sewered populations from 500 to several millions. For aeration purposes, horizontal rotors of about 2000 mm diameter are used; at each of the return bends in each trench a turbine rotor may also be inserted which helps to ensure an effective internal mixing of the sewage/sludge suspension.

Steel tanks (Plate 9)
Particularly where concentrated or malodorous industrial and agricultural effluents are concerned, and also in the case of constricted spatial circumstances and anaerobic activated sludge systems, circular, enclosed steel tanks with diameters of up to 18 m and a height of 10 m may be used. For aeration, intensive forced aeration equipment with oxygen input rotors of over 150 g/m^3 h, sometimes in conjunction with exhaust gas scrubbing, may be employed.

Deep shaft
The very deep tanks constructed with the aid of special equipment may be from 40 to 300 m deep. As a result of the evaluated hydrostatic pressure, the air bubbles are forced into the water at depths below about 25 m so that specific oxygen transfer coefficients of up to 4 kg/k Wh can be achieved. The gases which are flashed off when the sludge suspension is depressurised cause a partial flotation of the activated sludge flocs, which must be taken into consideration in the final settling stage [33]. Probably the organisms — like those in other very deep or tall reactors — acquire an enhanced level of activity as a result of the rhythmic pressure changes. The rates of elimination of organic substrate per unit volume and time are higher, the excess sludge production, that is, the synthesis/respiration ratio, is lower and the elimination of poorly degradable substances greater than in a conventional system. Owing to their high cost of construction these deep shaft systems are mainly used at present for congested sites, concentrated effluents and low volume flowrates.

Compact installations (Plate 12)
These comprise space and materials-saving installations combining aeration and final settling tanks, or primary settling, aeration and final settling compartments, in a single module (Randolf 1983). With such compact arrangements of aeration and final settling tanks, in order to avoid disruption of the sedimentation process and prevent scum formation on the clarifier, a degassing zone of suitable dimensions must be included in order to permit the escape of gas bubbles.

7
Final settling

Separation of the two phases composed of sludge biomass and treated effluent following the completion of the treatment processes in the aeration tank, constitutes the second decisive step for successful treatment performance. In this so-called final clarification step the sludge suspension, containing around 3500 mg/l of suspended sludge solids equivalent to about 2000 mg/l BOD_5 must be reduced to a sludge solids content of 20 mg/l or a maximum of 30 mg/l in the treated effluent. Too much carryover from the clarifier leads to a diminished sludge age or even to a lack of sludge biomass, and hence to an impaired treatment performance. During the separation process the sludge biomass thickens by a factor of 1.5–3.0 to a solids level corresponding to that in the recycled sludge. A further task of the clarifier, and one which is often overlooked, is to provide a buffer stock of activated sludge for use during periods of shock loading, stormwater inflows, and possibly ill-advised wasting of surplus sludge. The materials balance of an activated sludge plant showing the prominent role of the settling stage is indicated in Fig. 1.4. Clarifiers (settling tanks) which are undersized, incorrectly operated or subject to performance upsets, usually have the same adverse effects on the quality of the treated effluent as aeration tanks suffering from the same defects.

The clarification process will be considered here as concerned only with *sedimentation* on account of the exclusive use of this method in the GDR. *Flotation*, however, according to the literature, has certain beneficial effects with respect to water quality, particularly where bulking sludge is encountered, compared with gravity settling.

7.1 THEORETICAL PRINCIPLES OF ACTIVATED SLUDGE SEDIMENTATION

The settling process for particulates can be described with the aid of the following equation.

$$V_s = \sqrt{\frac{4gd\frac{\rho_F - \rho_w}{\rho_w}}{3\lambda_s}} \quad \text{(m/s)}$$

where: V_s = settling velocity of individual particles (m/s);
g = acceleration due to gravity
ρ_F = density of solid phase (kg/m³);
ρ_W = density of liquid (kg/m³);
d = particle diameter (m);
λ_s = drag coefficient of the settled particles $\lambda_s = f(Re)$.

If ρ_F is calculated from a settling velocity of 2 m/h for particles of 1 cm diameter assuming a value of $\lambda_s = 0.3$ for spherical floc particles, then we can arrive at approximate values for the settling rates of particles of different diameters:

— for 1 μm   : 0.02 m/h;
— for 10 μm   : 0.06 m/h;
— for 100 μm   : 0.2 m/h;
— for 1000 μm   : 0.62 m/h.

In the aeration tank the majority of floc particles have dimensions between 10 and 500 μm; in addition there are microflocs and also solitary bacteria in considerable numbers. The segregation of these small particles by sedimentation alone requires very long settling times, or very low upflow velocities in the final settling tank.

Fig. 7.1 shows some sedimentation curves for the same activated sludge at various solids contents in jar settling tests.

— At 0.57 g/l the flocs settle very rapidly without any mutual interference; from 4 min settling time upwards basically only thickening of the sludge was taking place, and after 30 min the maximum solids content (10.4 g/l) in the settled sludge had been reached).
— At 2.29 g/l a settling curve with three typical stages was obtained; the coagulation or flocculation phase, 0 to 1 minute, the linear or hindered settling phase, 2 to 4 minutes, and the thickening phase, 6 to 30 minutes. In the coagulation phase the microflocs coalesce on account of the surface properties of the sludge to form macroflocs. In principle the same physicochemical processes are involved as in the formation of activated sludge flocs in the aeration tank. During the agglomeration stage the bulk of the microflocs and solitary bacteria are incorporated into the macroflocs and hence induced to settle out. During the linear (hindered) settling phase the frictional resistance of the liquid must be overcome and the occluded water partly expelled; as a result pronounced turbulence is sometimes oberved in the settling chamber. The transition between settling and thickening is termed the *compression point*; the location of this point in time was used by Hackenberger (1968) for sizing the final settling tank. During the thickening stage of several hours' duration, the macrofloc structure of the activated sludge disappears. In the process of thickening sludge solids contents of 5–12 g/l are attained, or in the case of non-primary settled sewage, up to 25 g/l, but for bulking sludges the value may be only 1–2 g/l. The thickening process may also be disrupted by denitrification.
— At 4.29 and 7.15 g/l; under the conditions of the test only one thickening stage, of very low settling velocity, occurred. As a consequence of the high sludge

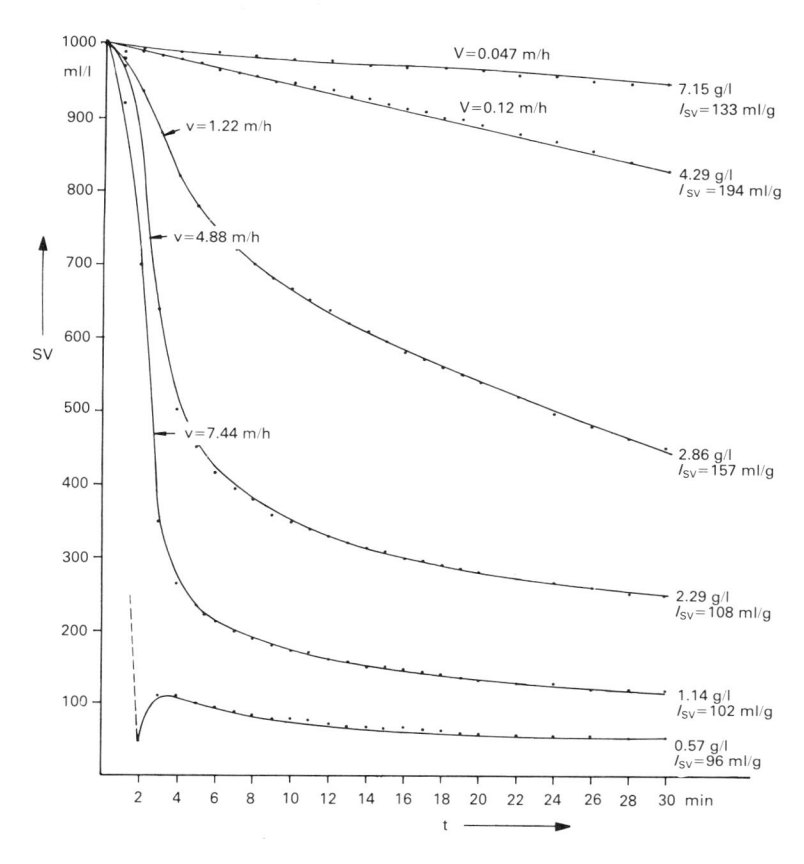

Theoretical principles of activated sludge sedimentation

Fig. 7.1 — Sedimentation profiles for a municipal activated sludge at different sludge concentrations. Measurements in a 1-litre measuring cylinder of d = 6.1 cm, h = 34.2 cm.

concentration there is a high filtering action with respect to the microflocs, so that the supernatant appears free of particulates.

From a comparison of the curves, it emerges that the *settling velocity* falls with increasing sludge solids content from a value of 7.44 m/h to 0.05 m/h. The sludge volume index rises from 96 to 194 ml/g as the solids rises to 4.3 g/l, while at 7.15 g/l sludge solids it was distinctly lower. The decrease in the sludge settling rate and the rise in the sludge volume index result from the mutual *hindrance* of the flocs during the settling process to elevated sludge solids contents. This hindrance effect is the basis for requiring less than 250 ml/l in the sample for determination of the sludge volume index (see Fig. 7.1 I_{sv} rises minimally from 96 to 108 ml/g as the sludge solids content increases from 0.57 to 2.29 g/l, but increases to 157 ml/g at 2.86 g/l). Under the conditions of the test in a measuring cylinder of 6 cm diameter there was also an effect due to wall friction which is hardly encountered in large-scale installations so that in the recycled sludge on the full scale, higher sludge solids concentrations may be realised than in the laboratory settling tests. According to Stobbe (1969) the wall friction effect does not arise in settling chambers of over 30 cm diameter, and hence is absent in all practical types of clarifier. On the other hand in the inlet and outlet regions of these vessels, there are flow patterns which adversely affect the settling process, but favour the flocculation and filtration effects; in addition, in the deeper types of settling tank a much stronger degree of compaction occurs by reason of the internal compression resulting form the increased density of the sludge relative to the aqueous phase.

The effects due to hindered settling on the optimisation of the aeration and settling tank volumes can be inferred from Fig. 7.2. According to Hörler (1968) several authors have devised similar diagrams with a trend towards further increases in the optimal sludge volume and hence a reduction in the overall tank capacity of an activated sludge installation. The greatest caution is called for in this connection, however, as the sludge volume index may influence the attainable sludge volume (and hence the solids concentration) adversely by reason of the poorly controllable settling properties of the sludge biomass. The Horler diagram represents a high degree of operational reliability for the final settling tank; the solids content indicated as optimal allows a high treatment performance to be achieved and ensures a relatively high α-value (Fig. 5.1). For the prolonged operation of existing activated sludge plants, the target value of sludge content may deviate substantially from the theoretical optimum for reasons of operating stability, the treatment performance or the energy consumption. In particular for plants with no primary settling facilities, values may often be set from 1 to 3 g/l higher on account of the high proportion of coarse solids.

7.2 FINAL SETTLING TANKS

For satisfactory functioning of the final settling tank, favourable hydraulic conditions are an essential prerequisite; these comprise:

settling time: ≥ 2 h, as a rule 3 h, for V_{max} always > 1.5 h, for bulking sludges 3–4 h, irrespective of V_{SR};

upflow velocity: < 1 m/h, for $V_{max} < 1.5$ m/h, for bulking sludges < 0.5 m/h, effect of aeration tank sludge volume (SV_{BB}) on upflow velocity given by Fig. 7.3;

outlet weir rating: 5–10 m³/m h; for bulking sludges 3–5 m³/m h and acccording to [38] up to 50 m³/m h for rectangular settling tanks;

recycled sludge flowrate: 0.3–1.0 V_h m³/h see Fig. 7.4.

According to [14] the following areal loading rates should not be exceeded:

— sludge content: 2.0–3.5 kg/m² h for $I_{SV} < 100$ ml/g, for bulking sludges < 2.0 kg/m² h;
— sludge volume: 0.2–0.4 m³/m² h, for bulking sludges < 0.2 m³/m² h.

Settling tanks may be funnel-shaped, circular or rectangular in design — see Fig. 7.5. For ensuring a high effluent quality the following additional requirements may be stressed.

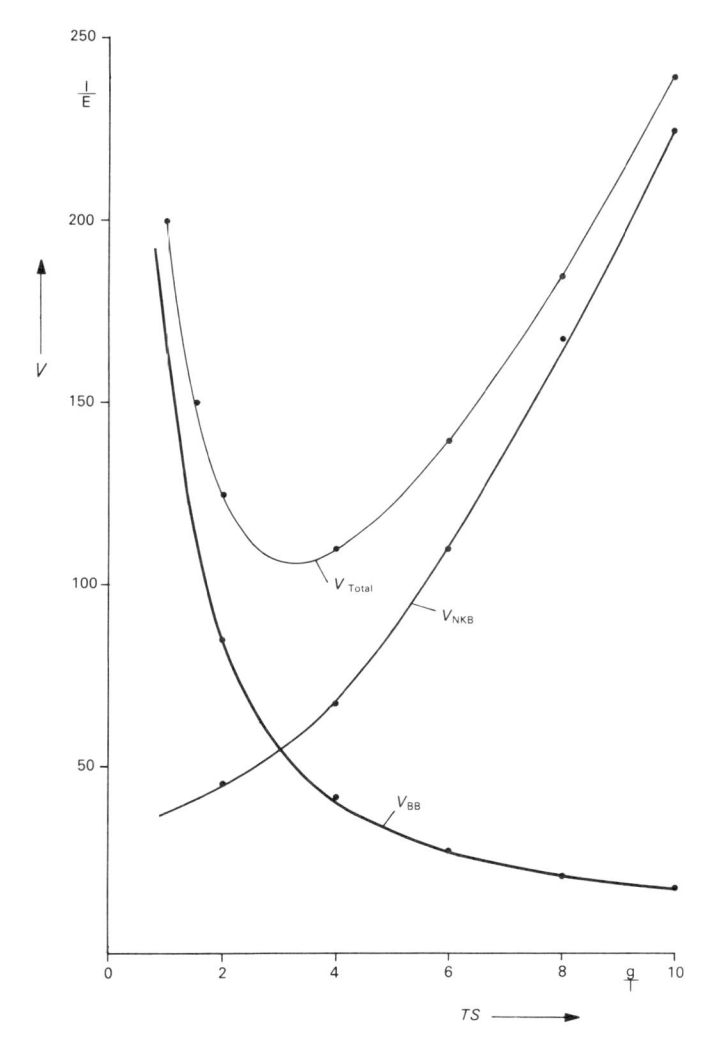

Fig. 7.2 — Working capacity of aeration tank and final settling tank as a function of the sludge suspended solids content in the aeration tank. Conditions: $n_{BOD} = 90\%$; $TS_a = 30$ mg/l; $V_{PE} = 21$ l/capita h. From Hörler (1968).

— *Scumboards* or baffles for retention of floating solids, which may be either carried into the tank or may be formed *in situ*; the action of such a submerged baffle may be seen in Plate 13.
— *Take-off device* for scum removal; the very wet scum may be returned to the primary settling tank, or possibly to the sludge treatment plant.

High-rate settling
In recent years exhaustive studies of methods of increasing the settling performance of clarifiers by means of tubular or parallel plate separators have been performed

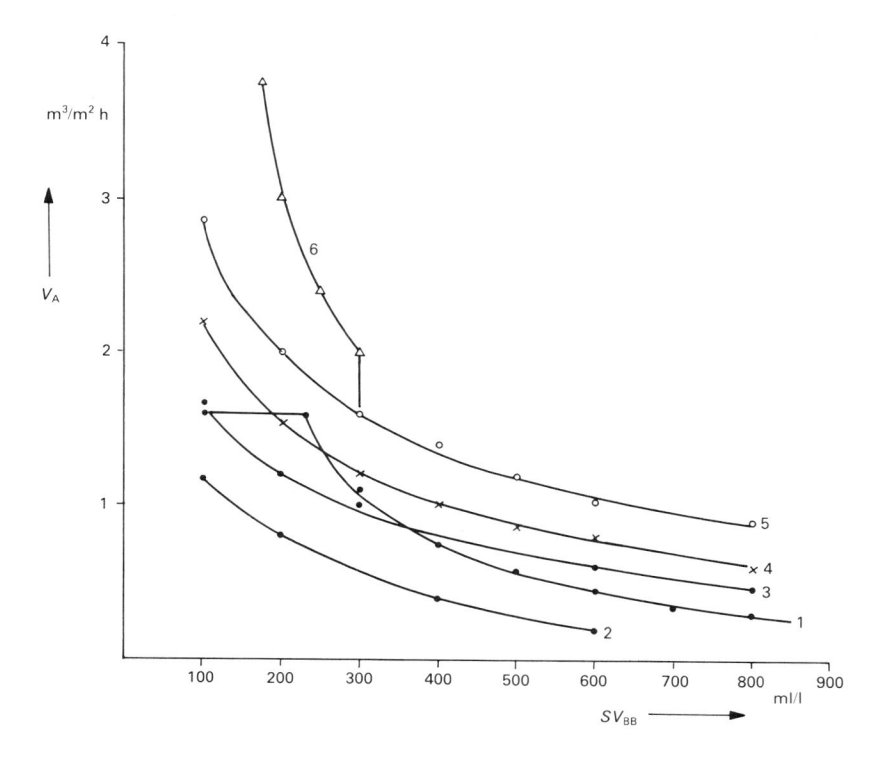

Fig. 7.3 — Dependence of the sludge volume on the areal loading rate of the final settling tank. (1) From [14] for $TS_a = 30$ mg/l; (2)–(5) After Ditsios (1984) for TS_a values of 10, 20, 30, and 50 mg/l; (6) From [31] for plug-flow tanks according to WAPRO 2.38/01 and for $TS_a \geq 30$ mg/l.

(Fig. 7.5B, [23], and Lützner 1979). The principle consists of inducing laminar flow conditions (Re ≤ 200, Froude, No. ≥ 10^{-5}) by improving the hydraulic conditions obtaining within the tank, so that the loading capacity can be increased (WAPRO 2.26: upflow velocity up to 4 m/h, sludge volume areal loading rate ≤ 0.6 m³/m² h).

Some studies on tubular separator systems showed that secondary filtration effects for microflocs occur at the tube walls, which have a positive effect on effluent quality. As, however, the incipient coating does not completely re-enter the settling compartment under the action of gravity, the extent of denitrification is enhanced and in the tubes suffering from the poorest transport rate, putrefaction reactions may be initiated. The floc coating in the tube walls must therefore be removed at regular intervals, usually about once a week.

Before installation of costly and maintenance-demanding high-rate settling facilities, tests should be undertaken in order to establish whether buffer storage of peak flows and their subsequent gradual introduction during periods of low sewage flow would enable hydraulic overloading of the final clarifier to be avoided.

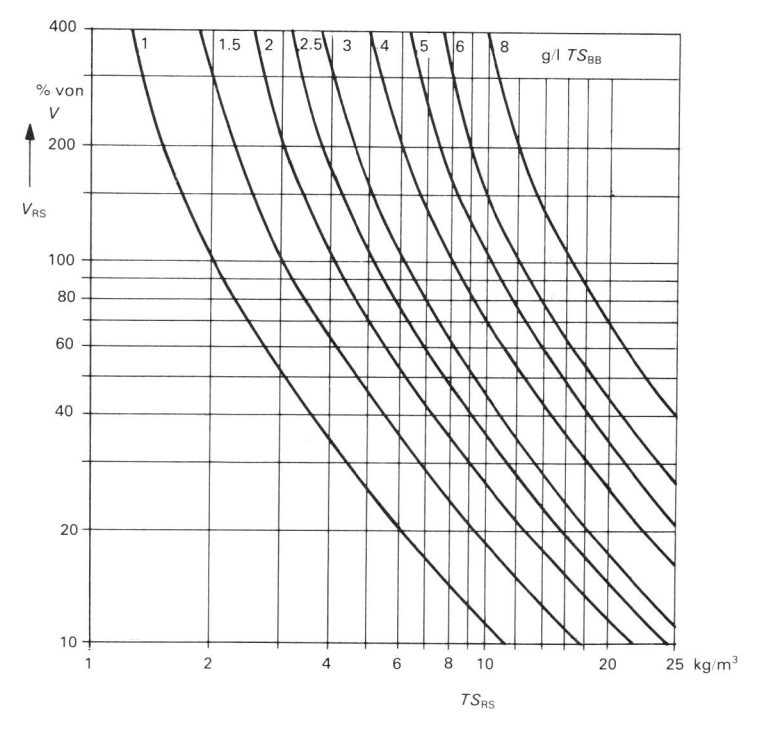

Fig. 7.4 — Recycled sludge flow-rates (V_{RS}) required to maintain various suspended solids contents in the aeration tank, (TS_{BB}) as a function of recycled sludge solids (TS_{RS}) at constant sewage inflow. From [14].

7.3 RECYCLED SLUDGE

The necessary recycled sludge flow can be read from Fig. 7.4; the curves are derived from the materials balance applicable to stationary conditions (that is, constant values of V_h, TS_{BB}, TS_{RS}, V_{RS}):

$$V_h\left(1 + \frac{V_{RS}}{V_h}\right) TS_{BB} = V_{RS} \cdot TS_{RS} \quad \text{(kg/h)}$$

(for symbols and abbreviations see Table 1.1).

The recycled sludge must be continuously removed in order to prevent digestion, phosphorus liberation and denitrification in the final settling tank. In the majority of final settling tanks the recycled sludge flowrate exerts only minimal influence in the settling process, up to a value of 2 V_h as it is immediately drawn off again at or just below the inlet zone and only the treated effluent passes through the interior of the tank.

The recycled sludge exhibits flow characteristics similar to those of water, so that the sludge recycle lines can be designed on this basis; for pumping the sludge, either

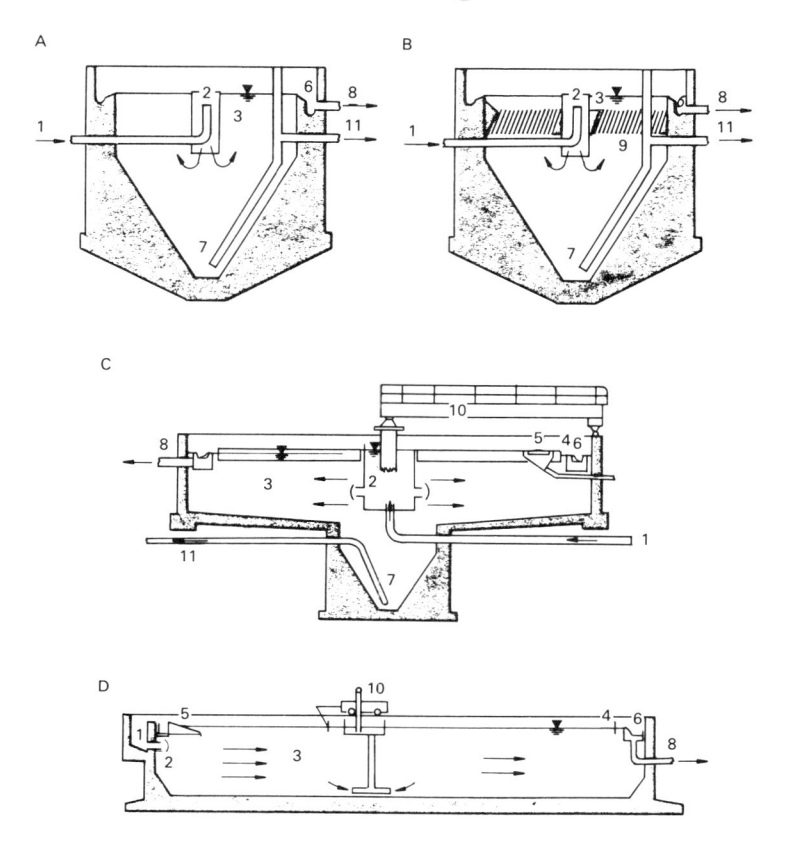

Fig. 7.5 — Principle features of settling tanks. 1: Inlet pipe. 2: Inlet aperture. 3: Settling zone. 4: Baffle. 5: Secure take-off. 6: Outlet duct. 7: Sludge thickening zone. 8: Outlet. 9: Lamellar inserts. 10: Sludge take-off point. 11: Discharge point for recycled and waste activated sludge.

screw or axial propeller-type pumps are suitable on account of the low hydrostatic head on the delivery side, while for small-scale plants mammoth-type rotors can be used. For very large plants several pumps of varying capacity are often employed.

An exact equivalence between the recycled sludge flowrate and the raw sewage flow is neither essential on biological grounds nor desirable from an operational viewpoint. During the entry of surface runoff from combined sewer networks, (frequently equal to $2\ V_h$) the final settling tanks will be subjected to inputs of very much greater quantities of sludge so that the sludge retention capacity under simple sludge recycle conditions is quickly exhausted and carryover of sludge into the outlet commences. Under such circumstances it is essential for a larger or additional sludge recycle pump to be brought into operation.

The wasted activated sludge may be tapped off from the delivery side of the sludge recycle pump; very often, however, separate waste activated sludge pumps are provided.

7.4 BIOCHEMICAL PROCESSES IN THE FINAL SETTLING TANK

Although clarifiers are equipment for mechanical settling, various biochemical processes take place inside, which are largely determined by the lack of oxygen. This oxygen deficiency results from the respiration of the activated sludge and even for plants with aerobic sludge stabilisation, the specific oxygen demand still amounts to 5 g/kg h, which means that at 4 g/l of activated sludge solid up to 20 g/m^3 h of oxygen may be called for. Hence a dissolved oxygen content of 1 mg/l in the outlet from the aeration tank is used up in about three minutes. If the sludge undergoes thickening to 12 g/l, then the oxygen demand rises to 1 mg/l min. The recycled sludge will therefore, even under the most favourable conditions, (for example, higher dissolved oxygen content in the aeration tank, shorter distance of travel to the clarifier, rapid take-off and return of the recycled sludge to the aeration tank) be free of oxygen. As a result of this oxygen deficiency, anoxic or anaerobic processes are set up, which lead especially to the formation of alchohols and organic acid (Fig. 4.20 and 4.25).

Should the anaerobic conditions be maintained for protracted periods, then in the presence of nitrate or nitrite, the denitrification reaction will commence. Also at the tank walls and on the surface of the submerged portions of the clarifier, flocs which are deposited there encounter ideal conditions for denitrification even where the theoretical retention time in the clarifier is quite short and where the nitrate concentration is low. From the reduction of only 1 mg NO$_3$-N, theoretically 0.83 ml of nitrogen gas is produced. According to Köhler (1975) 1 ml of gas is sufficient to cause from 53 to 125 mg of activated sludge to float. Floating activated sludge gives rise to scum formation at the surface of the clarifier and an increase in the filtrable and settleable solids in the final effluent.

It is also possible for phosphorus liberation to occur in the final settling tank in response to anaerobic conditions (see Fig. 4.28) and this is detrimental to water quality in the receiving stream.

In certain places such as the corners of the tank, at points of impact on the sides, and where the rubber scraper blades on the sludge removal device are defective, putrefaction of the sludge may occur. Similar pockets may also occur in the more quiescent zones of funnel-shaped tanks and in tubular separating equipment. The putrid sludge appears black owing to the formation of iron sulphide and comes to the surface from time to time in large buoyant flakes under the action of biogas.

8

Industrial effluents

Industrial effluents consist chiefly of process effluents from manufacturing operations. As long as they are of a non-toxic nature and of organic composition they may be treated by the activated sludge process. Owing to the multiplicity of the production processes and their operational peculiarities no generally applicable distinctions relative to municipal sewage are possible. With regard to the biochemical and biophysical processes occurring, their mode of purification does not differ from that of municipal sewage; however, the following particular aspects must be borne in mind.

8.1 CONTINUITY OF INCIDENCE OF THE EFFLUENT

Large installations with three-shift working usually generate a fairly even flow of effluent and a more or less constant BOD_5 loading over a 24 hour period so that the sludge biomass is subjected to an almost constant BOD_5 loading (see Fig. 5.3, Curve 1). By contrast, factories with single or two-shift operation, usually combined with weekend shut-down and holiday periods, are often associated with severe fluctuations in the BOD_5 loading during the course of a day. For factories with discontinuous operations like emptying reactors, washers and cooking retorts, very high shock loadings are created so that the loading conditions resemble those for a batch reactor with discontinuous feed arrangements. Often the reactors are emptied at the end of a shift, as for example in the food industry where sanitising agents may then be employed. In order to smooth out the hydraulic shock loads, which may cause problems due to sludge carryover from the final settling tank, equalisation or balancing tanks are provided in advance of the treatment plant. As a rule, however, it may be preferable to make the final clarifier larger. For pronounced shift-type operations the sludge loading (Table 10.1) should be expressed not in terms of daily amounts but with reference to the shift (e.g. instead of 1 kg/m³d a value of 1 kg/m³ 8 h, which is then equivalent to ~0.33 kg/m³ d). For seasonal operations, accidental shut-downs or maintenance operations on production lines, from time to time certain types of effluent or particular constituents may be lacking or may suddenly occur as shock loads. In such cases protracted plant upsets or start-up problems may arise.

8.2 ORGANIC SUBSTRATES

The organic constituents of effluents are entirely specific to a particular manufacturing process. In contrast to the numerous substrates present in municipal sewage, often only a few substances predominate. Where these are easily assimilable materials then populations of enhanced metabolic activity may be established. Very often, however, one is dealing with relatively poorly degradable substrates, for which specialist organisms with relatively tardy metabolic performance are required. Frequently this leads to large capacity aeration tanks with long retention times or to multistage plants. Because of the favourable physiological characteristics of municipal sewage as a nutrient solution, the foul sewage from any manufacturing plant should be combined with the process effluent from the same factory because it provides the necessary nutrient and trace elements needed for development of the bacterial population.

8.3 NUTRIENT ELEMENTS

Based on the classical studies involved in the development of the Magdeburg P-Process by Nolte *et al.* (1934) minimum nutrient contents are required in the proportion of $BOD_5:N \leqslant 100:7$ for nitrogen and $BOD_5:P \leqslant 100:1$ for phosphorus. This ratio can be obtained from the production of surplus sludge of empirical formula $C_{106}H_{180}O_{45}N_{16}P_1$. Where the content of nutrient elements in the activated sludge system is below 0.2 mg/l the synthesis of biomass is inhibited while the respiratory activity, that is the decomposition of organic matter by oxidative reactions as measured by the BOD_5 reduction per unit weight of sludge solids, remains virtually unaffected. Where the synthesis of surplus sludge is diminished on account of nutrient deficiency then the increase in aeration time necessary to maintain a constant effluent quality can be estimated as a function of the BOD_5 sludge loading, from Fig. 4.2. As nutrient-deficient effluents are often very rich in organic matter, it may be preferable, in order to achieve desirable effluent characteristics, to operate at an MLSS value of 8–12 g/l, possibly without any free dissolved oxygen, rather than with the usual MLSS content of 3 g/l and 1–2 mg/l of dissolved oxygen. Such facultatively anaerobic forms of sludge biomass usually exhibit SVI values of 50–80 mg/l so that the final settling tank is able to accommodate these high sludge concentrations.

As the surplus sludge production from many industrial effluents can be quite low, activated sludges of industrial origin are often characterised by lower concentrations of nutrient elements than those indicated earlier. For low ratios of surplus production the nutrients are generally preferentially incorporated in the biomass and repeatedly undergo successive endocellular biochemical reactions. With nutrient-deficient effluents sludge digestion facilities should be provided as a matter of course, so that the sludge liquor, which may contain 800–3000 mg/l of nitrogen and 50–800 mg/l of phosphorus can be used as a nutrient source in place of more expensive commercially available materials. In addition the supernatant liquid from the thickening tanks will be high in phosphorus.

The growth of organisms of the *Azotobacter* genus, which can fix nitrogen from the atmosphere and hence can rely on a self-contained source of nitrogen, has so far only been definitely observed in laboratory activated sludge systems.

8.4 OXYGEN DEMAND
Apart from those industrial effluents of very high BOD_5 content, the oxygen demand follows the pattern already indicated in Fig. 4.19.

8.5 BULKING SLUDGE
Many industrial effluents show a tendency to bulking sludge formation by reason of their composition, such as the presence of large amounts of low-molecular-weight dissolved substances, N- or P-deficiency, lack of trace elements or absence of inert weighting materials. Countermeasures may include larger final settling tanks and the process modifications indicated in section 3.7.2.3.

8.6 SURPLUS SLUDGE
Industrial effluents which consist principally of dissolved organic constituents with only very little inert suspended material usually generate only small quantities of surplus sludge, especially where low BOD_5 sludge loadings are adopted as a guarantee of a satisfactory effluent quality (see Fig. 4.17). Sludge yields of only 0.05–0.1 kg/kg are not an exceptional occurrence. Conversely liquid effluents high in coarse solids, such as liquid manure or slurry can produce several kilogrammes of waste activated sludge per kg BOD_5 eliminated.

8.7 TEMPERATURE
Many industrial effluents may be characterised either constantly or occasionally by distinctly elevated temperatures, which can lead to increased respiration and conversion rates in line with the Arrhenius Rule, but can also cause problems with the oxygen supply and enhanced denitrification. Also where the inflow is discontinuous, with no effluent being produced at week-ends, there may be other problems during the winter resulting from the pronounced temperature fluctuations, with impaired activity and flocculation behaviour in the biomass.

8.8 TOXIC AND ABIOTIC CONSTITUENTS
Many organically polluted effluents of industrial origin also contain toxic or inhibitory materials, such as fine coal solids, emulsified oils, etc. Besides the customary settling tanks it may thus be advisable to incorporate oil separators, physicochemical pretreatment with organic or inorganic coagulants, neutralisation facilities, ion exchangers, detoxification systems, and so on (Mangold *et al.* 1973).

9

Extended treatment

The activated sludge process removes from 90% to 98% of the organic contamination and pathogenic organisms as well as a certain portion of the adsorbable and absorbable inorganic compounds. This high degree of elimination, however, cannot obscure the fact that many water constituents, including noxious substances, pass through the system. Some of these water pollutants can accumulate in the fatty tissue of the aquatic organisms, while others accumulate in the sediments. Besides the nutrient elements nitrogen and phosphorus and those metal ions which are not retained, these comprise chiefly the refractory hydrocarbons, surfactants, aromatics, pesticides and organochlorine compounds. However, other substances such as the biologically stable products of metabolism, like humic and fulvic acids, must also be eliminated prior to the re-use of the water.

Riebhun and Manka (1971) cite the range of percentage composition of the dissolved organic fraction in 11 sewage works effluents as:

carboydrates	3.6–8%
proteins	18.9–24.8%
anionic surfactants	11.2–20.5%
tannins and lignins	0.8–2.4%
ether-extractable materials	10.3–19.9%
fulvic acids	16.8–30.2%
humic acids	3.6–15.7%
hymathomelanic acids	1.9–10.3%

Considering that a multiplicity of organic substrates together account for a BOD_5 of only 5–20 mg/l, each individual substrate must be present at an exceptionally low concentration that is far below the K_S value, so that according to the Michaelis-Menten equation its metabolism can hardly be achieved within a practically attainable period of time. Even readily-degradable substrates become poorly 'degradable' at this concentration level.

The frequency held view that the effluent quality improves with increasing sludge concentration, greater sludge age and longer aeration time only applies down to a BOD_5 sludge loading of 0.1–0.15 kg/kg d. Chudoba and Pokorny (1971) for example demonstrated an increase in COD from 20 to 40 mg/l, when the BOD_5 sludge loading was decreased from 0.2 to 0.082 kg/kg d. Especially at low sludge loadings, stable and sometimes strongly foaming organic products are generated by the organisms or liberated from them after death.

Both objectively and subjectively therefore the treated effluent can still be regarded as a contaminated liquid. As a result of the high content of *nutrient elements,* a well-treated transparent effluent becomes green after a few days due to the growth of algae. Owing to the continuing increase in the demand for water, multiple re-use is unavoidable, for which as a rule further treatment stages become necessary (Görbing 1981). Particularly stringent requirements are imposed if the receiving water body is to be used for the preparation of potable supplies. Then the treated effluent discharge must be of such a quality that, with due regard to the self-purification capacity of the receiving water, a safe potable supply can be obtained following soil passage and floc-filtration, possibly supplemented by active carbon filtration with or without ozonisation.

9.1 NUTRIENT REMOVAL

9.1.1 Importance of nutrient elements

Treated municipal effluents contain an average of 30 mg/l of nitrogen and 8 mg/l of phosphorus.

Primary production, which is the production of biomass consisting of chlorophyll-containing plants by means of photosynthesis, proceeds according to the following equation:

$$100\ CO_2 + 90\ H_2O + 16\ NO_3^- + PO_4^{3-} + \text{sunlight} \rightarrow C_{106}H_{180}O_{45}N_{16}P_1 + 154\tfrac{1}{2}\ O_2.$$

Just as for agriculture, the activity principle for growth factors applies in the case of natural waters, according to which that nutrient becomes limiting, which is present at the minimum concentration. Besides the starting materials indicated in the equation, primary production also requires the presence of potassium and trace elements such as Na, Ca, Mg, Co, Mn, Cu, Fe, Zn and Mo. Apart from sunlight, which fluctuates according to the season and particularly in winter becomes limiting on account of the screening effect resulting from a layer of ice, all other plant nutrients or trace elements not in the equation are usually present in sufficient concentration in the water body. Carbon dioxide becomes limiting only in the case of very strong photosynthesis, as it is continuously regenerated as a consequence of the respiration of aquatic organisms. Thus the nitrogen and phosphorus compounds remain as the decisive factors controlling the rate of growth. According to the above equation 1 g phosphorus yields 78.3 g plant biomass, and from 1 g nitrogen 10.7 g plant biomass is formed. Generally speaking, in the reservoirs in the central mountains of the GDR and in the lowland lakes, phosphorus is the growth-limiting factor.

As tolerable limiting values for the input of contamination to standing and impounded waters, Uhlmann (1982) quoted the following values, depending on the retention time and depth of the water body:

— phosphorus: 0.1–0.3 g/m² annum;
— nitrogen: 1–6 g/m² annum.

Standing water bodies are classified according to their content of nutrient elements in TGL 27886.

An accumulation of nutrients in natural waters, the process of eutrophication, leads to an increase in the production of algae and aquatic plants which can have the following adverse effects:

— formation of toxic materials — algal toxins;
— taste and odour problems;
— increased treatment costs;
— fish kills;
— depletion of the oxygen budget and of the self-purification capacity;
— impaired recreational potential;
— harmful effects on the health of man and animals due to nitrogen compounds.

The Drinking Water Quality Standard TGL 22433 specifies the following limiting values for nitrogen compounds in drinking water:

NH_4^+ 0.1 mg/l;
NO_2^- 0.2 mg/l;
NO_3^- 40 mg/l.

Persons specially at risk are small children on account of methaemoglobinaemia, while a further risk for people suffering from stomach disorders is assumed owing to the carcinogenic potential of nitrate in the event of its reaction with amines to produce nitrosamines with proven carcinogenic properties. Because of these potentially harmful effects of nitrate, the discharge of nitrate-rich biologically treated sewage effluents is harmful, as it may give rise to nitrate contamination and accumulation in the groundwater body. In this connection it is immaterial whether the nitrogen is discharged in the form of combined organic nitrogen, or as ammonia, nitrate or nitrite nitrogen, as given the necessary conditions the process of nitrification is initiated, with nitrate constituting the end product.

Ammonia also has a toxic effect on many aquatic organisms in the high pH region, and chlorination can give rise to chloramines, with similar toxic properties, from the reaction of chlorine with ammonia.

The *nitrite* formed as a result of nitrification and sometimes from denitrification is also toxic. Particular care should be exercised in respect of the effluent from activated sludge plants, in which the oxidation of nitrite to nitrate is blocked by inhibitory substances. According to Amlacher (1972) a nitrite content of only 3 mg/l of NO_2^--N can be toxic to fish. Wastewater left after cooking animal fodder and containing 0.4 mg/l NO_2^- killed over 48% of a herd of young pigs in 10 months (cited by Cena 1975).

In many publications reference is made to the role of nitrate as a source of oxygen in natural waters. Should the nitrate oxygen actually be utilised as an oxygen source in the water body, then natural denitrification is taking place, which has a beneficial effect on the state of the water. This nitrification process is, however, at the mercy of strong seasonal and climatic changes and hence cannot be relied on. It is, without any doubt, better to remove the nitrogen in the sewage treatment plant and to clean up the water body so that nitrate can be dispensed with as a source of oxygen. As denitrification is only initiated at very low oxygen concentrations the nitrate oxygen has no effect on the maintenance of fish populations in the water body.

The general quality of the receiving water and the uses to which it may be put have a decisive influence on the nature and extent of nutrient removal. For waters giving rise to plankton blooms phosphorus is generally the more critical pollutant, compared with nitrogen in those waters used for the preparation of potable supplies.

9.1.2 Origin of the nutrient elements

The ordinary human diet has a C:N:P ration of around 100:16:1. By the process of metabolism in the body a major portion of the carbon is converted into carbon dioxide and exhaled, while the nutrient elements enter the excreta. The C:N:P ratio in primary treated domestic sewage is then around 25:8:1. In the formation of the activated sludge biomass there is a further carbon loss as a result of CO_2 production, with the result that the elements N and P are present in excess. Neither CO_2 nor carbonate can be utilised as carbon sources by the heterotrophic organisms of the sludge biomass.

The mean incidence of nutrient elements per head of population is about 12 g/d of nitrogen and 2 g/d of phosphorus.

While the nitrogen content of municipal sewage has remained roughly constant during recent years, there has been a rapid increase in the amount of phosphorus as a consequence of the introduction of synthetic detergents. Household washing products contain 20–50% polyphosphate, while the 60°C formulations contain 40–70% pentasodium phosphate and special fabric washing products 10–25% polyphosphate. As a result the phosphorus content of domestic sewage usually consists of 60% derived from synthetic detergent products and only about 40% derived from dietary sources via the excreta. In the German Democratic Republic the specific phosphorus content of sewage rose from 0.75 to 2.21 g/head day between 1950 and 1967 (Heine 1969). For industrial effluents phosphorus concentrations may vary within wide limits according to the particular nature of the manufacturing process concerned.

9.1.3 Nitrogen removal

According to TGL 22764 (Table 2.3) biologically treated municipal sewage remains hypersaprobic in the absence of the oxidation of nitrogen, and polysaprobic without denitrification.

From a cost-effective angle nitrogen may be removed from sewage only by biological means. The two main removal mechanisms are the *incorporation* of nitrogen into the biomass and *denitrification*. For municipal sewage containing 50 mg/l N, heavily-loaded sludges will incorporate between 10 and 20% of the N into the surplus sludge. For BOD_5 sludge loadings less than 0.3 kg/kg d, the nitrogen removal is 30–50% however, because of the presence of nitrate or nitrite in certain portions of the flocs (see Fig. 4.26, the decrease of total nitrogen content in the aeration tank resembles that in the denitrification tank) and in anaerobic zones in the settling tank and sludge recycle lines which enable denitrification processes to occur. Just as for phosphorus, where an extended N-removal is required it is desirable to use the nitrogenous liquor for agricultural purposes.

In the extensive literature pertaining to denitrification, the use of methanol, acetate and molasses as non-nitrogenous hydrogen sources is described. For large-scale plants the use of such substrates incurs a heavy cost penalty where no other nitrogen-free wastes or by-products happen to be available. In order to remove 1 g of nitrite or nitrate oxygen, 1.5–2.0 g BOD_5 is required, or very much more than the theoretical equivalent. With a nitrate content of 30 mg/l NO_3^--N following nitrification, 133 mg/l of O_2 are present as nitrate and for denitrification of this amount of nitrate at least 200 mg/l BOD_5 is needed. Molasses has a BOD_5 of only 0.5 g/g [18], methanol 0.78 g/g and sodium acetate 0.39 g/g (Koumar 1964). However, up to 90% of the hydrogen needed for denitrification can be obtained from the organic substrates in the sewage. In this connection it is a point of economic significance that the BOD used for denitrification no longer imposes any oxygen demand in the aeration tank. With the use of the BOD already present in the sewage for the purpose of denitrification, therefore, a 20–30% saving in power costs may be achieved relative to an activated sludge plant without denitrification. For nitrite or nitrate-rich effluents the denitrification process should thus constitute the first stage of treatment.

Fig. 9.1 presents a selection of process variants for denitrification. The variants A and B and also E and F employ the incoming sewage as a hydrogen source, either in a preceding or (for variants E and F) a simultaneous denitrification stage. For variants A, B and F denitrification is essentially dependent on the recycled sludge flowrate.

For a sludge recycle ratio of 100%, η_N=50%, for a ratio of 200% η_N=67%. In addition there is the elimination of nitrogen by incorporation into the biomass and other random denitrification reactions, so that the total N-removal amounts to 70–80%.

By recycling 300% to 500% of the input to the final settling tank to the denitrification tank the nitrogen removal can be increased to 90% or possibly even 95%. Nowhere in the whole chain of water treatment operations can nitrite and nitrate be removed so simply and cheaply as in an activated sludge plant with a preliminary denitrification step. With variant B, from 90% to 95% of the nitrogen is removable, although 20–30% of the nitrogen is removed in the second tank where an artificial substrate is introduced as a hydrogen source. With variants C and D also it is possible for 90–95% of the nitrogen to be removed although at a high cost in terms of energy and substrate consumption, often accompanied by bulking sludge problems. If variant D is operated without any artificial substrate input the denitrification rate is low. According to Frangipane and Urbini (1978), denitrification rates of 2–3 g/kg h are achieved with preliminary denitrification, and only 0.3–0.5 g/kg h for post-denitrification. Variant E, termed simultaneous denitrification, can be successfully employed in some very large treatment installations. The principle lies in the repeated alternation between aerobic and anaerobic conditions along the length of the tank, so that optimal conditions for nitrification and denitrification are created in

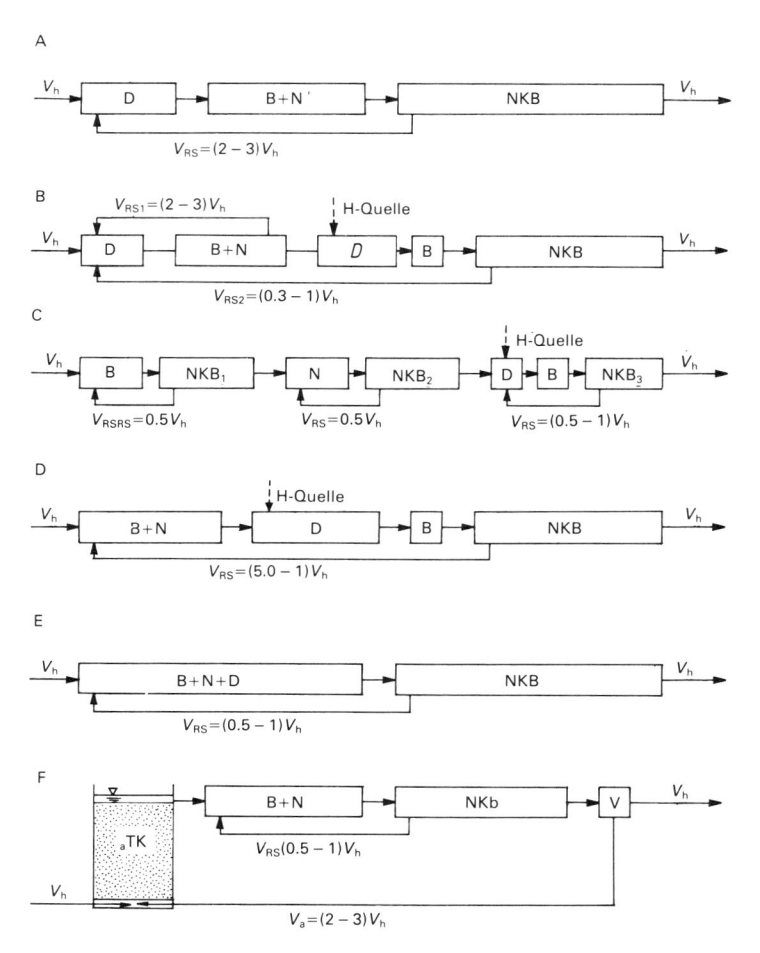

Fig. 9.1 — Process engineering options for denitrification. A: Preliminary denitrification. B: Preliminary denitrification and denitrification of the residual oxidised nitrogen compounds in an intermediate denitrification stage with substrate input. C: Three-stage activated sludge plant with the successive functions of (a) metabolic breakdown of organic matter, (b) nitrification and (c) denitrification with substrate addition. D: intermediate denitrification with substrate addition. E: Simultaneous and/or alternating denitrification. F: Preliminary denitrification employing an anaerobic trickling filter. Key: B=metabolic breakdown of organic substrate; D=denitrification; N=nitrification.

turn. As, however, the oxygen demand of the activated sludge undergoes appreciable daily, weekly and seasonal fluctuations, a close adjustment of oxygen input in keeping with the oxygen demand is necessary, which means that increased service and control expenditure is incurred. In stirred-tank systems denitrification can be realised as a consequence of intermittent aeration, for example 20 min aeration followed by 20 min denitrification. Both simultaneous and intermittent denitrification employ the hydrogen in the incoming sewage in a less efficient manner than in the case of preliminary denitrification so that extended reaction times are called for.

For variants B, C and D with denitrification stages in advance of the final settling tank, brief aerobic stages of 0.3–0.5 h duration should be inserted, in order to terminate the denitrification reaction, as otherwise it will lead to an upwelling of denitrifying sludge in the final settling tank, and hence to poorer effluent quality coupled with scum formation at the surface of the tank. In the aerobic stage also, those low-molecular weight substrates which are formed to some degree under anaerobic conditions, and which impose a significant BOD and COD (Table 4.7, condition 2) can be removed.

Under the climatic conditions prevailing in Central Europe, it is advisable [24], in the case of municipal sewage treatment and denitrification systems, to adopt BOD_5 sludge loadings of 0.1 kg/kg d with respect to the total biomass, or 0.15 kg/kg d for that in the denitrification tank, together with a denitrification rate of 1 g/kg h in the denitrification tank. For preliminary denitrification systems the anoxic phase for the sewage/sludge suspension should not last more than 2.5 h and make up not more than 30% of the total retention time, in order to avoid any injury to the aerobic biomass by reason of oxygen deficiency. For the same reasons, when designing and commissioning systems for preliminary denitrification according to the patterns indicated in diagrams A and B, care must be taken to ensure that under such initial conditions no denitrification occurs, but that the denitrification tank can either be by-passed during the start-up phase or can preferably be operated as an aerobic compartment. In order to prevent operational problems due to floating activated sludge, the outlet from the denitrification tank should be arranged at the same level as that of the liquid with no intervening underwater partition.

Conditions prejudicial to denitrification can arise as a result of substrate deficiency due to stormwater inputs, accentuated during snowmelt periods as a consequence of very low temperatures.

9.1.4 Phosphorus removal

Within the range of loadings cited in Table 10.1, from 2 to 4 mg/l P, or from 20 to 30% of the total P content of municipal sewage can be eliminated; owing to the hydrolysis of polyphosphates during biological treatment, however, the content of $\sigma\text{-}PO_4^{3-}$ in the effluent is often greater than that in the influent.

The phosphate may be removed, according to TGL 27886/01, by means of inorganic coagulants which may either be introduced into the inlet to the primary settling tank or in a separate physicochemical treatment stage connected in series with the final settling tank. The expensive *post-precipitation* method gives the lowest effluent concentrations, coupled with the additional elimination of filtrable solids, adsorbed compounds and end-products of bacterial metabolism, together with pathogenic organisms, etc. The P-content can be reduced to less than 0.5 mg/l. The use of *pre-precipitation* for municipal sewage results in effluent P-concentrations of 1–2 mg/l, but simultaneously removes up to 60% of the BOD and hence relieves the load on the activated sludge system and so reduces the energy consumption. With this method the P content is reduced to the critical range for P-demand of the microorganisms; however, no reductions in removal rates for organic substrates need be expected on account of phosphorus deficiency as the surplus biomass production from the pretreated sewage and hence the phosphorus demand is quite low. However, the inert ballast materials and other particulate solids helping the sludge settling properties are also eliminated, and in the liquid phase the low-molecular weight substrates are uppermost, so that the growth of thread-forming organisms in the biomass is encouraged and may lead to the formation of light, poorly sedimenting sludges.

Chemical phosphorus elimination has the following disadvantages:

— operating costs: added cost of chemicals, including transport, storage and dosage equipment, additional manpower costs and enhanced sludge production, and possible corrosion effects;
— capital costs: dosing plant, silo and pipelines;
— occupational safety: need to cater for aggressive and caustic substances;
— sludge utilisation: additional metals contained in sludge for disposal to land, with low availability to plants of phosphate combined with iron salts;
— water quality: increased salinity in the receiving water.

When considering the natural resources available on a global scale, there appears to be no long-term future for the widespread application of phosphate coagulation by means of iron or aluminium salts. As, however, the biological processes are still in the early stages of development, it is still the accepted practice to fall back on the use of chemical or combined methods for phosphate removal.

Several variants of the process of intensive phosphate removal using activated sludge are outlined in Fig. 9.2. Those exclusively concerned with biological mechanisms, such as over-compensation and luxury uptake, consist of the Anaerobic-Aerobic Process (Fig. 9.2(A)) and the Bardenpho Process (*Bar*nard *Den*itrification, *pho*sphate removal, Fig. 9.2(B), Barnard 1974). The residence times required for the individual stages should be established on the basis of long-term experiment. Under optimal conditions from time to time effluent concentrations of less than 1 mg/l total P may be obtained, with orthophosphate contents of less than 0.1 mg/l in the final aerobic compartment. Because of non-steady elimination rates, however, the biological phosphorus elimination must occasionally be supplemented by the addition of chemicals to the final aerobic compartment or the settling tank, or else a post-coagulation step introduced.

Among the combined chemical-biological methods, the Phostrip Process (Fig. 9.2(C)) and simultaneous coagulation (Fig. 9.2(D)) may be considered important. With the Phostrip process, from 15% to 30% of the recycled sludge is bled off and exposed to anaerobic conditions for 10–24 hour. The phosphate released by the sludge biomass during this period is then chemically precipitated after separation of the liquid and solid phases, while the phosphorus-depleted activated sludge is returned as recycled sludge to the aeration tank. The P-removal performance may from time to time be such that the effluent P-concentration is less than 1 mg/l. Advantages of this method include the lower chemical dosage, the separation of coagulants from the activated sludge and especially the possibility of using lime as a coagulant (150–300 mg/l CaO). According to Hagen (1980) the costs are around 55% lower than those of chemical coagulation alone. It is particularly effective when the phosphorus-rich liquor fraction, which may make up from 10% to 20% of the sewage flowrate, can be applied directly to land.

Simultaneous coagulation is the most widely-used method at the present time. Preference for this method is based on the fact that no additional tanks are required

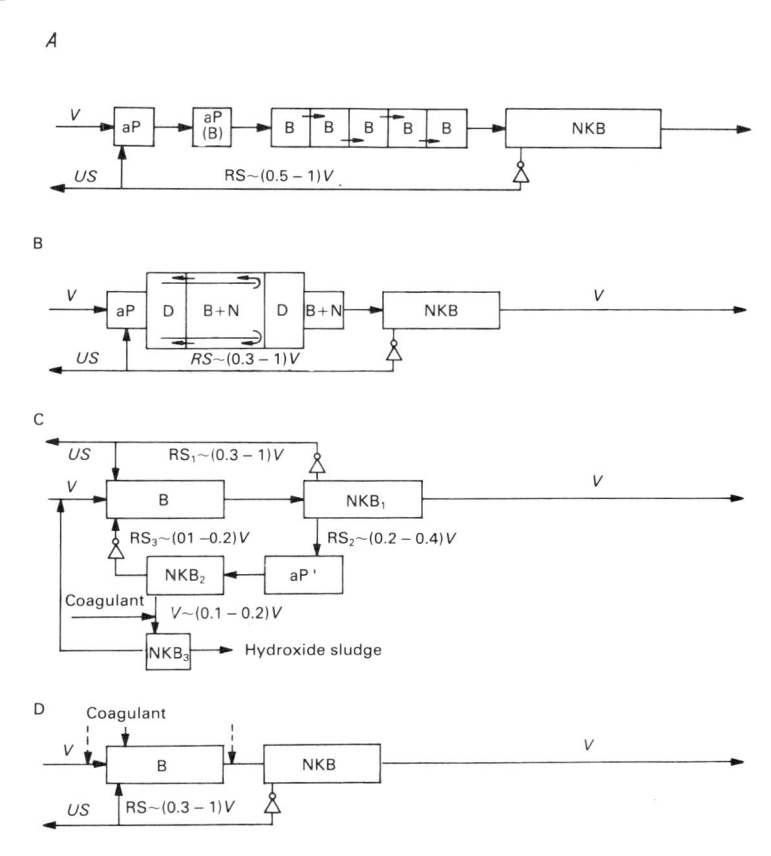

Fig. 9.2 — Process engineering options for biological and chemical phosphorus elimination in the activated sludge. A: anaerobic-aerobic process without nitrification. B: Bardenpho process from Osborn and Nicholls (1978). Key: B=metabolic breakdown of organic substrate; D=denitrification; N=nitrification; aP=anaerobic phosphorus liberation.

and that the coagulation process proceeds largely independent of fluctuations in input flowrate. The coagulants, usually ferrous salts, are introduced at the front of the aeration tank in the dissolved state or very often into the incoming sewage, and perhaps, as long as Fe^{2+} salts are not used, into the feed to the final settling tank. As the coagulants are continuously introduced into the cycle, their combining power is fully utilised; however, the high degree of turbulence in the aeration tank is detrimental to the coagulation process. After a running-in phase of several days or weeks, many plants produce very clear effluents, although in others the effluent may be very cloudy. P contents may be reduced to 0.5–1.5 mg/l, but often 2 to 3 moles of coagulant are required per mole of phosphate, and hence very much more than for other processes. The coagulant demand required to achieve a specified P-content is variable and not precisely calculable on stoichiometric grounds. For obtaining a residual P-content of 0.7 mg/l, Schuster (1971) recommends 40 mg/l of $FeCl_3$, 75 mg/l of $FeSO_4$ or 100 mg/l $Al_2(SO_4)_3$. The Soviet Planning Recommendations [15] envisage that the coagulant dosage, based on the relevant cation, should not exceed

25 mg/l if added prior to the aeration tank, but not be more than 10–15 mg/l if added subsequently. The sludge produced is relatively dense, it has a low sludge volume index, and hence settles well and is easily digested, despite its elevated metal content. The re-dissolution of phosphorus during digestion depends, according to Mosebach (1975) on the stoichiometric $Fe:PO_4^{3-}$ ratio. Where this ratio is greater than 2.0, then no re-dissolution of phosphorus occurs during continuous operation of the digester. On the contrary, the amount of dissolved total-P in the digester liquor is around 50–60% less than that of plants without simultaneous coagulation, as the excess combining power of iron available in the digestion vessel is used to immobilise the phosphate liberated from the sludge during the digestion process.

9.2 DISINFECTION

Even the very best standard of operation of activated sludge plants cannot achieve complete elimination of pathogenic organisms. For some types of sewage and with stringent water quality stipulations, for example for drinking water supply or beverages (TGL 22433), or for bathing waters (TGL 37780/01), either continuous disinfection or, in response to plagues or epidemics, intermittent disinfection may be necessary. However, disinfection should normally be avoided purely for water quality management purposes, because not only the pathogens, but also their natural enemies, the protozoa, are killed, and residual amounts of disinfectant may harm the natural biocoenosis in the water body.

Disinfection of a biologically-treated effluent, compared with untreated sewage, has the advantage that less disinfectant is required, and also that the coarse solids which shelter the pathogens are largely eliminated in the course of adequate final settling treatment.

The methods employed for destroying pathogens comprise chemical and physical processes, such as thermal treatment (TGL 22022), radioactive treatment or the application of ultrasonic, ultra-violet or ionising radiation (Table 9.1).

The most frequently-used disinfectant, namely chlorine, including its compounds such as chloramine, chloride of lime, sodium hypochlorite solution, etc., acts primarily on account of the atomic oxygen which is produced and is severely toxic. The reaction between atomic oxygen and the organic compounds remaining in solution also leads to a reduction in the BOD and COD of 10% .to 30%. The amount of chlorine required for disinfection is such that after a contact time of 30 minutes a residual of 0.1–0.2 mg/l is still detectable, which means that for well-treated effluents from 2–5 g/m^3 of chlorine is required.

Chlorine is a relatively weak disinfectant, so that many pathogens, particularly those in the interior of sludge flocs, together with bacterial spores and practically all helminth ova, are able to survive. The greatest drawback connected with the use of chlorine, however, lies in the formation of poorly biodegradable chlorinated hydrocarbons of which some may even have carcinogenic, mutagenic or teratogenic properties and hence must be regarded as injurious to health; on the other hand certain recognised carcinogens, such as 3,4-benzpyrene are rendered harmless by chlorination.

The other chemical disinfectants such as organic peracids, such as peracetic acid, hydrogen peroxide and ozone, likewise act by liberation of atomic oxygen. Ozone

Table 9.1 — Disinfectants active against pathogenic organisms from a summary by Clausing (1980)

Disinfectant	Dose g/m^3 for Cl$_2$·and O$_3$ K rad for radiation	Contact time (min)	Remarks
Chlorine	2–10	15–30	Biologically treated sewage effluent Residual Cl$_2$ 0.1–0.3 mg/l
Ozone	5–10	7	Destruction of Salmonellae
	5–10	30	Destruction of anthrax bacilli
	0.1–0.2	8	Used against coliforms
	1.6–3.2	8	Used against spore-formers
	20	12	Biologically treated sewage effluent
	1.24	0.003	Destruction of viruses
Irradiation	1000–15 000		Viruses
	12–20		Salmonellae
	200–280		Bacillus spores
	400–800		Spores of *Cl. botulinum*

may be regarded as the disinfectant of choice as its disinfectant activity is appreciably greater than that of chlorine and the resulting breakdown products are readily susceptible to subsequent biological decomposition. While the BOD$_5$ and the COD-Mn both increase slightly on account of the breakdown products the COD-Cr falls on account of a reduction in the chemically oxidisable, but biologically resistant, organic substrates. In the recent literature mention is made of a combination of two different disinfectants, as a result of which the disinfection rate was increased up to tenfold.

For the biological methods of advanced treatment referred to in section 9.3, the biological, chemical and physical mechanisms involved achieve a further reduction in the numbers of microorganisms of at least 98%. Polishing ponds with retention times of more than four days, which are associated with a massive development of filtering organisms such as water fleas, rotifers, etc., have removal performances of 3 to 4 powers of 10 for bacteria. Such a biological disinfection merits more widespread use in future.

9.3 OTHER METHODS

For extended treatment of biologically treated sewage effluents the following methods are widely used in practice:

— agricultural application according to TGL 6466/01 and 26567/01;
— soil filters according to TGL 26567/03;
— plant systems according to TGL 36430/03;
— rapid or slow sand filters;
— polishing lagoons according to TGL 28722;
— post-coagulation;
— floc-filtration.

The method to be employed is governed by the volumetric flowrate, location and quality required for the treated effluent.

Re-use of highly purified sewage effluent in the form of drinking water is not normally practised because persistent compounds may remain and, after accumulation or in conjunction with other toxins, may have adverse effects on health or impair the value of the recycled water. In certain arid districts, however, water re-use may be practised in such a manner that by a combination of very expensive chemical and physical processes the risk to health is practically eliminated.

10
Dimensional criteria

The starting point for the dimensional design of any sewage treatment plant consists of the treated effluent quality stipulations of the relevant Water Quality and Hygiene Inspectorate (Table 10.2). These requirements and the actual dimensional parameters of the activated sludge installation must be used in conjunction with statistically verified diurnal profiles for the volumetric flowrate of raw sewage and the most important sewage parameters such as BOD_5, COD, TS, pH-value, and sludge volumes, with some possible consideration of seasonal and climatic changes and holiday periods.

The Water Quality Inspectorate prescribes limiting values in respect of certain quality criteria for the treated effluent having regard to concentration and mass flow. As the maintenance of these limiting values affects the economic balance between sewage treatment and water quality protection, controversy has arisen, especially over the question of whether the limiting values ought not to be exceeded under any circumstances. However, every experienced operator knows that during cleaning operations on the final settling tank, the prescribed limits will be exceeded, and besides this, any biological system will from time to time and for no apparent reason 'go off' to a limited extent. It is not economically acceptable, and may even in some circumstances be detrimental to water quality, to over-design an activated sludge plant with the object of eliminating any risk of exceeding the limiting values. If such happenings must be excluded, it is more appropriate to add a further treatment stage.

For municipal plants with sewered populations of up to 50 000 PE, to which no harmful or reactive industrial effluents are discharged, the design of the activated sludge compartment can be based on the experience embodied in Fig. 10.1 or Table 10.1, while if nitrification and denitrification are called for, due regard should be paid to section 4.3.3. For the final settling tank the stipulations contained in sections 7.2 and 7.3 will apply. The aeration tank design should be based on the BOD_5 daily input, and the final settling tank should assume an effluent flowrate of 0.1–$0.2\ V_d$ (V_d flowrate per day).

Practical experience combined with empiricism provide a more reliable basis for

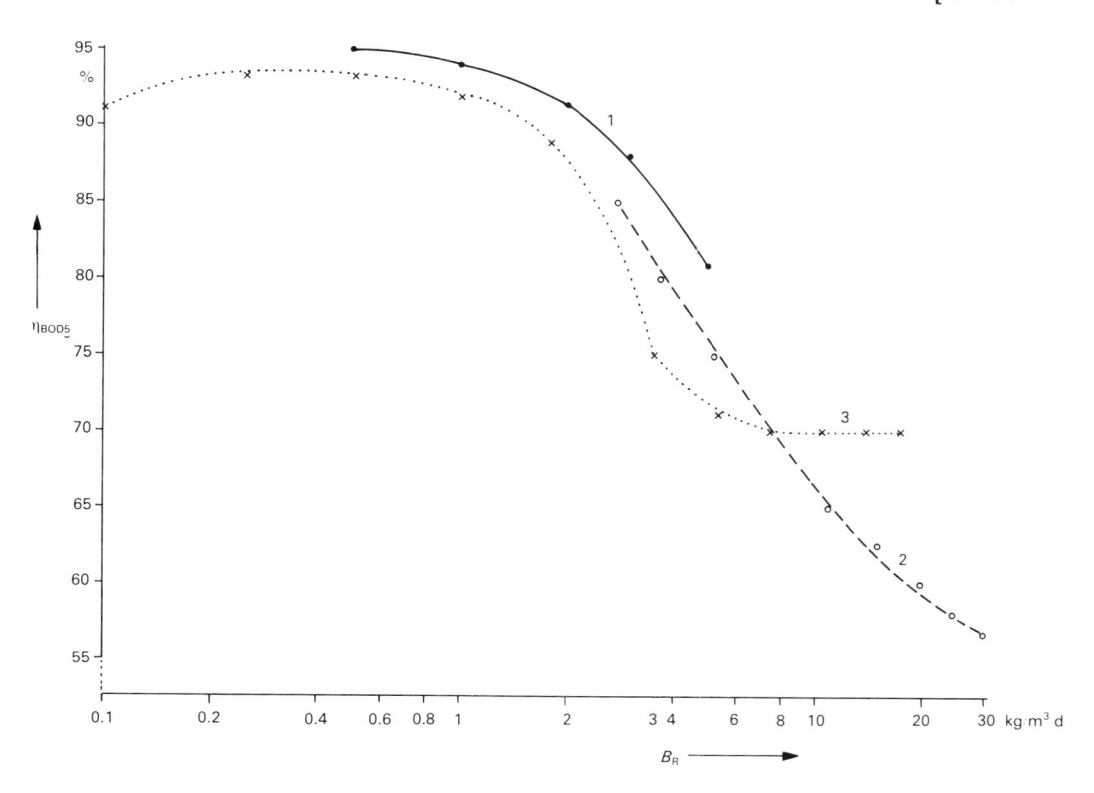

Fig. 10.1 — BOD$_5$ elimination as a function of the BOD$_5$ volumetric loading rate at a sludge suspended solids content of 3–4 g/l. Curve 1 from [19]; Curve 2: from [31]; Curve 3: recalculated from the BOD sludge loading rate according to Imhoff (1982).

the design of activated sludge plants than the theoretical models currently available. Even if the reaction constants for the metabolic decomposition of sewage constituents have previously been ascertained by experiment, this takes no account of the multiplicity of factors which can affect the operation of the plant in the long term. The time-related variability of the biocoenoses gives rise to continuous changes in the values of the reaction rate constants V_{max} and K_m. The current models (Vavilin 1982) accordingly do not provide a safe and at the same time economically acceptable method for design of activated sludge plants.

For larger catchments, or plants with purpose-designed nutrient removal, or where industrial or trade effluents are present in substantial amounts in municipal sewage, investigations in respect of degradability and conversion rate of specific constitutents may be required in addition to a knowledge of the parameters cited above. The nature and scope of the studies will necessarily be concerned with both the quantity and quality of the effluents to be treated.

For the simplest case, and for a low outlay in terms of effort and expense, batch or discontinuous degradation tests may be performed on a laboratory scale. For this an acclimatised or specially-adapted activated sludge will be instantaneously contacted

Table 10.1 — Dimensional parameters and the corresponding effluent quality criteria for BOD_5. Input BOD_5: 200 mg/l, or 300 mg/l for stabilisation without primary settling N = 40 mg/l. Stated effluent BOD_5 + 25% attained in 85% of cases. From [14]

	Aerobic sludge stabilisaton	Nitrification	Residual BOD_5 20 mg/l	Residual BOD_5 30 mg/l
BOD_5-volumetric loading (kg/m³ d)	0.25	0.5	1.0	2.0
BOD_5-sludge loading (kg/kg d)	0.05	0.15	0.3	0.6
Sludge content (g/l)	5	3.3	3.3	3.3
Recycled sludge flowrate (% of V_d)	100	100	100	100
Surplus sludge production (kg/kg)	0.8	0.75	0.85	0.95
Sludge age (d)	25	9	4	2
Effluent from final settling tank				
BOD_5 (mg/l)	12	15	20	30
NH_4^+-N (mg/l)	3	3	10	21
Org-N (mg/l)	0	1	1	2
NO_3^--N (mg/l)	2	16	12	5

Table 10.2 — Hygienic quality criteria for the effluent from municipal sewage treatment plants. Probability of exceedance — 20%. From Fiedler (1981) with modifications

| Parameter | Degree of biological treatment | |
	Partial	Complete
BOD_5 mg/l	75	25
COD mg/l	180	100
COD-Mn mg/l	40	20
TOC mg/l	60	20
DOC mg/l	50	15
SV ml/l	0.3	0.1
Colony count, per ml	10^6	10^4
Coliform count, per ml	10^4	10^3
Faecal coliforms, per ml	10^3	10^2
Enterococci, per ml	10^2	10
Helminth ova, per 1	—	1
Guppy test LC 96 h at 100% dilution	—	negative
Saprobic index	—	2.5

with a certain quantity of the raw sewage (ratios of 1:1, 1:0.5, 1:0.1, etc.) and the subsequent decrease in concentration of the target constituents, the oxygen demand and the sludge growth rate will be monitored while maintaining an adequate degree of aeration and circulation of the contents. From the rate of disappearance of the relevant constituents, which may follow zero, first, second or a mixed order of reaction following the initial adsorption phase, it is possible to determine the reaction velocities with respect to certain parameters as well as the value of the reaction constant K contained in the equations referred to in section 4.2.1. As a rule batch experiments of this type give values which are lower than those occurring in a full-scale continuous plant, so that the aeration tank dimensions will be found to be too large. Causes for this are:

— Frequently the process of adaptation of the biocoenosis, because it is difficult to achieve, remains incomplete. The prolonged operation of full-scale systems enables the adaptation as a result of induced acclimatisation and colonisation by selected strains to be performed over a period of several months, during which the value of the reaction constants may increase several-fold.
— The very high, and hence sometimes reaction-limiting, concentration (substrate-inhibition according to section 4.2.5) and also blocking of enzymes by a species-specific substrate (product inhibition).
— Toxic effects of degradable sewage constituents (pH, H_2S, etc., polyauxy) as a result of the sudden introduction of the sewage.

The two last-named effects may be largely overcome by a gradual addition of the sewage from a dropping funnel.

For more reliable dimensional parameter estimations, continuously-fed experimental systems operated on the flow-through principle for a period of eight to 12 weeks are necessary. Both laboratory and full-scale plants can be used for this purpose. The mobile test equipment described by Hänel and Schmidt (1975) (Plate 6), with its very considerable variety of operational conditions inclusive of the introduction of sewage or substrates which may only be expected to arise in the future, gives reliable and comprehensive data regarding dimensional parameters. However, all experimental systems which are less than full-scale exhibit the following deviations from the true situation:

— the recycled sludge flowrate must be greater as the sludge does not thicken to the same extent in small settling tanks, and hence the residence time of the sewage in the aeration tank is altered;
— where the test plant is set up in the open it often suffers from extremes of temperature;
— Experimental plants are usually controlled by highly qualified staff, so that very often exceptionally good quality effluents are produced.

The predicted residence times determined from the operation of experimental systems, particularly where there are imponderable dischargers, should be subjected to an incremental safety margin of 10–50%. However, too great a safety margin may have detrimental consequences for water quality, for example in the case of denitrification or the adsorptive elimination of constituents such as heavy metals or pesticides, on account of the much-reduced rate of waste activated sludge formation.

11

Treatment of surplus sludge

The waste activated sludge contains from 70% to 75% of organic matter with an energy content of 20.5 kJ/g organic solids. Many attempts have been made to make use of this material either for animal feeding or at least as a fertilizer on cultivated land, as a means of returning the biomass to the natural materials cycle. As a rule its use for animal feeding is ruled out on hygienic and toxicological grounds. Exceptions to this general rule are provided by the surplus sludge generated from effluents discharged by the food and beverage industries, and from dairies, abattoirs and even liquid manure which after autoclaving, for example, in animal by-product establishments, may be employed as additives in the preparation of protein meal or fortified silage (TGL 38867). A flotation thickening of the activated sludge prior to heat sterilisation is desirable as a means of raising the solids content to 50–100 g/l. Direct application of raw sludges to fields cultivated for agricultural or horticultural purposes is also unacceptable for hygienic reasons and also because putrefaction of the sludge is liable to cause the emission of foul odours.

The waste activated sludge is ordinarily recycled to the primary settling tank, and thickened together with the primary sludge before undergoing anaerobic digestion with the formation of biogas. Where the surplus sludge is introduced into the inlet to the primary settling tank the physicochemical and biochemical adsorptive properties can be put to good use. At some plants the surplus sludge is used for composting and in this case it may also be thickened by a flotation process.

Guidelines for sludge production relating to different methods of treatment will be found in the Standard WAPRO 2.19/01.

11.1 ANAEROBIC DIGESTION

The raw sludge, after thickening to a moisture content of 96–98% by settling or 90–95% by flotation, is subjected to a methanogenic digestion process in which the putrescible constituents of the sludge are converted into methane or other end-products of metabolism. From 0.3 to 0.5 m^3 of biogas is formed per kilogram of organic matter digested, and the organic solids in the sludge are thus diminished by

30% to 50%. Digestion may be performed under a variety of different temperature conditions namely:

— at 10°C, psychrophilic range, in open pits or Imhoff tanks with a digestion time of 60 to 90 days;
— at 30–33°C, mesophilic range, in enclosed tanks with a digestion time of 30 days;
— at about 56°C, thermophilic range, in closed vessels, digestion time five to six days.

The advantage of digestion from the water quality management aspect lies in the fact that some substrates which are resistant to decomposition under aerobic conditions, such as organochlorine compounds, are decomposed by the anaerobic treatment. The products of anaerobic metabolism, in so far as they are not converted entirely to methane, carbon dioxide and water, are usually well suited to act as substrates in a further aerobic treatment stage.

The aerobic, facultative and pathogenic bacteria and viruses which are exposed to the conditions of methanogenic fermentation are, apart from the spore-formers, inactivated. In addition a major proportion of the helminth ova is either killed or at least rendered non-viable.

The digested sludge is transferred in the wet state or after dewatering, to sludge drying beds or sludge lagoons, from which it may be utilised either for agricultural purposes, possibly after composting, or in the case of high levels of toxic materials (TGL 26056/02) disposed of by landfilling at a controlled site. Dewatering may also be performed by mechanical means [40]. The digester liquor (supernatant) is recycled to the primary settling tank.

11.2 AEROBIC STABILISATION

Since the introduction of oxidation ditches and compact activated sludge systems, aerobic sludge stabilisation has been carried out in those plants, and also in recent years in large-scale installations. The principle lies in aerating the sludge for just such a length of time that the sludge, according to Fig. 4.17, can utilise the putrescible organic matter by the process of endogenous respiration. The aerobically stabilised sludge may be dewatered on drying beds without fear of odour emissions. It consists principally of the cell-wall and envelope materials of bacteria, and hence of polysaccharides, hemicelluloses, polypetides, etc.

In the case of large-scale installations, aerobic stabilisation is performed in separate tanks, but for oxidation ditches and compact activated sludge plants it takes place simultaneously with the treatment of sewage at increased sludge solids suspension contents of 6–8 g/l. As the endogeneous respiration process is temperature-dependent there is a direct correlation between the sewage temperature and the stabilisation time (Table 11.1). Plants with simultaneous stabilisation are designed with organic sludge loadings of 0.03 to 0.06 kg/kg d.

For the oxidation of organic matter present in the waste activated and primary sludges, electrical energy is required in amounts represented by:

30–80 W/m^3 of aeration tank volume;

Table 11.1 — Dependence of stabilisation time and oxygen demand on temperature. From Müller-Neuhaus (1971)

Water temp (°C)	Stabilisation time (d)	Oxygen demand per kg oTS (kg/kg)
5	28	1.06
10	17	0.93
15	11	0.85
20	7.7	0.85
25	5.5	0.84
30	4.0	0.79
35	3.2	0.68

15 kWh/m³ of digested sludge;
10 kWh per capita per annum.

As the energy requirement for aerobic sludge stabilisation is almost as great as that of the activated sludge process, the method is not widely adopted. From the water quality management aspect aerobic stabilisation does not confer any benefits. The end-products of metabolism generated by the endogenous respiration of the sludge biomass pollute the receiving water by reason of their high nutrient content and the presence of poorly biodegradable or resistant compounds (Chudoba 1967). From the hygienic aspect there is only a negligible improvement, since the Salmonella, and helminth ova in particular are largely unaffected and retain their infective properties.

In recent times the process of *aerobic-thermophilic stabilisation* has received a good deal of attention. In this process the energy liberated by the exothermic processes of aerobic metabolism leads to a temperature increase in the sludge of 38–43°C, for a solids content in the raw sludge of 30–50 kg/m³ (Möller *et al.* 1984). The theoretical rise in temperature per kilogram of oxygen consumed of 1.7°C/m³ kg given in section 4.2.6 is, however, not quite reached under practical conditions despite the use of enclosed thermally insulated reactors, and a value of 1.3°C/m³ per kg of oxygen has been calculated for the 45 m³ reactor used by the above-named authors, while in [33] a value of 0.6°C/m³ per kg of solids decomposed is cited.

At these elevated temperatures, according to Table 11.1, the organic substrates are decomposed with a relatively low specific oxygen consumption. Depending on the effectiveness of the system of aeration employed, sludge stabilisation can be performed at the expense of a power consumption equal to 6–18 kWh/m³. The stabilisation time required on technical grounds is only 24 hours, but for hygienic reasons a period of two to three days is necessary. The exceptionally good disinfection performance, including destruction of the germination capability of resistant plant seeds, depends on the prolonged heat treatment and the rise in pH of about 2–3 units.

12

Operation of activated sludge plants

Operation comprises the monitoring and control of the treatment process. With the treatment capacities available nowadays, maximum treatment performance can be achieved at very low collective expense. For operation, specially trained workers are required who understand both the process itself and its effects on the other technological systems which make up the sewage treatment installation as a whole, and are also capable of maintaining the equipment in keeping with the manufacturer's instructions or plant requirements, and keeping it running. As routine duties carried out over long periods have a fatiguing effect, a high sense of responsibility is required in addition to the relevant technical qualifications, together with a certain enthusiasm for water quality protection. Even though the requirement for higher technical training on account of the use of more complex technical equipment during the last 10 or 20 years has been justified, nevertheless the personal commmitment of the staff still remains a decisive factor in achieving successful results.

The operation and maintenance of municipal activated sludge plants is compulsorily prescribed in TGL 26730/01 and 03. For industrial treatment plants the same measures are equally applicable on principle.

12.1 RUNNING-IN PHASE

The start-up of activated sludge plants, in particular the development of the sludge biomass, was regarded in the older literature as a particularly tricky problem, so that very many different, and sometimes highly original methods were described. It should, however, be assumed from the outset that the activated sludge biocoenosis does not consist of a 'special microflora'. The soil, the largest and most universal of all microbiological laboratories, contains up to 25 000 million organisms/g dry solids. The air is constantly inhabited by these organisms and in urban areas also with organisms of anthropogenic origin. Air contains between 100 and 500 bacteria/m^3 while 1 g faeces contains about 10^{11} organisms, and the number of microorganisms in domestic sewage is at least 10^8/ml. These sources are invariably adequate for the development in non-toxic effluents of a specific bacterial flora within a relatively short time.

A prerequisite for commissioning of an activated sludge plant is the serviceability of the aeration equipment and the sludge recycle facilities, as well as tanks which have been swept clean. For new plants and newly installed compressed air systems an initial filling with clear water is advisable as a means of checking the functioning of the mechanical equipment; non-uniform escape of air from compressed air operated systems becomes recognisable with the tank only partly filled. This water should be emptied out again before introducing either municipal sewage or non-toxic industrial effluent, in order to reduce the time required for running-in. For toxic effluents, however, the water may be used to dilute the effluent during the initial stages until the biological decomposition of the toxins has become fully established.

Once the aeration tank has been filled with sewage, the aerators may then be switched on, together with the sludge recycle pumps when the final settling tank is full. As long as the final settling tank is of sufficient dimensions, then for municipal sewage, contrary to the recommendations made in the older literature, the complete sewage input should be fed to the system. By this means the organisms receive the maximum nutrient supply, which in turn leads to a rapid rate of growth. Growth curves for mechanically pretreated municipal sewage are presented in Fig. 4.15. For achieving the desired volume fraction of activated sludge of 300 ml/l, a period of two to 14 days is usually required. For non-pretreated sewage this value may be reached in about a third of the time, although when dealing with industrial effluents largely devoid of microorganisms, the period may be increased severalfold.

For reasons of water quality protection it is very often necessary for the running-in time to be artificially curtailed; in extreme cases the discharge of partly treated effluent to the receiving water may even be forbidden. As a standard method for accelerating the growth of the activated sludge biomass, an *inoculum* consisting of sludge from another plant treating the same type of effluent may be introduced. The amount of this inoculum is usually about 20–50 l/m^3 of tank capacity, but may be as much as 300 l/m^3 where a fully-treated effluent is required immediately. However, such a sludge is never exactly equivalent to one which has developed at the point of use. Usually within the first 24 hours a considerable fraction is washed out, for which there may be a multitude of reasons: dilution effect, salt content, ionic strength, turbulence, damage in transit, etc. If the inoculated sludge readily accustoms itself to the new environment, a rapid growth to the intended concentration may occur, over two to four days.

In many cases there is no suitable sludge inoculum available, especially for industrial effluents, added to which the task of transport to the site may be problematic where large installations are concerned. In such situations it is advisable to introduce inert materials into the aeration tank as physical supports for biomass growth. Materials which have been tried include asbestos particles, sawdust, cellulose fibres, and also metal hydroxides and polymers. For municipal sewage the coarse solids removed at the primary settling stage may be suitable. The microorganisms are able to colonise these inert supports and then given suitable conditions may begin to multipy. Many of the newer activated sludge plants in the chemical industry were started up with the aid of $FeSO_4$ (e.g. spent acid) or $FeCl_3$ (40–150 mg/l in the aeration tank). As these metal salts exhibit a strongly acid reaction, pH adjustment to at least 6.0 is necessary. It is also recommended that about 1 ml/l of sludge from municipal treatment plants should be added to these hydroxide sludges as an inoculum, as by this means a large number of bacterial species and individual strains will becomes active. Of course the walls of the germination tank and final settling tank will also function as inert supports for growth of biomass. By brushing them down two or three times a day with a long-bristle brush the start-up time may be almost halved.

For special industrial effluents with poorly degradable substrates an inoculation with soil bacteria is often effective. From a suspension of 250 g soil/litre of tap water, amounts of supernatant equivalent to 1–10 l/m^3 of tank capacity may be added after settling of the suspension.

An activated sludge plant may be regarded as finally run-in when the entire sludge biomass has been bled off two or three times in the form of waste activated sludge.

The contents of TGL 26730/03 stipulate that following termination of the running in phase, a set of *operating instructions* is drawn up in which the mode of operation and the operating conditions are laid down for the benefit of the plant operators, and the extent of sampling and analysis is also prescribed for the benefit of the control laboratory. Where the proper functioning of the equipment cannot be guaranteed, running-in should not be attempted, for example, during periods of severe frost, for reasons of occupational health; no biological basis for this ancient ruling exists, however.

12.2 NORMAL OPERATION

Typical values of parameters pertaining to normal operation of activated sludge plants are:

Oxygen content in the aeration tank: 1–2 mg/l.
Sludge volume fraction in the aeration tank: ~300 ml/l.
  ($\leqslant$ 800 ml/l in compact systems and oxidation ditches).
Areal loading of the final settling tank: $\leqslant 0.8$ m^3/m^2/h.
Recycled sludge flowrate: 0.5 V_h.
Depth of visibility in the final settling tank: > 25 cm.
Settleable solids in effluent: < 0.1 ml/l.

Except for plants with simultaneous sludge stabilisation the recycled sludge is withdrawn continuously from the final settling tank in order to prevent the onset of putrefaction. For stabilisation plants at least enough sludge should be recycled each hour to maintain the final settling compartment free of sedimented sludge.

Where a plant is not receiving any highly-coloured influents then the colour of the activated sludge allows conclusions to be drawn regarding the quality of the sludge biomass.

— Brownish: light-to-moderate loading with an adequate supply of oxygen.
— Light to dark grey: heavily-loaded sludge or oxygen deficiency.
— Black: extreme lack of oxygen with formation of hydrogen sulphide leading to deposition of FeS, or else prior introduction of large amounts of digester liquor.
— Whitish–light grey: extreme cases of bulking sludge with very low solids contents or sulphur deposition in the biomass following introduction of digester liquor.

— Yellow, orange or reddish colour: growth of special bacterial or fungal strains.

For the smaller plants the following tasks should be performed once a day:

— visual inspection of the state of operation;
— measurement of sludge volume in the aeration tank and settleable solids in the effluent;
— inspection of the sewage and recycled sludge distributors in the case of multi-compartment systems;
— measurement of depth of visibility in the final settling tank;
— wastage of surplus sludge;
— inspection of measuring instruments;
— entry of data into the plant log book as specified in TGL 26730/03.

For large plants with their own laboratory facilities, in addition to the above process control variables, other determinations should be carried out daily, consisting of the oxygen content, sludge solids content, and as a precaution against bulking sludge formation, the sludge volume index. For the biological analysis of the sludge biomass from one to three analyses per week will usually suffice.

From the water quality management aspect the maintenance of an optimum sludge volume fraction is the most important task. Where continuous wastage of the surplus sludge is either difficult or impossible, for example, in the case of stormwater inflows or severe scum formation in the secondary settling tank, then the following would apply for the frequency of sludge take-off:

— for $B_R > 3$ kg/m^3 d — several times a day;
— for $B_R = 0.5$–3 kg/m^3 d — one to three times a day;
— for $B_R < 0.5$ kg/m^3 d — every two to ten days.

Owing to the oxygen requirement and the formation of stable organic compounds, it is desirable to withdraw the excess sludge after the peak loading period has passed. As the waste activated sludge from municipal sewage treatment plants must be regarded as an infective material, the plant operators should wear protective clothing and pay special attention to personal hygiene.

12.2.1 Treatment performance

BOD_5

Fig. 10.1 and Table 10.1 present the BOD_5 elimination as a function of BOD_5 volumetric loading for correctly operated activated sludge plants. As the BOD_5 is used, irrespective of the type of sewage, as a measure of the treatment performance no very great effluent-dependent differences arise such as are characteristic of COD removal.

As a consequence of the diurnal fluctuations in BOD_5 volumetric and sludge loading rates, the BOD_5 removal and concentration of BOD_5 in the plant effluent exhibit similar systematic fluctuations. Maximum percentage removal effects, with respect to the BOD_5 eliminated, occur between 7.00 am and 11.00 am for municipal sewage, and the lowest between 1.00 am and 6.00 am. The lowest BOD_5 concentrations in the final effluent are observed between 2.00 am and 8.00 am and the highest between 3.00 am and 8.00 pm. The arithmetic mean of the treated effluent concentration, which usually differs little from the weighted mean, exhibits a normal distribution pattern:

> 68.3% of values in the range $\pm 1.0\sigma$
> 95.5% of values in the range $\pm 2.0\sigma$
> 99.73% of values in the range $\pm 3.0\sigma$

Very often the treated effluent concentration values more clearly reflect a logarithmic than a normal distribution, so that they can be more satisfactorily represented with the aid of a cumulative frequency distribution curve.

The breadth of variation in the values is dependent on the actual magnitude of the BOD_5 sludge loading; lightly-loaded plants exhibit a relatively narrow range of variation while the more heavily-loaded plants show a greater degree of scatter.

The BOD_5 of a filtered sample, even with a low sludge loading and a healthy biocoenosis, only rarely falls below 5–10 mg/l. This *residual BOD* is primarily occasioned by the endogenous respiration of solitary bacteria and microflocs carried over into the final effluent, but is sometimes also naturally induced by the utilisation of residual substrate. Also it is not possible, within an acceptable period of time, to reduce the treated effluent concentrations to the desired level where the plant is overloaded or the biocoenosis injured, even though the input of raw sewage or substrate may be interrupted or curtailed. While the BOD_5 in Fig. 4.9, Curve 2, had fallen to 144 mg/l at 11.00 h, at 6.00 h it was only 12 mg/l; in many cases the treated effluent concentration at that time may even be higher than the influent concentration. The fact that the sludge biomass is unable to utilise these low substrate concentrations is accounted for on reaction kinetic grounds by the high K_m-value under such conditions (see Fig. 4.5, Curve 1).

COD

The attainable percentage and absolute elimination rates for COD are always lower than those for BOD_5; the same applies to the DOC (dissolved oxygen concentration). The principal reason for this is that the COD is in part the outcome of the presence of non-biodegradable compounds, which cannot be utilised by the biomass. Many organic substrates exhibit no BOD_5 and can at best be adsorptively bound to the activated sludge flocs and hence withdrawn along with the waste activated sludge.

The absolute level of the COD in the final effluent and its percentage decrease do not allow for any conclusions regarding the function of an activated sludge system. Low COD-values in biologically treated municipal sewage effluents are of the order of 15 mg/l for COD-Mn and 60 mg/l for COD-Cr. For many industrial effluents however, values several times these may be obtained, depending on the properties of the effluent being treated.

Filtrable and particulate solids

With adequate final settling and well-flocculating activated sludges the filtrable solids content will be less than 20 mg/l, or 30 mg/l in the case of heavily-loaded sludges. Higher values than these generally point to disturbances in the final settling or sludge recycle operations. For many industrial effluents the content of filtrable solids may of course be very much greater where the materials concerned fail to settle out within the mean retention time in the final settling tank. In such cases a *post-coagulation* step (and perhaps also a preliminary precipitation in the primary settling tank) with the aid of cationic flocculating agents, in particular polyacrylamides, may be beneficial.

12.2.2 Effect of other technological processes
Treatment of stormwater flows
Municipal activated sludge plants are as a rule designed to accept a proportion of the wet weather flow equal to 1.5–2.0 dry weather flow (DWF). This has the advantage that diurnal peaks in the DWF can be readily accommodated, and there is also a certain amount of reserve capacity for new connections. When stormwater enters the combined sewer system the sludge which has deposited is resuspended and flushed out of the sewer. The first flush is thus of exceptionally high concentration, although the actual runoff or snowmelt from the street gullies may be only lightly polluted. Concentrated discharges may also originate from the yards and paved areas of stables, agricultural and industrial premises. As the sewer network is usually designed only for 1 + 4 DWF, and the entry of runoff from a design storm may give rise to roughly 1 + 200 DWF, the concentrated first flush will largely escape via the stormwater overflows. As a rainstorm usually affects the whole of a given sewer catchment at the same time, this means that only a small fraction of the sludge deposits will actually reach the treatment plant. Following the first-flush action, only a very dilute sewage with a BOD_5 of 40–100 mg/l is left. For the design and operation of aeration tanks it is hardly necessary as a rule to take into account the effect of the stormwater fraction of the BOD_5 as the BOD_5 load is less than that for DWF. However, transfer pipelines, final settling tanks and sludge recycle pumps must be able to accommodate the increased flows. During periods of wet weather flow a very clear treated effluent from the final settling tank is often obtained; this is due less to the lower BOD_5 volumetric or sludge loading rate than to the entrainment of fine particulate matter which assists the flocculation of the sludge biomass.

Screens
Intermittent cleaning operations may give rise to hydraulic shock loading effects, with consequent carryover of sludge, in cases where the final settling tank is too small or contains too much sludge.

Primary settling tanks
According to Fig. 1.2, primary settling tanks with mean retention times of 1–1.5 h are connected in advance of the activated sludge installation. Their function is to remove the settleable solids (about 5 ml/l consisting of detritus, helminth ova, paper, soap, metal hydroxides, etc.) as well as the more buoyant materials (oils, fats, corks, faecal balls, food residues etc.). Where the weste activated sludge is recycled to the primary settling tank, a mean retention time of at least 30 min will be necessary for it to settle out, otherwise proper management of the sludge content of the activated sludge system will be impracticable.

For oxidation ditches, compact activated sludge plants and also for a few large-scale plants, the primary settling tank may be dispensed with. According to the existing literature [26] and also the author's own results, this practice is not recommended for municipal sewage treatment plants for a variety of reasons, especially mechanical problems such as blockage of sludge recycle lines and pumps, failure of compressed-air diffusers, poor operation of scum removal facilities in the final settling tank, deposition of coarse solids in the oxygen electrodes and hence breakdown of oxygen consumption measurement and control facilities, and deposition and accumulation of grit in the aeration tank. In addition the biological treatment performance may be impaired by the inhibitory substances or toxins which are removable at the primary settling stage, for example oils or heavy metal salts, while the buffering capacity of the primary settling tank is also lost. The settleable solids removable in the primary treatment stage, which may constitute from 20% to 35% of the total BOD_5 load, can also create the following problems in the aeration tank:

— Surplus sludge production: Fig. 4.17.
— Oxygen demand: as course solids, from Fig. 4.17, are only broken down at a sludge loading of <0.2 kg/kg d, for more heavily loaded sludges there is a relatively minor increase in oxygen demand of only about 2–4% despite the 20–35% increase in the BOD_5 load.
— Settleability of sludge: the coarse particulates carried into the aeration tank in the absence of a primary settling stage enhance the settleability of the sludge and reduce the sludge volume index (Hackenberger 1968). Any trend to bulking sludge formation is thus diminished, partly aided by the more favourable ratio of coarse solids to dissolved organic matter.
— Treatment effect: owing to the accumulation of biologically inactive coarse solids in the sludge, the proportion of active organisms in the sludge is depressed. According to Veits (1977) the sludge loading should be reduced in order to achieve the same treated effluent quality as follows:

for nitrifying systems, from 0.15 to 0.12;
for a final BOD_5 of 20 mg/l, from 0.30 to 0.24;
for partial BOD_5 removal, from 0.70 to 0.55 kg/kg d.

The treated effluent quality may still be adversely effected by the presence of unpleasant floating solids, high levels of filtrable solids and their accompanying helminth ova.

Preliminary coagulation — see section 9.1.4.
Input of digester liquor
The digester liquor is introduced into the primary settling tank from which it passes to the activated sludge system. Possible adverse effects can occur due to the intermittent nature of digester liquor take-off, and on acccount of its high content of sulphide and organic acids and occasionally also due to sludge particles present in suspension; the Kraus process described in section 6.2 affords a reliable countermeasure. As a rule, however, it is possible to cope with a controlled take-off of digester liquor during the daily low-load periods without any deleterious effects on the activated sludge process. In cases of severe bulking sludge formation liable to affect

the treatment performance, the sudden entry of digester liquor into the aeration tank may even suppress the thread-forming bacteria which are the cause of the sludge bulking effect.

12.3 OPERATIONAL DISTURBANCES

In every activated sludge plant, occasional plant operating disturbances may be expected. Such disturbances do not occur only in industrial effluent plants or in municipal plants treating a high proportion of industrial discharges. For compact activated sludge plants faulty handling of chemicals in a school or photographic laboratory may be the cause of damage to the biocoenosis. The majority of plant upsets, however, are caused not by chemicals or toxic substances, but by mechanical failures, hydraulic deficiencies, lack of maintenance, and unwelcome biological processes. Operational failures are as a rule first detectable by the appearance of turbidity in the treated effluent. By daily measurement and recording of the depth of visibility in the final settling tank, such interferences in the treatment performance can become apparent even to unskilled operators.

Sludge carryover from the final settling tank
Properly functioning final settling tanks deliver effluents containing less than 20 mg/l of total solids and 0.1 ml/l of settleable solids. Where a level of 0.1 ml/l settleable solids is clearly exceeded, then, assuming proper sludge recycling, there is either too much sludge biomass in the system or the hydraulic loading of the final settling tank is excessive. The activated sludge volume in the aeration tank and the upward velocity of the treated effluent in the final settling tank are related according to Fig. 7.3. As a rule sludge carryover can be prevented by increased sludge wastage. In the case of excess hydraulic loading, before attempting an extension or possibly the insertion of a plate separator unit, the effect of an automatic sludge wasting device should be examined.

In cases where there is a sudden drop in atmospheric pressure, for example just prior to storms, there may be carryover of sludge into the outfall. The causes for this phenomenon are associated with microscopic gas bubbles, primarily consisting of nitrogen, or possibly air, which begin to float toward the surface as a result of the fall in ambient pressure.

The clearly recognisable 'pinhead flocs', which are dispersed throughout the upwelling liquid and are visible in direct sunlight, should not be considered to represent 'sludge carryover' in the sense of an operational disturbance; they are still resistant to sedimentation even after a 2 h settling time in a conical funnel.

Many activated sludge plants from time to time suffer from a 'deflocculant growth' of activated sludge. In place of the customary activated sludge flocs, only microflocs are formed which hardly settle out in the final settling tank. The cause is mostly due to the presence of toxic substances. Countermeasures may often take the form of adding inorganic and/or organic flocculating agents.

Scum on the final settling tank
Scum occurs as a result of the lack of a primary settling stage, or the formation of buoyant activated sludge. The often greasy, viscous layer of scum is only removed

from conical and circular final settling tanks with difficulty, the use of scum take-off devices being only partly effective. It is possible to break it up by direct impingement from a jet of water, so that it begins to settle, but jetting of this kind should only be attempted after lowering the liquid level, otherwise too much sludge will be carried over into the outlet. After jetting of a denitrification scum, a foamy layer will persist at the surface of the liquid which consists of stable cream-coloured gas bubbles, formed from gaseous nitrogen and adhering microfloc particles.

The flotation of activated sludge can be induced by means of air bubbles which originate from the aeration tank and are unable to disperse for reasons of inadequate gas-separation space or are entrained because of excessive mixed liquor input velocities into the final settling tank (Table 13). In such cases modifications to the tank are indicated.

The most common cause of flotation is the process of denitrification, occurring in nitrate or nitrite-containing liquids. Very severe denitrification can lead to the formation of scum layers at the rate of 20–30 cm/day, so that sludge carryover is observed. The denitrification reaction may be transferred to the aeration tank by means of a 10–20% reduction in the normal air supply to the tank, so that both nitrite and nitrate are reduced in a few hours and hence the cause of floating sludge formation is eliminated. With a very low dissolved oxygen content of 0.2–0.5 mg/l there is usually only a low level of nitrification, while dentrification is initiated. As, however, nitrification is a desirable process the control of denitrification without precise adjustment of the O_2–NO_3^-/NO_2^- regime is tricky, and in the case of oxygen-deficiency the effluent quality deteriorates; it is thus preferable to increase the rate of sludge recycle so that the sludge retention time in the final settling tank under anaerobic conditions is reduced and hence denitrification is suppressed. Intensive denitrification reactions in the settling tank usually occur only over a period of two to three days at the start of nitrification in the aeration tank. Critical conditions may also occur in those settling tanks receiving nitrate-rich mixed liquors with coincidental high BOD_5 values, that is, with the occurrence of nitrate-rich sewage and heavily-loaded sludge biomass simultaneously.

Causes of final effluent turbidity
The turbidity of the treated effluent is a definite indication of the operational status of the activated sludge plant. It undergoes a systematic variation as a result of periodic changes in hydraulic and BOD_5 sludge loadings, which should not be interpreted as a sign of abnormality. A number of factors may give rise to a sudden abnormal deterioration, however, such as an increase in the saprobity of the sludge biomass, partial carryover of solitary bacteria and microflocs in response to severe shock loading, toxic injury to the biomass from sewage constituents, or possibly the presence of non-biodegradable dyes or colloidal matter. A biological analyses of the sludge biomass provides the safest indication of the cause in such cases as it will enable the technician to differentiate between toxic injury, increased saprobity and floc destruction. Measurements of respiratory activity can also supply supplementary information concerning the physiological state of the biomass.

An increase in saprobity may occur as a result of an increased sewage loading, possibly with consequent oxygen deficiency in the aeration tank. Before extensive process analyses and laboratory tests are undertaken, however, it is advisable to check the operating condition of the activated sludge system itself and that of the ancillary equipment. The search for possible causes should be initiated in the following order: sludge recycle rate; dissolved oxygen and sludge content in the aeration tank; sludge content of the recycled sludge and upflow velocities in the primary settling and final settling tanks; input of digester liquor; complete and uniform sludge take-off from the final settling tank; persistent sludge deposition in the aeration tank, final settling tank or sludge recycle line.

Cloudy effluents occur spasmodically in many systems as a result of breakdown of the sludge flocs without any accompanying injury to the biocoenosis; it often disappears after only a week, or perhaps even two to three days from its first occurrence. Apart from the natural floc disintegration as a result of nutrient deficiency, for example at $B_{TS} \leqslant = 0.03$ kg/kg d or an interruption lasting several days in the sewage inflow, the causes are usually apparent without the need for any very penetrating investigations. Possible causes may include the presence of toxins with a selective action on the surface structure of the bacteria, the death of a bacterial species on account of nutrient deficiency, the growth of poorly flocculating bacteria or the effects of viral infection. If floc breakdown is the cause of the effluent turbidity, then the BOD_5 of a filtered sample is generally no greater than during normal operation of the plant. A distinct increase in the effluent turbidity also occurs as a result of the carryover of microflocs just prior to sludge carryover caused by an excess of sludge in the final settling tank.

Injury to the sludge biomass as a result of the presence of toxic substances manifests itself as follows:

operationally by:
— increased turbidity in the treated effluent;
— elevated dissolved oxygen contents in the aeration tank owing to depressed respiratory activity;
— lower rates of surplus sludge production owing to increased wash-out of floc-forming particles;
— increased foam formation in the aeration tank;

under the microscope:
— increased propositions of solitary bacteria and microflocs;
— reduced motility of ciliates and flagellates;
— contraction of individuals or penduncles and the swarming of peritrichal ciliates;
— rounding of cells;
— formation of large vacuoles in cells;
— inactivity, die-off and cyst-formation among the microorganisms.

The identification of the toxic substance and the institution of measures by the discharger in order to prevent a repetition are urgently called for. The elimination of the toxic effect will depend on both the nature and amount of the toxin concerned. For biologically degradable substances and those which can be neutralised, and where the biomass has been only partially damaged, the ordinary mode of operation may be resumed with the biomass fully regenerated after only a few days. For more severe injuries, however, the sludge content should be reduced by wasting of surplus sludge to around 50 ml/l. The remaining sludge biomass then serves as a support for organisms which are still healthy or are able to colonise the system afresh. Where the damage is caused by heavy metal ions, these ions are largely bound to the activated sludge and in such cases it is desirable to waste the entire sludge biomass. On no account should the wasted sludge be introduced to the digester as the methanogenic bacteria are even more susceptible to heavy metals than the aerobic organisms. The sludge must therefore be collected in special tanks or in a primary settling tank, and ultimately incinerated or disposed of by landfilling, and a new biomass must be started from scratch.

In the event of the ingress of mineral oils and organic solvents, for example, in plants with no primary settling stage, the sludge content should be reduced by controlled wastage to around 50–100 ml/l so that the offending material is flushed out, and relatively insensitive bacteria are then encouraged to multiply as a consequence of the increased BOD_5 sludge loading. Oil floating on the surface of the aeration and final settling tanks should be removed with the aid of one or other of the oil-collecting agents prescribed in TGL 22213/06; the aeration equipment should be switched off before the collecting agent is introduced.

Non-metabolisable constituents and those which are poorly adsorbed onto the sludge biomass may give rise to highly turbid or coloured final effluents. Such materials are usually of industrial origin, scuh as drilling and cutting oil emulsions, synthetic dyes, etc. They should be removed by physicochemical treatment before discharging the effluent to sewer, although they do not normally harm the biocoecnoses, nor increase the BOD_5 of the treated effluent. Because the COD of the effluent is increased, however, the BOD_5: COD ratio becomes very small. Some effluents, such as those from fermentation processes, tanneries, flax retting, the pulp and paper industry and many others, remain very cloudy despite very low BOD_5 sludge loadings.

Where a sludge biomass has been damaged by the effect of toxins, it is of no avail to reduce the input of sewage and hence the BOD loading, that is to diminish the food supply. As the curve for 6 mg/l $HgCl_2$ in Fig. 2.10 shows, a high level of substrate input enables the toxic effect to be overcome.

Failure of aeration equipment and sludge recycle pumps
Mechanical failures of this nature must be notified immediately to the relevant members of the Water Quality Inspectorate and possibly also to the Sanitary Inspectorate, since only a few hours later it will become necessary to discharge untreated or only partially treated sewage effluents to the environment.

Short-term failures, such as power cuts lasting only a few hours, can be coped with and no long-term effects will ensue; between 24 and 48 hours after reconnection of the supply the orginal treatment performance has usually been regained. Now and again problems arise when restarting fine-bubble aeration systems, as the air may no longer issue uniformly from the surface of the ceramic diffusers. In difficult cases the level in the aeration tank should be reduced by half and then the air supply resumed using all the available fans in order to apply the maximum pressure as a means of unblocking the diffusers. Longer-term failure of individual diffusers in aeration tanks with several such devices will lead to only partial treatment as a result of a lack of oxygen even though the necessary turbulence is still produced. For municipal sewage and aeration times of three to five hours, the effluent BOD_5 may increase to 30–50 mg/l, or possibly to 100 mg/l. In order to prevent the onset of putrefactive changes, it may be desirable to reduce the sludge content far enough to ensure that aerobic conditions are maintained.

Failure of sludge recycle pumps and of the sludge scrapers in the final settling tank lead to a complete breakdown of the activated sludge process. By means of appropriate stocking and availability of replacement parts, repairs may be effected in a short space of time. In the case of breakdown in the sludge recycle system, a decision must be made first of all on how the plant may be enabled to continue operating. One method which often works is to discharge the primary treated sewage and to continue aeration in the aeration tank. Following successful repair of the sludge recycle system the entire quantity of sludge in the final settling tank, which will have begun to putrefy, is wasted in the form of surplus sludge, and the supply of sewage to the aeration tank resumed, where the biomass remaining in the tank will have been kept healthy in the meantime. Another method is to flush the biomass from the aeration tank into the final settling tank and hold it there, while the primary treated sewage is fed into the aeration tank and aerated before being discharged to the receiving water body. In this case a BOD_5 reduction of 30–50 mg/l may be obtained with retention times of three to five hours.

Where the sludge recycle system failure may go undetected, the biomass from the aeration tank accumulates in the final settling tank. As this is usually incapable of accommodating the entire quantity of biomass, a pronounced carryover of sludge into the final effluent is the result. In this case too it is necessary immediately the situation is discovered, to discharge the sewage immediately after the primary settling stage. After carrying out the repairs, the sludge biomass should be regenerated by returning it to the aeration tank for a period of two to four hours before any further primary sewage is introduced. The same course should be adopted when one is confronted with very putrid black or acrid-smelling sludge from the sludge recycle line, for example from blocked pipelines or badly fouled filters in Dortmund tanks, as long as it is impossible to discard the sludge completely, which is the better alternative.

Where a plant has to be shut down for several days, it is also desirable to re-aerate the sludge biomass for a period of two to four hours under turbulent conditions. By this means any hydrogen sulphide which may have formed due to putrefaction will be either stripped or oxidised, thus removing a potential hazard or occupational health risk.

Oxygen deficiency in the aeration tank
Oxygen deficiency manifests itself in highly turbid effluents and increased saprobity of the sludge biomass; the effluent BOD_5 will as a rule be several times the value given in Table 10.1 or Fig. 4.9. If the cause lies in failure of aerators or defective compressed air lines then immediate repairs are called for. If, however, it is the result of such high BOD loading rates that the oxygen supply is inadequate for the normal sludge biomass content, then some method of increasing the oxygen supply must be found as an interim measure. According to the literature this may involve chemical coagulation prior to or in the primary settling tank (see section 9.1.4) or the addition

of hydrogen peroxide at the head of the aeration tank. Both these alternatives are costly and very often give rise to secondary problems.

Based on the prediction in Fig. 4.9 that the treatment performance falls only slightly in response to elevated BOD_5 sludge loading, and from the level of maximum oxygen demand derived from section 4.2.4, it may be desirable in such situations to lower the MLSS content from the optimum of 3 g/l to 2 g/l or even 1 g/l. A minimum sludge volume of 60–80 ml/l should, however, be retained in the interests of good sedimentation in the final settling tank.

Severe foaming in the aeration tank
Foam on the surface of the liquid is primarily caused by the proteinaceous constituents of sewage (whipped cream effect) by glutinous polysaccharides, and some industrial chemicals, as well as certain metabolic intermediates with surface-active properties. The 'soft' detergents in current use are adsorptively taken up by the sludge biomass on account of their surface active characteristics and are then very largely metabolised, so that only in cases of insufficient sludge supply or injured sludge biocoenoses will there be any tendency to foam-formation. Where the sludge volume has reached 100 ml/l, hardly any detergent foam will be observed.

Severe foam-formation upsets the operation of activated sludge plants. The foam overflows from the aeration tank onto footpaths and access ways, and may even be carried away in the wind. When certain types of fungi (Fig. 3.1(21)) develop in the sludge biomass of a foaming tank, they stabilise the foam to such an extent that visco-gelatinous foam layers of considerable thickness may develop (Plate 8). In such plants there is usually a stable layer of foam at the surface of the final settling tank.

Some industrial effluents, for example from the pulp and paper industry or timber processing, may give rise to foaming to such an extent that the foam must be destroyed. For this it may be sufficient to spray the foam with treated effluent, although this leads to a marked increase in aerosol emission. In extreme cases an *anti-foam agent* may be necessary such as Elvaumi from VEB Polychemie, Velten, or the Defoaming Compound 3577 or 7800 from VEB CKB Wolfen or VEB Fettchemie Karl-Marx Stadt, usually in amounts of 1 g/m³. These anti-foam agents are biodegradable; they thus give rise to an increase in the BOD_5- volumetric loading and to all the consequent secondary effects.

Sludge deposits
Small sludge deposits may form in the corners of the aeration tank, in the bends of oxidation ditches and in the wet well of pumps. In addition deposits of coarse solids which occur in tanks and channels in the absence of primary settling facilities, and are usually unavoidable. Such deposits do not give rise to serious plant upsets and may be removed when there is the opportunity.

Where the level of turbulence in the aeration tank is so low that an activated sludge biomass is not produced, or where localised deposits of activated sludge occur, the entire biological process is affected and the treatment performance reduced. Deposits of sludge may result in the onset of putrefaction with the formation of hydrogen sulphide, which will interfere with the development of a healthy biocoenosis. Such sludge deposits are frequently observed in plants with compressed air or submerged-jet aeration systems, in oxidation ditches and in compact activated sludge plants with self-induced sludge recycle. With horizontal rotors as the means of aeration the level of turbulence may be so low as a consequence of the small depth of immersion that insufficient aeration occurs at times of low flow, especially at night. Polysaprobic or hypersaprobic sludges, with poor effluent quality despite good sludge and dissolved oxygen levels during the daytime, are the first signs of this happening, which at first sight seems inexplicable.

Bulking sludge
The most dreaded form of plant is that due to the occurrence of bulking sludge. In qualification of this statement, however, it may be noted that the best effluent sludge quality is obtained from a combination of a slight bulking slude formation (150–200 ml/g) and a final settling tank of sufficient dimensions (Table 3.4).

Bulking sludges which occur due to a deficiency of nutrient or trace elements may be rectified by the addition of these elements (see Chapter 8). The cheapest, most reliable and, in the case of well-run-in activated sludge plants, most readily available source of these nutrients, is the digester liquor. Where septic sewage is the cause of bulking sludge, then very often a preliminary aeration and perhaps the elimination of septic tanks in the catchment may be beneficial. In order to limit the formation of bulking sludge the process configurations indicated in section 3.7.2.3 may be adopted. Where such measures are ineffective, and where sludge carryover from the final settling tank still occurs despite reduction of the sludge volume to ≤ 200 ml/l, normal sewage input and adequate tank dimensions, then countermeasures must be introduced. These may have the effect of either weighting the flocs or of physiologically inhibiting or weakening the thread-forming organisms.

For weighting of thread-like organisms sometimes accompanied by growth inhibition, inorganic coagulants such as $FeSO_4$, $FeCl_3$ and $Al_2(SO_4)_3$ may be effective. For many types of sludge, milk of lime, kieselguhr, alumina, activated carbon, bentonite, etc., may be successfully employed. Recommended dosages are 50 g/m^3 for $FeCl_3$, 80 g/m^3 for $FeSO_4$, and 100 g/m^3 for $Al_2(SO_4)_3$. The chemicals are introduced into the aeration tank just ahead of the point of entry of the recycled sludge, in order to keep the pH shift as low as possible. Nevertheless on commencing to dose these chemicals there is usually a noticeable deteriation in effluent quality. For some activated sludge plants the coagulants must be dosed continuously; very often waterworks sludges can be used for this purpose [25]. The chemicals referred to are relatively slow-acting. Cationic flocculating agents on the basis of polyaminocarboxylic acid and polyacrylamides, however, when added to the inlet to the final settling tank, are effective within one to three hours. They are, however, biologically degradable, and also the macromolecules may be disrupted by the intensive local turbulence in the aeration tank. When using Stipix KMN, the effect had completely disappeared only 24 hours after dosing the chemical.

The selective physiological impairment of thread-like organisms is possible because the threads projecting from the flocs or floating in the liquid are more readily attacked by toxins than are the bacteria within the protective influence of the flocs. Most frequently either chlorine or hydrogen peroxide are used; both cold water and digester liquor were successful in particular situations. In order to minimise the consumption of chemicals they may be introduced into the sludge recycle line, preferably near the point of take-off from the final settling tank, so as to allow the maximum contact time with the sludge. The dosages must be determined empirically with due regard to the retention time in the aeration tank; typical quantities are:

Cl_2:	5–20 g/m³ of sludge	for 6–12 h;
H_2O_2:	80–200 g/m³ of sludge	for 6–12 h.

Following the initial shock dose, repeat treatments at half the original dose should be given for two to three days, and even longer in the case of stubborn bulking sludges; in many plants it is impossible to manage without continuous use of these chemicals. At the same time as preventive treatment is commenced, surplus sludge should be wasted in order to remove the dead filiform organisms; otherwise they will remain behind and impede the reduction of the sludge volume index on account of their unwieldy structure.

Chemical addition, especially where chlorine is used, generally leads to a deterioration in effluent quality, and usually bulking sludge reappears about two to three weeks after it has been successfully suppressed.

12.4 AUTOMATED OPERATION

The current state of the art of microprocessors and process control computers permits a large degree of automation to be applied to activated sludge plants. Automatic feedback control enables maximum treatment performance to be realised for the lowest possible expenditure of energy. Hardly any savings in operator working time are possible, however, as servicing and repair of sensing and control components is very time consuming and demands skilled personnel.

Depending on the desired result, the following sub-processes may be selected for automated operation.

— Control of the oxygen input as a function of the oxygen demand, or maybe a time-related programming of the aeration equipment.
— Wasting of activated sludge as a function of the sludge volume in the final settling tank, or the sludge solids content in the aeration tank and/or effluent turbidity.
— Phased input of digester liquor, for example between 22.00 h and 04.00 h.
— Control of sludge recycle rate as a function of the sewage inflow.
— Dosage of chemicals as a function of flow rate and composition of sewage.

In selecting the particular operation to be automated, equal weight must be attached to the water quality management and economic criteria. The number of processes forming part of an activated sludge installation which can be cost-effectively adapted for automatic control is often overestimated.

In addition, continuous monitoring of plant operation and treated effluent quality is only recommended for a limited number of plants because of the profusion of data and the problems of worthwhile data utilisation.

13

Future prospects

In order to estimate the future of a particular process a knowledge of its advantages and disadvantages is required, and in particular those which are liable to change with the state of science and technology, or with movements in the price and availability of materials, compared with the effect of similar changes on alternative processes. As physicochemical treatment of organically polluted sewage is less effective than biological treatment, and hence can only be applied as a preliminary treatment with specific objectives in view, only the alternative biological methods will be considered here.

From the water quality and socio-economic aspect, agricultural utilisation is the most important method. For agricultural activities in a Central European climate however, this method, [4] with operating costs of 0.19–0.40 M/m^3, only pays during periods of drought in the growth season. Hygienic and toxicological factors further limit the application of the method.

Accordingly, for all-the-year round operation, this leaves the soil infiltration method according to TGL 7762, use of sewage lagoons according to TGL 28722, and the trickling filter. Septic tanks of the single or multi-compartment type according to TGL 7762 for so-called partial treatment are of such poor efficiency relative to the activated sludge process, that they do not qualify.

Soil infiltration has an upper limit of 50 PE for the population served. In situations with sandy soil this method is usually preferable for isolated dwellings, guesthouses, etc., on account of its lower maintenance and energy requirements relative to compact activated sludge systems. The variants of the infiltration method comprising the soil filter and the irrigation of poplar plantations can provide complete biological treatment of sewage from sewered populations of up to 10 000 PE, including a large degree of nutrient removal. Where a sufficient area of suitable land is available the poplar plantation method in particular presents a energy-saving alternative to the activated sludge process.

Sewage lagoons, on account of their high space requirement of 5–10 m^2/PE and an upper limit for population served of about 2000 PE overlap the capacity range of compact activated sludge plants and oxidation ditches. Despite its poorer treatment performance a lagoon system is often the preferred choice, because it uses less energy and has a relatively low maintenance requirement.

The trickling filter, which dominated sewage treatment in the GDR up until 1960, has scarcely been built at all during the last 15 years. Reasons for this include its inferior and rather less stable treatment performance, particularly in winter. Lightly-loaded trickling filters with $B_R=0.175$ kg/m^3 d often have to be taken out of service from November to March because of the danger of icing up. Other contributory reasons are their high construction cost, problems with the packing material, and the lack of control possibilities. However, their much lower energy consumption, of about 10 kWh/PE per annum gives them one decided advantage over activated sludge systems. Their operational drawbacks may be partly overcome by the use of synthetic packing materials consisting of plastics such as foamed polystyrene, rigid PVC sheet, etc., while the treatment effect can be considerably improved by adding a biological post-treatment step consisting of a lagoon with a residence time of two to five days or an activated sludge tank in series between the trickling filter and the final settling tank, with aeration times of 10–30 min. Despite all these modifications the upper limit in terms of population served remains about 50 000 PE.

The activated sludge process may be used for any size of sewered population up to several million. The space-saving form of construction, and in addition the special versions consisting of the deep-shaft and tower systems also enable the method to be used in constricted areas. The same high level of treatment, coupled with the biological nutrient elimination, cannot be achieved all the year round by any other method. Consequently, despite the changed situation regarding energy supplies, the activated sludge process remains the treatment method of the future.

The disadvantages of the activated sludge method which still remain are a challenge for scientists and technologists as well as for the plant operators. For biologists, biochemists, process engineers and technologists the most pressing problem is the optimisation of biological methods of nutrient removal. The use of the organic substrates present in the sewage as a source of hydrogen for denitrification and the optimisation of conditions for those organisms with the capacity for luxury phosphorus uptake are the most urgent problems for water quality management. Research concerning the occurrence, prevention and counteraction of bulking sludge must also be taken beyond its present largely empirical basis towards a precise scheme of identification for the filiform organisms responsible, and a scientific understanding of the conditions favouring their growth. In addition there is the problem of appropriate post-treatment methods for the elimination of the organic compounds resistant to degradation in the activated sludge system.

A deliberate control of the aquatic biotope, already achievable with the conventional activated sludge process according to the text-books, will become a major task for future generations of sewage biologists and a principal aim of limnological research. A step in this direction is the adoption of two-stage and multistage activated sludge treatment systems. Increasing the biomass in the aeration tank by means of suspended growth supports, rotating discs or foam inserts, leads to a reduction in the sludge loading and possibly also to a selection of certain organisms with particular metabolic capabilities.

The trend in recent years towards sewage treatment solely by aerobic methods is being called into question for those industrial effluents in particular which give rise to numerous persistent organic residues in the treated effluent. The current developments along the lines of anaerobic activated sludge processes are a useful amplification of the method for use with highly concentrated effluents, as a means of reducing energy consumption and of enabling a high level of treated effluent quality to be realised through the use of a terminal aerobic treatment stage.

Attempts to inoculate certain bacterial strains with specific metabolic properties into the sludge biomass, and to enable them to become stable members of the biocoenosis, have so far been of no avail. The activated sludge is a wild, spontaneously-emerging biocoenosis. In the laboratory and in small-scale pilot plants it has proved possible to cultivate certain artificial strains successfully, which are capable of performing the reduction reactions referred to in sections 4.3.2 and 6.2 with respect to oxygen-containing and strongly toxic anions. Certainly the future will see an extension of the *immobilisation technique* as a consequence of further microbiological research, and further water quality improvements should be possible as a result.

Irrespective of all the efforts towards raising the treatment performance of activated sludge installations, further solutions must be devised in conjunction with agricultural and water quality specialists for disposal of the nutrient-rich fractions, such as digester liquors and liquid sludge, which will allow the agricultural utilisation of these liquids to continue all the year round.

In recent years there has been a certain stagnation in the development of more effective *aeration systems*. Better aeration mthods are needed (all the more urgently because of the energy situation) which will allow specific oxygen transfer efficiencies of over 2 kg/kWh to be realised with oxygen transfer rates of 40–100 g/m^3 h. A further condition is that such oxygen transfer rates must be almost continuously variable in line with the oxygen demand prevailing at any time. In order to achieve a constant high level of treated effluent quality, the discriminating application of automated control to certain sub-processes of the activated sludge plant is a current necessity. By means of further improvements in measurement and control technology a marked reduction in the cost of servicing and routine maintenance of such apparatus should be achieved.

For continous *monitoring of final effluent quality* for activated sludge plants, further parameters must become routinely measurable which are of decisive importance for water quality production protection, such as oxygen demand and ammonium, nitrite, nitrate and phosphate concentrations.

14

References

ABBREVIATIONS

Ahh	Acta hydrochimica et hydrobiologica; Berlin
GWA	Gewässerschutz — Wasser — Abwasser; Aachen
GWF	Gas- und Wasserfach, Ausgabe: Wasser, Abwasser; München
JWPCF	Journal of the Water Pollution Control Federation; Washington
MB	Münchner Beiträge zur Abwasser-, Fischerei- und Flußbiologie; München, Wien
Sbornik	Sbornik Vysoké Školy Chemicko-Technologické, Technologie vody a prostředí; Praha
vh	vodní hsopodářstvi; Praha
WAF	Zeitschrift für Wasser- und Abwasserforschung; München
WWT	Wasserwirtschaft — Wassertechnik; Berlin

REFERENCES

1. Wasserwirtschaftlicher Entwicklungsplan, Inst. f. Wasserwirtschaft, Berlin 1979.
2. *Ausgewählte Methoden der Wasseruntersuchung. Vol. 1: Chemische, physikalische und physikalisch-chemische Methoden.* 2nd edn. Fischer, Jena 1986.
3. *Ausgewählte Methoden der Wasseruntersuchung. Vol. 2: Biologische, mikrobiologische und toxikologische Methoden.* 2nd edn. Fischer, Jena 1982.
4. *Schutz der Trinkwasserressourcen,* 2nd edn. Ministerium für Umweltschutz und Wasserwirtschaft. Berlin 1984.
5. *Infektionsschutz.* 4th edn. Thieme, Leipzig 1979.
6. Bestimmungtabelle für fadenbildende Mikroorganismen im Belebtschlamm. *Korr. Abwasser* **24** (1977) (1) 12–15.
7. *Wasserschadstoffkatalog.* Inst. f. Wasserwirtschaft. Verlag Bauwesen, Berlin 1975, 1979, 1981.
8. *Künstliche Belüftung von Oberflächengewässern, industriellen und kommunalen Abwässern. Technik und Umweltschultz,* Vol. 20. Leipzig 1978.
9. Issledovanija raboty aérotenkov s primeneniem techničeskogo kisloroda (oksitenkov) na obektach chimičeskoj promyšlennosti. SÉV, Svodnyi otčet po pod-

teme 3 'Metody i sooruženija biochimičeskoj očistki stočnych vod' temy II-01 'Issledovanie metodov intensifikacij i povyšenija effektivnosti očistki i doočistki stočnych vod', Moskow (1980): 197–226 (russ.).
10. Vlijanie vysokich koncentracij mineralnyeh veščestv na processy biologičeskoj očistki. *Wie* **9** (1980): 252–258 (russ.).
11. Kombinirovannoe sooruženie s protivotočnoj aéraciej (KSPA-21). *Wie* **9** (1980): 4–5 (russ.).
12. Issledovanija biochemičeskoj očistki vysokomineralizirovannych stočnych vod. *Wie* **9** (1980): 138–158 (russ.).
13. Informationskatalog Abwasser. Angebotsunterlagen, VEB Projektierung Wasserwirtschaft, Halle 1983.
14. *Lehr- und Handbuch der Abwassertechnik*. Vol. II, 2nd edn. Verlag W. Ernst und Sohn, Berlin 1975.
15. *Rekomendaeii po proektirovaniju sooruženij biologo-chimičeskoj očistki gorodskich stočnych vod*. Akademija kommunal'nogo chozjajstva im. K. D. Pamfilova, Moskow 1980 (russ.).
16. *Bergey's manual of determinative bacteriology*. 8th end. Williams and Wilkins, Baltimore 1974.
17. *Katalog: Ausrüstungen und Apparaturen für Wasserbehandlungsanlagen*. Rat für Gegenseitige Wirtschaftshilfe, Dresden 1976.

Research reports from the Forschungszentrum Wassertechnik, Dresden
18. *Richtlinie: Anwendung von Belüftungskreiseln für die biologische Reinigung von Abwässern nach dem Belebtschlammverfahren.* 1971.
19. *Optimierung des Betriebes von Belebtschlammanlagen mit Belüftungskreiseln.* 1973.
20. *Messung und Regelung des Sauerstoffeintrages.* 1976.
21. *Optimierung der organischen und hydraulischen Raumbelastung von Abwasserteichen durch künstliche Belüftung.* 1976.
22. *Optimierung von Kleinbelebungsanlagen.* 1977.
23. *Anwendung der Röhrensedimentation zur Nachklärung in der Abwasserbehandlung in der Kläranlage Leipzig-Wahren.* 1978.
24. *Untersuchungen zur weitergehenden Abwasserbehandlung.* 1978.
25. *Einsatz von Wasserwerksschlämmen zur Verbesserung der Absetzleistung in der Vorklärung.* 1980.
26. *Verfahrensoptimierung zur Steigerung der Belastung von Vorklär- und Belebtschlammbecken.* 1980.
27. *Untersuchungsmethode für biologisch schwer abbaubare Substanzen im Abwasser.* 1982.
28. *Optimierung einer Kleinbelebungsanlage für den VEB Grosswäscherei Geithain.* 1982.
29. *Empfehlungen für Lärmschutzmassnahmen der Belüftungskreisel in Kläranlagen zur Minderung von Nachbarschaftslärm.* 1982.
30. *Inaktivierung PSM-haltiger Abwässer in kommunalen bilogischen Abwasserbehandlungsanlagen.* 1982.
31. *Verfahrenskombination zur Leistungssteigerung von mechanischen Kläranlagen.* 1982.

32. *Biologische Phosphoreliminierung mit Hilfe des Belebtschlammverfahrens.* 1983.
33. *Belebtschlammverfahren nach dem Tiefschachtprinzip.* 1983.
34. *Sauerstoffbegasung von Abwasserschlamm.* 1983.
35. *Verfahrenskombination zur Leistungssteigerung von biologischen Kläranlagen.* 1983.
36. *Verfahren zur enzymatischen Schlammstasbilisierung unter Nutzung exothermer Prozesse.* 1984.
37. *Untersuchungsbericht zur Messung des Sauerstoffeintrages und zur Durchmischung der Belüftungsbecken mit Tauchstrahlbelüftung in der Abwasserbehandlungsanlage des VEB PCK Schwedt im Kominationsbetrieb 'Otto Grotewohl' Böhlen.* 1984.
38. *Entwicklung eines Saugräumers nach dem Heberprinzip für rechteckige Nachklärbecken.* 1984.
39. *Bericht zu Untersuchungen zur biologischen Reinigung von Abwässern der Lederfabrik 'August Apfelbaum' Neustadt-Glewe.* 1984.
40 *Maschinelle Entwässerung bei chemischer P-Elimination.* 1983.

Amlacher, E. (1972) *Taschenbuch der Fischkrankheiten.* Fischer, Jena.
Bagby, M. M. and Sherrard, J. H. (1981) Combined effects of cadmium and nickel on the activated sludge process. *JWPCF* **53** (11) 1609–1619.
Bargel, U. (1965) Über die Lebensfähigkeit der Erreger der Tuberkulose. Thesis, Karl-Marx-Universität, Leipzig.
Barnard, J. L. (1974) Cut P and N without chemicals. *Water and Wastes Engn.* **11** (41) (7): 33–36 and 43–44.
Barth, E. F., English, J. N., Salotto, B. V., Jackson, B. N., and Ettinger, M. B. (1965) Field survey of four municipal wastewater treatment plants receiving metallic wastes. *JWPCF* **37** (8) 1101–1126.
Barth, E. F., Ettinger, M. B., and Salotto, B. V. (1965) Summary report on the effects of heavy metals on the biological treatment processes. *JWPCF* **37** (1) 86–99.
Bartos, E. (1959) Viřníci — Rotatoria. In *Fauna ČSR,* Vol. 15, Nakladatelství československé adkademie ved., Prague.
Batek, J. (1979) Vliv některých technologických parametrü aktivačniho čištění na biocenózu aktivovaného kalu. *vodní hosp.* B **29** (8): 142–147.
Behrens, U. and Hannes, J. (1984) Zum Abbau von Formalehyd durch adaptierte Bakterien. *Ahh* **12** (1) 39–43.
Berg, G. (1965) Die Virusübertragung auf dem Wasserweg. *Archiv f. Hygiene und Bakteriologie* **149** (3/4) 310–335.
Bernsen, H. (1983) *Rechenprogramm zur Wärmebilanz von Kleinbelebtschlammanlagen.* VEB Abwasserbehandlungsanlagen, Merseburg.
Beuthe, C.-G. (1967) Verfahrenstechnische Grundlagen des Belebtschlammverfahrens. *GWF* **108** 143–149.
Billmeier, E. (1985) Der Sauerstoffeintrag in der Abwasserreinigung nach dem Belebungsverfahren. *Korr. Abwasser* **30** (4) 244–248.
Billmeier, E. (1983) Bewertung der bei Betrieb massgebenden parameter für den Sauerstoffeintrag. *Korr. Abwasser* **30** (5) 334–339.

Birr, R., and Mrugowski, F. (1981) *Respirometer*. Forschungszentrum Wassertechnik, Aussenstelle, Leipzig.

Blaszcyk, M. (1978) The influence of some carbon substrates on the efficiency of denitrification of high concentrations of nitogen. *Acta Microbiol. Pol.* **28** 145–156.

Böhnke, B. (1985) Vergleichende Betrachtungen von Versuchs- und Betriebsergebnissen der zweistugigen AB-Technik unter besonderer Berücksichtigung mikrobiologischer Reaktionsmechanismen. *Korr. Abwasser* **30** (7) 452–461 and (8) 530–536.

Bonomo, L. (1974) Einfluss von Zink in einem Belebtschlammverfahren. *Ingegneria Ambientale inquinamento e depurazione* **3** (6) 538–547.

Borneff, J., and Kunte, H. (1967) Kanzerogene Substanzen in Wasser und Boden. XIX. Wirkung der Abwasserreinigung auf polyzyklische *Aromaten. Arch. f. Hyg. u. Bakteriol.* **151** 202–210.

Braha, A. (1983) Über das kinetische und hydraulische Verhalten mehrstufiger Belebungsanalagen. *Korr. Abwasser* **30** (2) 92–99.

Brix, J., Imhoff, K., and Weldert, R. (1934) *Die Stadtentwässerung in Deutschland*. Fischer, Jena.

Brown, P., and Andrew, P. R. (1972) Some effects of zinc on the performance of laboratory-scale activated-sludge units. *Water Pollut. Control* **71** (5) 549–554.

Buck, H. (1968) Die Ciliaten des Belbtschlammes in ihrer Abhängigkeit vom Klärverfahren. *MB* **5** 206–222.

Bucksteeg, W., and Thiele, H. (1959) Die Beurteilung des Schlammes mittels TTC (2,3,5-Triphenyltetrazoliumchlorid). *GWF* **100** 916–920.

Burdick, C. R. (1982) Advanced biological treatment to achieve nutrient removal. JWPCF **54** (7) 1078–1085.

Busse, W., Schuster, H., Jagusch, L., and Püschel, S. (1970) Verfahren zur biochemischen Reinigung konzentrierter tensidhaltiger Abwässer. WP 77466,85c, 1, C02e.

Cena, M. (1975) *Wasser und Tierproduktion*. Fischer, Jena.

Chaudhuri, N., Engelbrecht, R. T., and Austin, J. H. (1965) Nematodes in an aerobic waste treatment plant. *Jour. Amer. Water Works Assoc.* **57** 1561–1571.

Chudoba, J., Blaha, J., and Madera, V. (1974) Control of activated sludge filamentous bulking. III. Effect of sludge loading. *Water Res.* **8** (4) 231–237.

Chudoba, J., and Porkorny, S. (1971) Residual organic matter in activated sludge process effluents. VI. Effect of sludge age and sludge load. *Sbornik F* **16** 5–17.

Clarke, N. A., Stevenson, R. E., Chang, S. L., and Kabler, P. W. (1961) *Amer. J. Publ. Health* **51** 1118 (cited by Berg 1965).

Clausing, D. (1980) Möglichkeiten und Grenzen der Desinfektion von Abwässern, besonders aus Schlachtbetrieben. *Z. ges. hyg.* **25** (7) 486–491.

Cooke, W. B., Bridge, A. F., and Pipes, W. O. (1970) The occurrence of fungi in activated sludge. *Mycopathol. Mycol. Appl.* **40** (3/4) 249–270.

Cooke, W. B., and Ludzack, F. J. (1958) Predacious fungi behavior in activated sludge. *Sewage and Ind. Wastes* **30** (12) 1490–1495.

Cooper, D. G., and Zajic, J. E. (1980) Surface-active compounds from microorganisms. *Advances in Applied Microbiol.* **26** 229–253.

Curds, C. R., Cockburn, A., and Vandyke, J. M. (1967) An experimental study of the role of the ciliated protozoa in the activated-sludge process. *Water Pollut. Control* **67** 312–329.

Curds, C. R. and Cockburn, A. (1970) Protozoa in biological sewage treatment processes. II. Protozoa as indicators in the activated-sludge process. *Water Res.* **4** 237–249.

Cyrus, Z., and Sladeček, V. (1973) Určovaci atlas organismü z čistiren odpadnich vod. Výzkumný Ustav vodohospodářský, Práce a studie, 133, Prague.

Dautermann, J. (1969) Technologishce Kennziffern für das Belebtschlammverfahren durch den Betrieb einer halbtechnischen Versuchsanlage. *WTZ-Mitteilung* **1** 10–20.

Decloitre, L. (1962) Le genre Euglypha Dujardin. *Arch. Protistenk.* **106** 51–100.

Dias, F. F., and Bhat, J. V. (1964) Microbial ecology of activated sludge. I. Dominant bacteria. *Appl. Microbiol.* **12** (5) 412–417.

Ditsios, M. (1984) Horizontal durchströmte Nachklärbecken von Belebungsanlagen — Festsotffgehalt im Ablauf und Zulässige Flächenbeschickung. Österr. Wasserwirtsch. **36** (5/6) 85–89.

Dobberkau, H.-J., and Walter, R. (1979) Abwasser. In Horn, K. *Grundlagen der Kommunalhygiene*. Volk und Gesundheit, Berlin.

Dohanyos, M., Kosova, B., and Grau, P. (1982) *Možnosti intenzifikace metanizačniho procesu*. vh **32** (7) 179–186.

Eckenfelder, W. W. (1966) Principles of biological oxidation. In *Industrial water pollution control,* pp. 134–187. McGraw-Hill, New York.

Eden, G. E. (1972) Effect of temperature on the removal of NTA during sewage treatment. *Wat. Res.,* **6** 877–883.

Eickhoff, F. F. (1969) Aerobe Schlammstabilisierung. *Veröffentl des I. für Siedlungswasserwirtschaft, TU Hannover* **31** 179–197.

Eikelboom, D. H. (1975) Filamentous organisms observed in activated sludge. *Water Res.* **9** (4) 365–388.

Elmore, H. L., and Hayes, T. W. Solubility of atmospheric oxygen in water. *Proc. Amer. Soc. Civil Eng.* **86** 41–53.

Engelhardt, H. (1978) Untersuchungen und Versuche zur Geruchsbekämpfung in Kläranlagen. *Forum Städte-Hygiene* **29** 268–269.

Erman, L. A. (1963) Nutrition and reproduction of Brachionus calyciflorus in mass cultures. *Dokl. Akad. nauk SSSR* **144** 926–929.

Fair, G. M., and Geyer, J. C. (1961) *Wasserversorgung und Abwasserbehandlung*. R. Oldenbourg, Munich.

Fenchel, T. (1980) Suspension feeding in ciliated Protozoa: functional response and particle size selection. *Microb. Ecol.* **6** 1–11.

Fenchel, T. (1980) Feeding rates and their ecological significance. *Microb. Ecol.* **6** 13–25.

Fiedler, W. (1981) *Technologisches Projekt: Schaumstoffbelüfter,* VEB Zellstoff- und Papierfabrik Blankenstein.

Fischer, E. (1894) Einfluss der Configuration auf die Wirkung der Enzyme. *Ber. dtsch. chem. Ges., Berlin* **27** (3) 2985–2993.

Fleissig, (1960) Die Lebensdauer und das Zusammenleben von Mikroorganismen im Wasser. Thesis, TU Munich.

Frangipane, F., and Urbini, G. (1978) Modern trends in plants design for the

removal of nutritive substances from sewage. *Poltechnika Krakowska* **1** 127–149.
Franz, B. (1968) Einsatz von Saugkreiselbelüftern bei der biologischen Reinigung von Gaswässern. *WWT* **18** (5) 165–170.
Fuka, T., and Pitter, P. (1981) Stanoveni biologické rezoložitelnosti běžných pracich prostředků. *Sbornik F* **24** 5–47.
Fush, G. W. (1961) Der mikrobielle Abbau von Kohlenwasserstoffen (Sammelbericht). *Arch. f. Mikrobiol.* **39** 374–422.
Fash, G. W., and Chen, M. (1975)Microbial basis of phosphate removal in the activated sludge process for the treatment of waste waters. *Microbial Ecol.* **2** 119–138.
Gassen, M. (1982) *Nitrifikation und Denitrifikation bei sauerstoffbegasten ein- und zweistufigen Belebungsanlagen.* Institut zur Förderung der Wassergüte- und Wassermengenwirtschaft, Essen, 19. Kurzbericht.
Gassen, M. (1980) Sauerstoffkonzentration, Schlammbelastung und Blähschlammbildung. *Wasserwirtschaft* **70** (4) 164–168.
Gilsenbach, R. (1971) *Wasser. Probleme — Projekte — Perspektiven.* Urania, Leipzig.
Görbing, F. (1981) Grundsatzuntersuchung zur Kreislaufnutzung und Wiederverwendung von Wasser. B. thesis, TU Dresden, Sekt. Wasserwesen.
Gottschaldt, N. (1968) Biochemische Mmethoden in der Abwasseranalytik. Fortschr. *Wasserchemie* **10** 123–132.
Grosse, G., Hänel, K., Keller, H., and Poschke, H.-J. (1978) Belüftungsverfahren der biologischen Abwasserreinigung. *WWT* **28** 1 15–18.
Grünwald, A. (1980) Biologická rozložitelnost uhlovodiků aktivovaným kalem. *Sbornik F* 23 65–78.
Grüwald, A., Koller, J., and Krnáčová, L. (1983) Vliv sulfidů a thiosiranů na aktivační proces. *vh. B* **33** (9) 231–235.
Grutsch, J. F. (1980) The γ-β-γ's of wasterwater treatment. *Environmental Science and Technology* **14** 276–281.
Gudernatsch, H. (1970) Verhalten von Nitriloessigsäure im Klärprozess und im Abwasser. *GWF* **111** (9) 511–516.
Gruz, J. (1981) Aerobni rozklad suspenzi na čistírnách prasečí kejdy. *vh B* **31** (9) 245–247.
Guhl, W. (1979) *Opercularia coarctata,* ein variables Peritrich. *Arch. Protistenk.* **121**: 3208–346.
Hackenberger, J. (1968) Untersuchungen über den Einfluss der Absetzvorgänge bei Belebungsanlagen mit und ohne Schlammstabilisierung zur Reinigung von Abwässern. Thesis, TU Dresden, Sekt. Wasserwesen.
Hagen, H. (1980) Phosphatentfernung aus dem Abwasser. *WWT* **30** (1) 24–26.
Hale, J. M. (1983) Factors influencing the stability of polarographic oxygen sensors. in: Gnaiger, E., and Forstner, H. *Polarographic oxygen sensors.* Springer-Verlag, Berlin.
Halle, K., and Trautsch, J. (1977) Abwasserprobleme in Wäschereien. *Textilreinigung* **24** (5) 146–149.
Hamm, A. (1963) Untersuchungen über die Ökologie und Variabilität von *Aspidisca costata* (Hypotricha) im Belebtschlamm. First thesis TU Munich.

Hänel, K. (1963) Biologische, chemische und physikalische Untersuchungen an einem Oxydationsgraben unter besonderer Berücksichtigung der Nährstoffelimination. Diploma study, Karl-Marx-Universität Leipzig.
Hänel, K. (1980) Weiterentwicklung von Kleinbelebungsanlagen — Effektivitätssteigerung und Verbesserung der arbeitshygienischen Bedingungen. *WWT* **30** (6) 202–205.
Hänel, K. (1979) Systematik und Ökologie der farblosen Flagellaten des Abwassers. *Arch. Protistenk.* **121** 73–137.
Hänel. K., and Schmidt, E. (1975) Ermittlung von Bemessungsparametern der Abwasserreinigung in einer fahrbaren Mess- und Versuchsanlage. *WWT* **25** (3) 95–98.
Hänel, K., Liedke, P., Gutendorf, W., and Anton, Ch. (1981) Kontinuierliche Kontrolle der Ablaufgüte einer Belebtschlammanlage mittels Trübungsmessung. *WWT* **31** (10) 337–340 and 342.
Häusler, J. (1982) Schizomycetes (Bakterien). In *Süsswasserflora von Mitteleuropa*. Vol. 20, 1st edn, Fischer, Jena.
Hardin, G. (1943) Flocculation of bacteria by protozoa. *Nature* **151** 643.
Harnisch, O. (1963) Wurzelfüssler, Rhizopoda. In *Die Tierwelt Mitteleuropas*. Leipzig vol. 1 (1b) 1–75.
Hashimoto, A. G. (1975) Turbine-air aeration system for poultry wastes. *Proceedings of Third International Conference on Livestock Wastes,* pp. 530–534.
Haupt, R. (1978) Über die lärmdämpfende Wirkung der Vegetation, untersucht am Beispiel von Waldrändern und Waldbeständen, und Möglichkeiten ihrer Anwendung. Thesis, TU Dresden, Sekt. Fortwirtsch.
Heine, A. (1968) *Technisch-ökonomische Untersuchungen zur III, Reinigungsstufe unter besonderer Berücksichtigung der Schlammbehandlung.* TU Dresden, Sekt. Wasserwesen.
Heyden, W. (1982) Durch Mutation gewonnene Mikroorganismen verstärken den biologischen Abbau von chemisch stark verunreinigten Abwässern. *Chemie-Technik* **11** (7) 863–865.
Hofman, T., and Lees, H. (1952) *Biochem. J.* 52 140–142 (cited by Thimann, 1964).
Hörler, A. (1968) Probleme der Abwassertechnik und des Gewässerschutzes. *GWF* **109** 1337–1340.
Horvath, R. S. (1972) Microbial Co-Metabolism and the Degradation of Organic Compounds in Nature, Bacteriol. Rev. 2 146–155.
Huber, L. (1968) Erfahrungen bei der Abwasserbeseitigung in vier neuen Erdölraffinerien. *WAF* (5) 180–187.
Hunken, K.-H. (1960) Untersuchungen über den Reinigungsverlauf und den Sauerstoffverbrauch bei der Abwasserreiningung durch das Belebtschlammverfahren. *Struttgarter Berichte zur Siedlungswasserwirtschaft,* vol. 4 Verlag R. Oldenbourg, Munich.
Hurwitz, E., Beck, A. J., Sakellariov, E., and Krup, M. (1961) Degradation of cellulose by activated sludge treatment. *JWPCF* **33** 1070–1075.
Hussel, L. (1963) *Lehrbuch der Veteriñarhygiene.* S. Hirzel Verlag, Leipzig.
Ilič, P. (1977) Untersuchungen zum Verhalten schwer abbaubarer Modellsubstanzen bei der biologischen Reinigung. Thesis, TU Darmstadt.

Imhoff, K. (1949) Die Gould-Stufenbelüftung bei der Abwasserreinigung mit belebtem Schlamm. *Gesundheitsingenieur* **70** 393–395.

Imhoff, K., and Imhoff, K. R. (1972) *Taschenbuch der Stadtentwässerung*. R. Oldenbourg, Munich.

Imhoff, K. R. (1982) *Leistungsvergleich ein- und zweistufiger biologischer Abwasserreinigungsverfahren,* pp. 1–8. Institut zur Förderung der Wassergüte- und Wassermengenwirtschaft, Essen, 15. Essener Tagung.

Jähnig, C. (1980) Neue leistungsfähige Mikroorganismen optimieren biologische Kläranlagen. *Chemie-Technik* **9** (10) 497–500.

Jakobczyk, J., Kwiatkowski, M., and Ostrowska, J. (1984) Untersuchung der biologischen Zersetaung aromatischer Nitroverbindungen in Abwässern. *Chem. Techn.* **36** (3) 112–116.

Jarnych, V. S. (1976) *Aerosolve in der Veterinärmedizin*. Beutscher Landwirtschaftsverlag, Berlin.

Jones, H. (1964) The effect of temperature and oxygen tension on one of the microorganisms responsible for sludge bulking. *Ind. Wast. Conf.* 902–904.

Jones, P. H., and Sabra, H. M. (1980) Effect of systems solids retention time (SSRT or Sludge Age) on nitrogen removal from activated sludge systems. *Water Pollut. Control* **79** (1) 106–123.

Kahl, A. (1935) Urtiere oder Protozoa. I. Wimpertiere oder Ciliata (Infusoira). In *Die Tierwelt Deutschlands*. Dahl, F. (ed.) Parts 18, 21, 25 and 30. Fischer, Jena.

Kalinske, A. A. (1978) Verwendung von Luft oder technisch reinem Sauerstoff in Belebtschlamm-Anlagen. *Wasser-Luft-Betrieb* **22** 24–26.

Kapp, H. (1983) Zur Interpretation der 'Säurekapazität des Abwassers'. *GWF* **124** (3) 127–130.

Kayser, R. (1967) Ermittlung der Sauerstoffzufuhr von Abwasserbelüftern unter Betriebsbedigungen. Thesis, TH Braunschweig, Fakuiltät für Bauwesen.

Kienzle, K. H. (1972) Die beeinflussung des BSB_5-Abbaus durch die Nitrifikation in schwachbelasteten Belebungsanlagen. *GWF* **113** (6) 261–263.

Kienzle, K. H. (1974) Der Einfluss der Temperatur auf die Reinigungsleistung und den Betrieb von Oxydationsgräben. *Österr. Abwasserrundsch.* **19** (2) 17–22.

Klimowicz, H. (1974) Przewodnik do szybkiejoceny przebiegu procesu oxyszczania ścieków osadem ezynnym. Część I — Orzeski. *Gaz, woda i technika sanitarna* **40** (1) 278–283.

Klimowicz, H. (1976) Część II. Wrotki. *Gaz, woda i technika sanitarna* **12** Tom L 331–335.

Klimpel, H. (1976) Kleinbelebungsanlage. Patentschrift 124793, WPC 02c/19190, 18.3.

Knaack, J., and Ritschel, H. (1975) Zur Eliminierung von exogenen Helminthenstadien aus dem kommunalen Abwasser durch verschiedene Abwasserreinigungsverfahren. *Zeitschrift ges. hygiene und ihre Grenzgebiete* **21** 746–750.

Kohler, R. (1975) Technologie und Anwendung der Entspannungs-Flotation in der Abwasserreinigung. *Wasser, Luft, Betrieb* **19** (2) 72–77.

Kolkwitz, R., and Marsson, M. (1908 Ökologie der pflanzlichen Saprobien. *Ber. dtsch. Bot. Ges.* **26** 505–519.

Kolkwitz, R., and Marsson, M. (1909) Ökologie der tierischen Saprobien. *Intern. Rev. Hydrobiol.* **2** 126–152.

Kone, S., and Behrens, U. (1981) Zur Kinetik der Denitrifikation. Part 1: Mischpopulation und Acetat als Kohlenstoffquelle. *Ahh* **9** (5) 523–533.
Koska, H. (1983) *Lärmschutzhaube*. VEB Wasserversorgung und Abwasserbehandlung, Gera.
Koumar, L. (1964) Biochemické odbourávání čistých organichkých látek. *vh* **14** (4) 143–145.
Kovacs, E., Bucz, B., Kolompar, G. (1969) Propagation of mammalian viruses in protista and visualisation of fluorochrome labelled EMC-virus in yeast and tetrahymena. part IV. *Proc. Soc. Exp. Biol. (N.Y.)* **32** 971 (cited by Möse *et al.* 1970).
Kraus, L. S. (1945) The use of digested sludge and digester overflow to control bulking activated sludge. *Sew. Works Journ.* **17** 1177–1190.
Kroiss, H. (1981) Anaerobe Abwasserreinigung. *Österr. Wasserwirtsch.* **33** (3/4) 56–64.
Kroiss, H. (1981) Adsorptions- und Speichervorgänge beim Belebungsverfahren. *GWF, Schriftenreihe Wasser — Abwasser* **19** (1) 101–122.
Krul, J. (1976) Untersuchungen über die Milieubedingungen innerhalb von Belebtschlammflocken. *Stuttgarter Berichte zur Siedlungswasserwirtschaft* **54** 51–66.
Kurnilovič, O. B. (1972) Očistka slabokoncentrirovannych stočnich vod v aerotenkach. *Vodosn. i sanitarnaja technika* **9** 18–19.
Lautenbach, E. (1966) Untersuchungen über die Lebensbedingungen der Organismen im Tropfkörper und Belebungsbecken, die zum spezifischen Abbau von chemischen Stoffen in industriellen Abwässern befähigt sind. Part 1: Cyanide. *Schriftenreihe des Deutschen Arbeitskreises Wasserforschung e. V. (DAW)*, Erich Schmidt Verlag, Berlin.
Lautenbach, E. (1966) Part 3: Mineralöle. *Schriftenreihe des Deutschen Arbeitskreises Wasserforschung* pp. 153–257.
Levin, G. V., and Shapiro, J. (1965) Metabolic uptake of phosphorus by waste water organisms. *JWPCF* **37** 800–821.
Liebmann, H. (1962) *Handbuch der Frischwasser- und Abwasserbiologie*. vol. I, 2. edn. Oldenbourg Verlag, Munich.
Liebmann, H. (1965) Mikroorganismen als Indikatoren bei der Abwasserbelüftung. *MB* **12** 367–378.
Lienig, D. (1979) *Wasserinhaltsstoffe*. Akademie-Verlag, Berlin.
Lindner, K. A. (1978) *Milliarden Mikroben*. Urania-Verlag, Leipzig.
Lineweaver, H., and Burk, D. (1934) Determination of enzyme dissociation constants. *J. Amer. Chem. Soc.* **56** 654 (cited by Ruttloff *et al.* 1978).
Ludwig, C., Matsche, N., and Streichsbier, F. (1982) Beeinflussende Faktoren der biologischen Phosphorentfernung mit dem Belebungsverfahren. *Forum Städte-Hygiene* **33** (5) 259–262.
Lützner, K. (1979) Leistungssteigerung biologischer Reinigungsanlagen durch Einsatz der Röhrensedimentation. B thesis, TU Dresden, Sekt. Wasserwesen.
Mangold, K.-H., Kaeding, J., Lorenz, K., and Schuster, H. (1973) *Abwasserreinigung in der chemischen und artverwandten Industrie*. Deutscher Verlag f. Grundstoffindustrie, Leipzig.

Marvan, P., and Zelinka, M. (1961) Zur Präzisierung der biologischen Klassifizierung der Reinheit fliessender Gewässer. *Arch. Hydrobiol.* **57** 389–407.

Matsché, N., Usrael, G., and Ludwig, C. (1982) Die biologische Phosphorentfernung mit dem Belebungsverfahren am Beispiel der Kläranlagen im Einzugsgebiet des Neusiedler Sees. *Österr. Wasserwirtsch.* **34** (9/10) 219–227.

Matthes, D., and Wenzel, F. (1960) Wimpertiere (Ciliaten). Kosmos-Verlag, Stuttgart.

McDermott, G. N., Moore, W. A., Post, M. A., and Ettinger, M. B. (1963) Effects of copper on aerobic biological sewage treatment. *JWPCF* **35** (2) 227–241.

McKeinney, R. E. (1962) Microbiology for sanitary engineers. McGraw-Hill.

Meissner, M., Birr, R., and Hänel, K. (1983) *Bekämpfung von Blähschlamm in einer Kleinbelebungsanlage KBA 100.* Technische Hochschule Merseburg, Sekt. Verfahrenstechnik.

Michaelis, L., and Menten, M. L. (1913) Die Kinetik der Inverteinwirkung. *Biochem. Z.* 333–369.

Möller, F.-W., Kuhles, H., and Hattenbach, J. (1984) Kläranlage Friedrichroda — Ergebnisse eines Experimentes. *WWT* **34** 3 58–64.

Möse, J. R., Dostal, V., and Wege, H. (1970) Inaktivierung von Viren durch Abwasserprotozoen. *Arch. f. Hyg.* **154** (4) 319–330.

Moore, A. W., McDermott, G. N., Post, M. A., Mandia, J. W., and Ettinger, M. B. (1961) Effects of chromium on the activated sludge process. *JWPCF* **33** (1) 54–72.

Mortimer, C. H. (1942) The exchange of dissolved substances between mud and water in lakes. *J. Ecol.* **29** 280–329.

Mosebach, K.-H. (1975) Phosphatrücklösung bei der Ausfalung von Simultanschlämmen. *Wasser und Abwasser in Forschung und Praxis,* vol. 11. E. Schmidt Verlag, Berlin.

Müller, C., and Bartocha, W. (1978) Zur Frage der Verbreitung von Bakterien über Belebungsanlagen auf Kläranlagen. *Bundesges.-Bl.* **21** (23) 21–33.

Müller, H. E. (1981) Abschätzung des Infektioinsrisikos durch Abwasser und Klärschlamm. *Forum Städte-Hygiene* **32** 146–152.

Müller-Neuhaus, G. (1971) Untersuchungen über getrennte Schlammstabilisierung und Folgerungen für die Praxis. *Abwassertechnik* **22** 1.

Nähle, C. (1980) *Uber den Einfluss biogener und anthropogener Komplexbildner auf die Eliminierung von Schwermetallen bei der Landsamsanfiltration.* Veröffentl. Inst. Wasserforschung GmbH, Dortmund 32.

Nechvatal, J., and Sladka, A. (1981) Zkušenosti s čištěnim odpadnich vod z pražske aglomerace kyslíkovou aktivaci. *vh B* **31** (2) 48–52.

Neufeld, R. D. (1976) Heavy metals-induced deflocculation of activated sludge. *JWPCF* **48** (8) 1940–1948.

Neafeld, R. D., Guttierrez, J., and Novak, R. A. (1977) A kinetic model and equilibrium relationship for heavy metal accumulation. *JWPCF* **49** (3) 489–499.

Neufeld, R. D., and Hermann, E. R. (1975) Heavy metals removal by activated sludge. *JWPCF* **47** 310–329.

Nicholls, H. A., and Osborn, D. W. (1979) Bacterial stress: prerequisite for biological removal of phosphorus. *JWPCF* **51** (3) 557–568.

Niethammer, A. (1947) *Technische Mykologie.* F. Enke Verlag, Stuttgart.

Nikaido, H. (1979) Die Permeabilität der äusseren Bakterienmembran. *Angew. Chem.* **91** 394–407.

Nihaido, H., and Rosenberg, E. Y. (1981) Effect of solute size on diffusion rates through the transmembrane pores of the outer membrane of Escherichia coli. *J. Gen. Physiol.* **77** 121–135.

Nolte, E., Meyer, H. J., and Fromke, E. (1934) Versuche zur Durchführung des Belebtschlammverfahrens bei gewerblichen Abwässern. *Vom Wasser* **8** 126–147.

Nösel, H. (1978) Über die Zuverlässigkeit von O_2-Sättigungstabellen. *Wasser, Luft und Betrieb* **22** (5) 176–180.

Offhaus, K. (1965) Die Bewertung eines Abwassers unter besonderer Berücksichtigung des biologisch abbaubaren Anteils und der Toxizität. *Die Wasserwirtschaft* **55** 220–229.

Offhaus, K. (1968) Zinkgelhalt und Toxizität in den Abwässern der Chemiefaserindustrie. *WAF* **1** 7–21.

Osborn, D. W., and Nicholls, H. A. (1978) Optimisation of the activated sludge process for biological removal of phosphorus. *Progress in Water Technology* **10** (1/2) 261–277.

Ostermann, G. (1971) Selbsttätige Belebtschlammanreicherung im Oxydationsgraben durch eine Schlammtasche. *Rationalisierung in der Abwassertechnik* 73–82.

Ottova-Svobodova, V. (1962) Biocenosa plisní a hub v městských odpadních vodach. *Sbornik* **6** (1) 563–572.

Page, F. C. (1976) An illustrated key to freshwater and soil amoebae. Freshwater Biological Association, Scientific Publication 34. Ambleside, Cumbria.

Parkin, G. F., and McCarty, P. L. (1981) Sources of soluble organic nitrogen in activated sludge effluents. *JWPCF* **53** (1) 89–98.

Parkin, G. F., and McCarty, P. L. (1981) Production of soluble organic nitrogen during activated sludge treatment. *JWPCF* **53** (1) 99–112.

Pasveer, A. (1958) Über die Theorie des Sauerstoffeintrages und des Sauerstoffverbrauches beim Belebtschlammverfahren. *MB* **5** 152–163.

Paucke, H., and Bauer, A. (1979) *Umweltprobleme — Herausforderung der Menschheit*. Dietz, Berlin.

Peters, H. (1976) Einfluss der Temperatur auf die BSB-Kinetik, gezeight am Beispiel *Escherichia coli*. *Karlsruher Berichte zur Ingenieurbiologie* 9.

Peukert, V., and Thomsch, U. (1981) Automatisch arbeitende Messeinrichtung zur Bestimmung des biochemischen Sauerstoffverbrauchs von wässruigen Medien. *Ahh* **9** (4) 463–465.

Pike, E. B. (1975) *The organisms and their ecology*. Academic Press, New York.

Pipes, W. O. (1967) Bulking of activated sludge. *Advanc. Apl. Microbiol.* **9** 185–235.

Pöpel, J. (1971) Schwankungen von Klärenlagenabläufen und ihre Folgen für Grenzwerte und Gewässerschutz. *GWF, Schriftenreihe Wasser — Abwasser*. Oldenbourg, Munchen.

Popp, W. (1979) Neuere Erkenntnisse über Blähschlamm. *MB* **31** 139–151.

Pöpopinghaus, K. (1975) Chemisch-physikalishe und -biologische Massnahmen zur Stick stoffeliminierung aus dem Abwasser. *GWA* **17** 77–106.

Port, E. (1978) Verwendung von Belebtschlamm zur selektiven Adsoprtion toxischer Abwasserinhaltsstoffe. Thesis. Fachbereich 13 — Wasser und Verkehr, D 17, Eigenverlag TH, Darmstadt.

Randolf, R. (1983) Studien zur Entwicklung von Bauwerken auf kommunalen Kläranlagen der DDR. B. thesis TU Dresden, Sektion Wasserwesen, 1983.

Rebhun, T., and Manka, I. (1971) Classification of organics in secondary effluents. *Environ. Sci. Technol.* **5** (cited in [27]).

Reichelt, H. (1981) Rationelle Wasserverwendung — zentrale Aufgabe zur Verwirklichung der ökonomischen Strategie in der Wasserwirtshaft. *WWT* **31** (11) 363–365.

Reichert, J., Kunte, H., Engelhardt, K., and Borneff, J. (1971) Kanzerogene Substanzen in Wasser und Boden. XXVII. Weitere Untersuchungen zur Eliminierung kanzerogener, polyzyklischer Aromaten aus Abwasser. *Zbl. Bakt. Hyg., I. Abt., Orig. B.* **155** 18–40.

Reimann, K. (1969) Untersuchungen über den Mechanismus der toxischen Hemmung des Belebtschlammes durch ein Schwermetallion, dargestellt am Beispiel des Kobaltes. *WAF* **1** 25–35.

Reinbothe, H. (1975) *Einführung in die Biochemie.* Fischer, Jena.

Richarme, G. (1982) Interaction of the maltose — binding protein with membrane vesicles of *Escherichia coli. J. Bacteriol.* **149** 662–667.

Riedel, W. (1971) Spezifische Aktivität des Belebtschammes in einem Oxydationsgraben in Abhängigkeit von Belastung und Turbulenz. Diplomarbeit, TU Dresden, Sekt. Wasserwesen.

Rienzo, J. M. DI, Nakamura, K., and Inouke, M. (1978) The outer membrane proteins of gram-negative bacteria: biosynthesis, assembly, and functions. *Ann. Rev. Biochem.* **47** 481–532.

Roske, J., Hackenberger, J. and Uhlmann, D. (1982) Untersuchungen zur Ermittlung der Überschussschlammproduktion in Abhängigkeit von der Schlammbelastung. *Ahh* **10** (5) 497–503.

Ruf, M. (1958) Das Verhalten radioaktiver Stoffe in Tropfkörpern und Belebungsbecken. *MB* **5** 56–66.

Ruttloff, H., Huber, J., Zickler, E., and Mangold, K.-H. (1978) *Industrielle Enzyme.* Fachbuchverlag, Leipzig.

Sarfert, F. (1978) Bestimmung von fadenförmigen Mikroorganismen im Belebtschlamm. *WAF* **11** (5) 176–178.

Sawyer, C. N. (1964) Advanced water pollution research, vol. 2, p. 508. Pergamon Press, Oxford.

Schäfer, E. (1978) Virusnachweis im Bereich der Wasserhygiene. *Arch. Hyg.* **154** (4) 299–313.

Scherb, K. (1979) Abwasserreinigung mit Hilfe con kleinen Kläranlagen. *MB* **31** 61–76.

Scherb, K. (1965) Verleichende Untersuchungen über das Sauerstoffeintragsvermögen verschiedener Belüftungssysteme auf dem Münchner Abwasserversuchsfeld. *MB* **12** 330–350.

Scherb, K. (1979) Verfahren zur Begasung von belebtem Schlamm mit technischem Sauerstoff. *MB* **28** 99–119.

Scheuring, L., and Höhnl, G. (1959) *Sphaerotilus natans.* Seine Ökologie und

Physiologie. *Schriften des Vereins d. Zellstoff- und Papier-Chemiker und -Ingenieure* **26**.

Schlegel, H. G. (1969) *Allgemeine Mikrobiologie.* Thieme, Stuttgart.

Schmidt, E. (1977) *Die Behandlung vom Schlammfugat in einer Versuchsbelebtschlamm anlage.* Ingenieurarbeit, Ingenieurschule für Wasserwirtschaft, Magdeburg.

Schneider, W. (1939) Freilebende und pflanzenparasitische Nematoden. In *Die Tierwelt Deutschlands.* vol. 36, Fischer, Jena.

Schoenichen, W. (1927) *Einfachste Lebensformen des Tier- und Pflanzenreiches. Vol. II: Urtiere, Rädertiere.* Hugo Bermühler, Berlin.

Schönborn, W., and Beyer, H. (1966) Untersuchungen über die Lebensbedingungen der Organismen im Tropfköper und Belebungsbecken, die zum spezifischen Abbau von chemischen Stoffen in industriellen Abwässern befähight sind. Part IV: Phenole mit Cyaniden und Rhodaniden. *Schriftenreihe des Deutschen Arbeitskreises Wasser-forschung e. V. (DAW)* vol. 13, pp. 259–341. E. Schmidt, Berlin.

Schöborn, W., and Lautenbach, E. (1966) Part 2: Rhodanide *Schriftenreihe des Deutschen Arbeitskreises Wasserforschung e. V. (DAW)* vol. 13, pp.99–152. E. Schmidt Verlag, Berlin.

Schuster, G. (1971) Entwicklung eines wirtschaftlichen Verfahrens zur Elimination der Phosphorverbindungen aus häuslichem Abwasser. *Wasserwirtschaftsdirektion Saale — Weisse Elster,* Magdeberg.

Schwägeer, U. (1980) Vergleichende Untersuchungen über das Absetzverhalten des Belebtschlammes in volldurchmischten Becken und beim Kaskadenbetrieb. *Stuttgarter Berichte zur Siedlungswasserwirtschaft* **67**.

Schwartz, H. G., Porowchak, T., and Becker, K. (1980) Control of sludge bulking in the brewing industry. *JWPCF* **52** (12) 2977–2994.

Seeliger, H. P. R. (1978) *Taschenbuch der medizinischen Bakteriologie.* Johann Ambrosius Barth, Leipzig.

Seltmann, G. (1982) *Die bakterille Zellwand.* Fischer, Jena.

Shamat, N. A. and Maier, W. J. (1980) Kinetics of biodegradation of chlorinated organics. *JWPCF* **52** (8) 2158–2166.

Sixt, H. (1979) Reinigung organisch hochverschmutzter Abwässer mit dem anaeroben Belebungsverfahren am Beispiel von Abässern der Nahrungsmittelherestellung. *Veröffentl. des Inst. f. Siedlungswasserwirtschaft d. Universität Hannover* **50**.

Skidrov, J. V., et al. (1980) Isseldovanie kinetiki biochimičeskogo razloženija organičeskich veščesty promyšlennych stočnich vod. Vodgeo, Moskva.

Skuja, H. (1956) Taxonomische und biologische Studien über das Phytoplankton schwedischer Binnengewässer. *Nova Acta Regiae Societatis Scientiarum Upsaliensis, Ser. 4* **16** 3.

Sládeček, V. (1963) A guide to limnosaprobical organisms. *Sbornik* **7** (2) 543–612.

Sládeček, V. (1955) Vysledky výzkumu o biologii aktivační čistírny v Hostívaři. *vh* **5** 132–133.

Sládeček, V. (1973) System of water quality from the biological point of view. *Arch. f. Hydrobiol., Beiheft: Ergebnisse der Limnologie* **7** 1–218.

Sladká, A. (1975) Biocenóza a morfologie aktivovaného kalu. Výzkumný ústav vodohospodářský, Práce a Studie, Praha 139.

Sladká, A., and Hänel, K. (1965) Kvasinky v akitvačním čistícím procesu. *vh* **15** (10) 457–460.

Sprenger, F.-J. (1982) *Einfluss erhöhter Schwermetallgehalte und Organohalogenverbindungen auf den Faulprozess und die weitere Schlammverwertung*. Institut zur Förderung der Wassergüte- und Wassermengenwirtschaft, 15. Essener Tagung, 16. Kurzbericht.

Stikute, J. A., Kuzmina, L. F., Pimenova, L. J., and Bolochovec, V. G. (1979) Effektivnost biologičeskogo okislenija proizvodstvennych stočnych vod pri raznych temperaturnych režimach v periodach aeracii. *Vodosn. i sanitarnaja technika* **11** 6–8.

Stobbe, G. (1969) Über den Schlammindex. *Städtehygiene* **20** 222.

Straub, F. B. (1972) Enzyme, Moleküle, Lebenserscheinungen. Akademische Verlagsgesellschaft Geest & Partig KG, Leipzig.

Streeter, H. W., and Phelps, E. B. *U.S. Pub. Health Bull.* **146** (1925) (cited by Fair, and Geyer, 1961).

Streichsbier, F., and Washüttl, J. (1980) Untersuchungen über die biologische Reinigung von peroxidhaltigen Industrieabwässern nach dem Belebungsverfahren. *Österr. Abwasserrundsch.* **25** 24–28.

Sudo, R., and Shuichi, A. (1973) Mass and monoxide culture of *Vorticella microstoma*, isolated from activated sludge. *Water Res.* **7** (4) 615–621.

Sutton, B. M., Murphy, K. L., and Dawson, R. N. (1974) *Continuous biological denitrification of wastewater*. Environment, Canada.

Takahashi, S., Toshihiko, F., and Saiko, T. (1968) Metabolism of suspended matter in activated sludge treatment. *Fourth Int. Conf. on Water Pollution Research*, Prague 1968, Sect. II, Paper 3.

Tischler, L. F., and Eckenfelder, W. W. (1968) Linear substrate removal in the activated sludge progress. *Fourth Int. Conf. on Water Pollution Research*, Prague 1968.

Tomlinson, T. G., and Suaddon, D. H. M. (1966) Biological oxidation of sewage by films of micro-organisms. *Int. Air. Wat. Pollut.* **10** 865–881.

Uhlmann, D. (1982) *Hydrobiologie*. 2nd edn. Fischer, Jena.

Vavilin, V. A. (1982) Obobśćennaja model' aerobnoj biologičeskoj očistki. Vodnye resursy **10** (4) 136–148.

Veits, G. (1977) Einfluss der Vorklärung auf die biologische Stufe und die Wirtschaftlichkeit von Belebungsanlagen. *Berichte aus Wassergütewirtschaft und Gesundheitsingenieurwesen der TU München* **18**.

Ventz, D. (1971) Einige Aspekte zur Abwasserbehandlung in der fischverarbeitenden Industrie. *Fortschr. Wasserchemie* **13** 81–86.

Vogel, G., and Angermann, H. (1972) *dtv.-Atlas zur Biologie*, 5th edn. Deutscher Taschenbuchverlag, München.

Voigt, M. (1955) *Rotatoria. Die Rädertiere Mitteleuropas*. Bornträger, Berlin.

Wagner, F. (1982) Ursachen, Verhinderung und Bekämpfung der Blähschlammbildung in Belebungsanlagen. *Stuttgarter Berichte zur Siedlungswasserwirtschaft* **76**.

Wanner, H. U. (1975) Mikrobielle Verunreinigungen der Luft über Belebtschlammbecken. *Zhl. Bakt. Hyg., I. Abt., Orig., B* **161** 46–53.

Weber, A. S., and Sherrard, J. H. (1980) Effects on cadmium on the completely mixed activated sludge process. *JWPCF* **52** (9) 2378–2389.

Weide, H., and Aurich, H. (1979) *Allgemeine Mikrobiologie*. Fischer, Jena.

Wilcke, D. E. (1957) Oligochaeta. In: *Die Tierwelt Mitteleuropas* 1, 7a. Leipzig.

Wildmoser, A. (1978) Abwässerreinigung mit technischem Sauerstoff. *Wasser, Luft und Betrien* **22** 26–39.

Winter, W., and Erbert, L. (1981) Anwendung von Textilhilfsmitteln unter dem Aspekt einer minimalen Abwasserbelastung. *Textiltechnik* **31** (9) 575–579.

Wolf, B. (1979) Inbetriebnahme der ersten kommunalen Kläranlage in Deutschland nach dem Unox-Verfahren. *Wasserwirtschaft* **69** (4) 133–139.

Wolf, P., and Nordmann, W. (1983) Untersuchungen zur Verbesserung der Leistung und Wirtschaftlichkeit kommunaler Abwasserreinigungsanlagen. Bayer. Landesamt für Wasserwirtschaft 1979, 3 (cited by Billmeier, 1983).

Wollenweber, H. W., and Reinking, O. A. (1935) *Die Fausarien*. P. Parey, Berlin.

Wood, D. K., and Tchobanoglous, G. (1975) Trace elements in biological waste treatment. *JWPCF* **47** 1933–1945.

Wuhrmann, K., Beust, F. v. and Ghose, T. K. (1958) Zur Theorie des Belebtschlammverfahrens. *Schweizer Z. f. Hydrol.* **20** 284–330.

Wullenweber, M., and Joret, J.-C. (1983) Zur Problematik viraler Aerosole in der Umgebung von biologischen Kläranlagen und Abwasserverragnungsgebieten. *Gesundheitsingenieur* **104** 254–258.

Zelinka, M., and Marvan, P. (1961) Zur Präzisierung der biologischen Klassifikation der Reinheit fliessender Gewässer. *Arch. f. Hydrobiol.* **57** 389–407.

Zimmermann, G., Tiedemann, E., and Ludwig, H. (1977) Zum Problem der Redoxpotentialmessung in Abwäsern und Schlämmen. *Ahh.* **5** (3) 299–307.

Zlokarnik, M. (1975) Einfluss einiger stofflicher und verfahrenstechnischer Parameter auf den Sauerstoff-Eintrag bei der-Abwasserbelüftung. *Chemie-Ing.-Techn.* **47** (7) 281–282.

Zülke, H.-J. (1967) Kontinuierliche technische Nachreinigung von Braunkohlen-Kokereiabwässern nach dem biologischen Intensiv-Verfahren. *Fortschr. Wasserchemie* **5** 172–188.

Standard	Date	Title
TGL 6466/01	Dec 1977	Soil enrichment: irrigation of cultivated land: basic requirements in respect of quality of irrigation water
TGL 7762	Jul 1979	Small-scale sewage treatment systems: application, design, installation and operation
TGL 22022	Jun 1967	Sewage disinfection; continuous three-stage thermal disinfection procedure
TGL 22213/06	Jan 1977	Agriculture and environmental protection; protection of natural waters, and countering the effects of oil spills
TGL 22433	Apl 1971	Drinking water; quality requirements
TGL 22764	Mar 1981	Use and protection of natural waters; classification of natural properties of flowing water
TGL 22767	Nov 1984	Compact activated sludge plants; application, design and construction features
TGL 24350/01	Apl 1971	Oxidation ditch sysems — general principles
TGL 26056/02	May 1972	Sewage treatment residues — agricultural and horticultural utilisation
TGL 26567/01	Jan 1974	Sewage treatment — land treatment methods, process principles
TGL 26567/03	Mar 1975	As above — plants for populations greater than 2000 PE
TGL 26730/01	Mar 1975	Sewage treatment — operation and maintenance of municipal treatment facilities; general requirements
TGL 26730/03	Mar 1975	As above — biological treatment
TGL 27796/01	Nov 1972	Residues from plant protection chemicals and growth regulators
–/32	– Feb 80	
TGL 27885/01	Apr 1982	Use and protection of natural waters; static water bodies — classification
TGL 27886/01	Dec 1974	Phosphate removal from sewage — chemical coagulation
TGL 28400	Nov 1983	Water investigations; principles of experimental methodology
TGL 28722/01	Feb 1982	Sewage treatment, naturally aerated lagoons — application and design
TGL 33408/01	Nov 1980	Concrete structures — corrosion and corrosion protection; severity of attack
TGL 36430/03	Nov 1979	Water supply; groundwater recharge — technology and design of planted beds
TGL 36763/02	Sep 1979	Sewage treatment; centrifugal aerator Type B — application, sizing
TGL 37780/01	Aug 1980	Use and protection of waters; bathing waters — hygienic requirements
TGL 38867	Dec 1981	Feedstuffs; animal protein feeds, mixed protein silage
TGL 55032/04	Jan 1984	Terminology for water management; water supply, biol. sewage treatment
WAPRO 2.19/01	Jan 1982	Sludge treatment processes; general remarks
WAPRO 2.26	Jan 1980	Sewage treatment plants — settling tanks; principles comments on dimensional design
WAPRO 7.45	Oct 1977	Instrumentation and control systems for water management; regulation and control of dissolved oxygen — time-based control systems

Plates

Plate 1 — Compact activated sludge flocs of heterogeneous composition with pronounced development of *Sphaerotilus natans* as a consequence of inputs of sugar-refining liquors. Municipal sewage. Magnification 200×.

Plates

Plate 2 — Activated sludge flocs with threadlike fungal mycelia and yeast-like cells in the interior of the flocs. Effluent from the wood pulp industry. Magnification 300×.

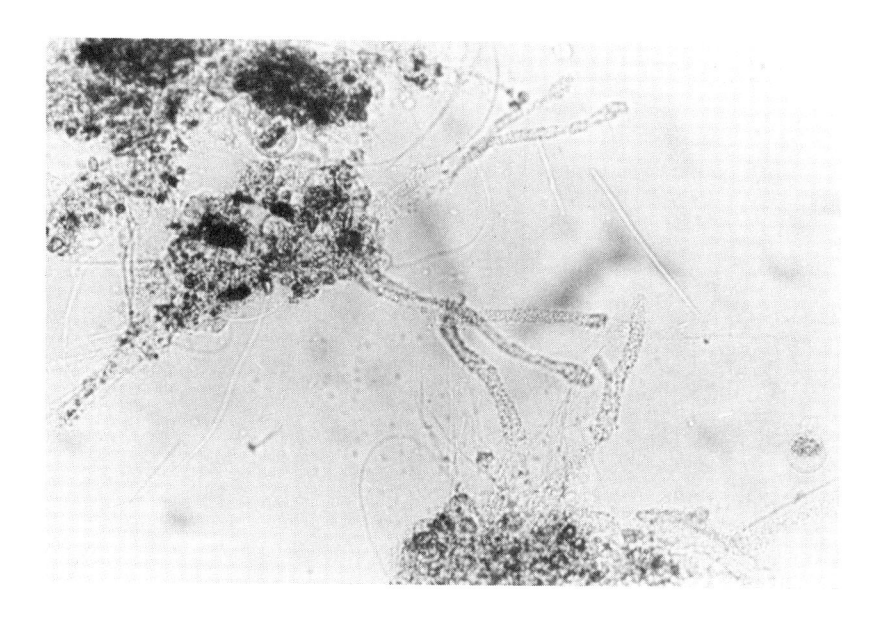

Plate 3 — Activated sludge flocs with pronounced development of *Zooglea ramigera*, *Thiothrix nivea* and *Haliscomenobacter hydrosses*, together with two specimens of *Chamydophrys stercocea*. Effluent from the textile industry. Magnification 300×.

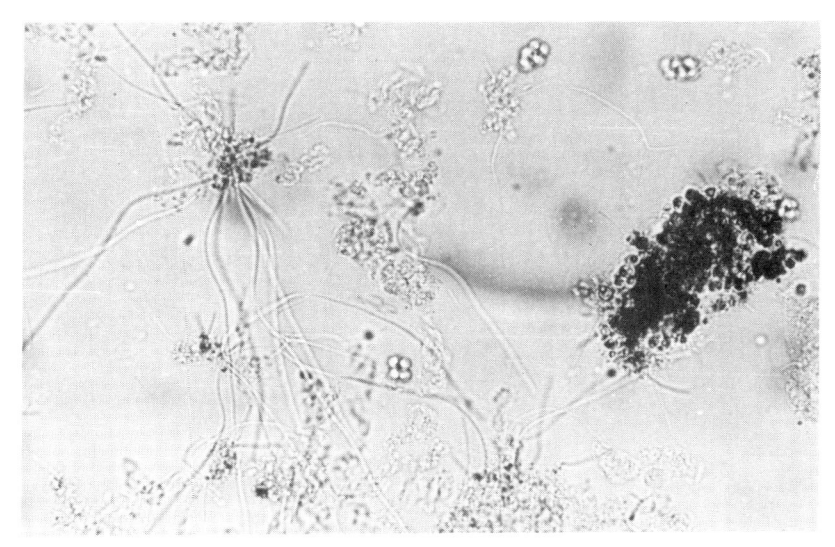

Plate 4 — Activated sludge flocs with pronounced development of *Thiothrix nivea* as a consequence of putrefaction of the biomass caused by discontinuous and hence incomplete delivery of recycled sludge.

Sulphur granules are no longer detectable as the sulphur in the aeration tank was already oxidised to sulphuric acid, while in the compact flocs to the right deposits of black iron sulphide are visible in the floc interior. Oxidation of this FeS is inhibited as a result of the screening of the oxygen supply by the bacteria and colonies of *Diplophrys archeri* around the margin. Effluent from animal by-product plant. Magnification 300×.

Plate 5 — Respirometer for physiological studies. Principal components: thermostat enclosure with input and output connections, motor and gear box for stirring gear, thermometer, dosing points for substrate addition or input of effluent or toxic constitutents; aeration equipment; O_2-recorder and sensing apparatus. Prototype devised by Birr and Mirugowski at the Water Engineering Research Centre (1981).

Plate 6 — Mobile experimental plant belonging to the Water Engineering Research Centre Technical components: primary settling compartment; dosing controls for sewage flowrate with variable speed motors; two aeration chambers with infinitely variable speed drive for B 200 aerators; hose connections for input of a prospective industrial effluent; O_2-electrodes from the Mv. Ardenne Research Institute; two Dortmund tanks as final settling compartments. From Hänel and Schmidt (1975).

Plate 7 — Aeration tanks with compressed air aeration facilities. Photograph taken from the outlet end; air input lines to the right; on the left stepped sewage feed arrangements in the front, O_2-electrode from Meinsberg Research Institute at the rear.

Plate 8 — Aeration tanks with battered sides and centrifugal aerators of the BK B 2400 type according to TGL 36763 for treatment of warm wastewater from an industrial fermentation plant. Severe aerosol formation in winter when the difference between ambient and effluent temperature exceeds 15°C. Exceptionally pronounced foam formation due to fungal growths of *Nocardia* spp. As a result of coalescence inhibition in the mixed liquor, $OC/P = 2.6$ kg/kWh (see Table 5.3).

Plate 9 — Aeration tanks of enclosed reactor design with submerged jet aeration. Reaction tank equipped with four immersed jets and four circulating pumps. Picture — Weber.

Plate 10 — Small-scale activated sludge plant ref. TGL 22767. Aeration tank with compressed air injection, final settling compartment (showing scum due to denitrification) sludge recycle and wastage connections. At bottom right: beakers for collecting floating sludge due to denitrification.

Plates 287

Plate 11 — Oxidation ditch with cage rotor. The rotor is covered by a hood as a means of enhancing the oxygen input and reducing aerosol emissions. The greenery to the left of the picture helps to minimise aerosol dispersion in the atmosphere.

Plate 12 — Compact tanks. Typical solution to the problem of increasing the capacity of a plant for primary treatment by the construction of a heavily loaded activated sludge system in two existing settling tanks [31].
Left: primary settling tank showing severe scum formation; activated sludge tanks with $3 \times BK$ B 900 aerators.
Right: final settling tank with suction scraper.

Plate 13 — Showing the effect of a submerged baffle on the quality of the treated effluent from a final settling tank fitted with a mechanical scraper.
For the effluent discharging on the right hand side of the picture the outlet weir assumes the function of the submerged baffle. Settleable solids: 14 mg/l with baffle, 74 mg/l without baffle.

Plate 14 — Surface-attached growth in a ponded outlet duct beyond the final settling tank. Bright zone to upper right of picture — duct wall; black zone — growth of blue algae; cloud-like zone — whitish surface colonies of the bacteria-consuming bell-shaped protozoan *Carchesium polypinum*.

Plate 15 — Trickling filter plant.
Left: Section of the trickling filter surface with rotary spreaders; surface wetted and with the foamed polystyrene packing covered with a biological film. Black surface growth on the internal wall surface of the filter composed of blue algae.
Right: general view of filter showing rotating spreader arms, outer casing containing air inlet apertures and outlet duct.

Plate 16 — Sewage lagoon treatment system ref TGL 28722.
Right hand smaller lagoon: anaerobic primary tank; when used in series following an activated sludge plant it functions as a sedimentation chamber.
Middle lagoon: facultative aerobic tank; algal treatment stage; when used following an activated sludge plant acts as an aerobic lagoon with algal and/or filtering organisms.
Rear lagoon: aerobic tank with filter-feeding zooplankton.

Index

absorption, 183
abundance rule (ecology), 76
acclimatisation (adaptation), 16, 42, 130, 149
   induced adaptation, 37, 101
   to salt, 85, 165
Acetobacter spp., 113
acid-forming bacteria (acetogenic), 140
acidic conditions
   acidity measurement (*see also* pH), 30–32
   organisms found, 78, 84, 170
   pulp mill effluent, 89, 90, 91
Acinetobacter spp., 23, 83, 136, 151, 169
Actinomycetales (bacteria), 59
Actinomycetes (bacteria), 62, 165
Action Law (ecology), 76, 85
activated sludge (*see also* biomass; BOD_5 sludge loading; bulking sludge; flocs; sludge age; waste)
   carry over from final settling tank, 252, 254
   characterisation of properties, 24–27
   deposit putrefaction, 259
   flotation, 55, 255
   heavy metal ion contamination, 257
   regeneration, *see* recyled sludge
   respiration (OV), 35–38
   sludge level, 27
   sludge solids content (TS), 15, 25–26, 29, 251–252
   sludge volume (SV), 15, 24–25
   sludge volume index (SVI), 15, 26–27, 89, 250
activated sludge plants (*see also* activated sludge process; reactor types)
   anaerobic, 210–212
   compact or small, 213, 214, 245, 286
   dimensional design, 239–242
   operation of, *see* plant operation
   on ships, 85
   two-stage, 203, 206, 263
   typical layout, 18
activated sludge process (*see also* activated sludge plants; final settling), 13, 17–20
   advantages and disadvantages, 44–45
   future prospects, 263
   history, 21–23

   standards (TGL and WAPRO), 280
   terms and symbols used, 13–17
   variations (*see also under* aeration systems), 80, 83, 200–214
active transport through cell wall, 96
acyclic compounds, 134
adaptation, *see* acclimatisation
adenosine triphosphate, *see* ATP
adenoviruses, 55, 174
adsorption, 43, 109, 112, 183
   adsorption isotherms, 153
   special variant processes, 208–209
aeration systems (*see also* aeration tanks; centrifuge systems; compressed air; oxygen input)
   as cause of pollution, 195–199
   equipment failure, 257–258
   future prospects, 264
   gases given off, 183
   history, 22, 23
   illustrations, 285–287
   reaeration of recycled sludge, 14, 91, 205–206
   variations, 22, 189–192, 203, 212–214
   submerged jet aerators, 192, 260, 286
aeration tanks (*see also* aeration systems), 14, 23, 210, 285–287
   alternate operation of two, 207
   anaerobic metabolism, 138
   covered, 198
   Crabtree effect, 127
   enzyme activity, 101
   foaming, 157, 259
   living communities, 77–91
   oxygen deficiency, 78, 258–259
Aerobacter spp. 140, 175
aerobic bacteria, 54, 123
aerobic processes, 16
   carbon and hydrogen metabolism, 134, 137
   stabilisation of surplus sludge, 245–246
Aeromonas punctata, 176
aerosols, 192, 196, 197–199
agricultural effluent, 177, 238, 262, 264
Alcaligenes spp., 116–117
alcohol-forming bacteria, 140

Index

alcohol industry effluent, 89, 90, 184, 208
alcohols, 138
aldehydes, 134
algae, 57, 58, 62, 74, 76, 92
aliphatic compounds, 134
alkaline disinfectants, 167
alkalinity (*see also* pH), 30, 31, 32, 210
alkylbenzenesulphonate, 137
allergens, 153
allyl thiourea, 33, 35
amoebae, 63, 175
ammonia
   oxidation, 29, 30, 34–45
   as atoxin, 229
   as volatile gas, 183
ammonification, 141–142
amoebiasis, 177
anabolic processes, 102
anaerobic bacteria, 54, 123, 140
   facultative, 123, 140, 210
   methanogenic, 137, 140, 257
anaerobic processes, 16
   anaerobic sludge plants, 210–212, 264
   detoxification, special methods, 209
   digestion of waste sludge, 244–245
   fermentation, 16, 139–140, 210
   final settling tank, 223
   floc interior, 116
   metabolism, of carbon and hydrogen, 137–140
Ancylostoma duodenal, 75
animal diseases, 174, 175, 176, 177, 178
animal feed, waste sludge as, 244
anoxia, 16, 79–80, 138
Anthophysa spp., 83, 84, 85
anthrax, 176, 177
anthrobacter (bacteria), 136
antibiotics, 97, 98
anticoagulants, 98
antisaprobic state, 50, 77
apoenzymes, 99, 100
aromatic hydrocarbons, polycyclic, 166–167
Arrhenius Rule, 84, 125, 128–129, 150
arsenate, 127
Ascaris spp., 75, 178, 179
Aspidisca spp., 79–80, 83
atomic absorption spectroscopy, 33
ATP method of biomass measurement, 42–43
ATP production, *see* Krebs cycle
autoecology, 76
automated operation, 261
aztobacter, 113, 135, 146, 226

Bacillus spp., 136, 146, 171, 176, 177
bacteria (*see also* anaerobic bacteria; solitary bacteria)
   in aerosols, 197–198
   bacteriological investigations, 52–55
   cell components, 93–97
   communities in municipal plants, 76
   counting, 53–54
   culturing, 169, 264
   flocs, 78–87, 98–99, 116

   growth, 118–121, 209–210
   help to eliminate pathogens, 175, 177
   metabolism (*see also* anaerobic bacteria, denitrification; nitrification), 115, 117, 135, 136, 168–169, 195
   numbers in sludge and sewage, 113
   pathogenic, 171, 175–177
   pleomorphy, 59
   polysaprobic, 79
   salinity adaptation, 85
   staining methods, 48
   taxonomic identification, 56, 57–60
   temperature tolerance, 129–130
   thread-forming, 56, 58, 59–60, 61, 62, 87, 207, 254
bactericides, 163
bacteriophages, 57, 175
Balantidium coli, 177
Balantiosphorus minutes, 171
basal respiration, *see* endogenous respiration
beef tapeworm, 75
Beggiatoa (bacteria), 58, 62, 135, 151
benzenes, 157, 184, 196
biochemical methods, 33–44
biochemical oxygen demand (BOD) (*see also* BOD_5 sludge loading; respiration), 33–35
   high for industrial effluent, 191–192
   high for regeneration systems, 205
   as toxin indicator, 39
biocide elimination, 208
biological methods, 13, 44–55, 250
   history of, 20–23
   terminology, 16
biomass
   frequency of analysis, 250
   growth, 43, 118–121
   measurement, 42–43
   sludge colour, 249–250
biosorption, 183
biosurfactants, 137
Boda caudatus (bacteria), 76, 79, 81, 83
   found during start-up, 87, 88
   taxonomy, 64
BOD_5 sludge loading (*see also* biochemical oxygen demand), 14, 27, 33, 35
   anaerobic metabolism, 138
   determination by ciliates, 52, 53, 81
   diurnal fluctuations, 11, 12, 192, 224, 250–251
   living communities, 80, 81
   municipal effluent, 19
   reactor type, effects of, 131–133, 201
   substrate–enzyme ratio, 102–110
   temperature effects, 131–133
   toxins, effects of, 128
   treatment performance, 250–251
BOD–Warburg Methods for toxin analysis, 39–40
Brachionus calyciflorus, 83, 171
brewery effluent, 89
Brucella spp., 176
bulking sludge
   analysis, 87, 250
   causes, 44, 60, 79, 80, 87–91, 98, 170, 226, 260

292 Index

elimination methods, 260–261
prevention, 91, 182, 205, 206, 207, 208, 211, 226, 254
research required, 263

CH_4 (methane), 138, 140, 183, 211
cadmium, 152, 153, 154, 156
Campylobacter fetus, 176
Candida, 66
carbocyclic compounds, 134
carbohydrates, 16, 78, 89
carbon, activated, 183
carbon and hydrogen metabolism (*see also* Krebs cycle), 33–34, 134, 1437–140
carbon, organic, 28–29, 184
carbon tetrachloride, 159
Carchesium spp., 77, 80, 83, 288
carcinogens, 152, 153, 158
chlorine as, 236
polycyclic aromatic hydrocarbons, 166–167
carrousel systems, 201, 205, 214
cascade systems (*see also* multistage systems), 77, 205
catabolic processes, definition, 102
caustic solutions, 167
cell (bacterial) structure and function, 93–97
centrifuge systems, 22, 191, 192, 285, 287, 303–305
biochemical oxygen demand, 35
worm egg separation, 55
Cephalosporium spp. (Fungi), 60
Chamydophyrs stercocea, 282
chemical factory effluent, 85, 131, 133, 184
chemical oxygen demand (COD), 28, 128, 251
chemolithotrophic bacteria, 168
chemoorganotrophic bacteria, 168
Chilodonella (rotifer), 84
Chironomid larvae, 74, 83
chlorinated hydrocarbons, 137, 167
chlorine
as distinfectant, 167, 236, 237
impairs thread-forming, 260–261
chloroform, 159
cholera, 176
chromates, detoxification of, 208
chromatography, thin-layer, 33
chromium, 153, 156
cilia, 59
Ciliata, 67
ciliates
BOD_5 estimates, 52, 53, 81
effect on treated effluent quality, 174
food chain, 171, 172
found during start-up, 87
found when oxygen deficiency, 79
indicators of saprobic index, 52
in intake channel, 76
Cinetochilum margazritaceum, 84
citric acid cycle, *see* Krebs cycle
Cladosporium cladosporioides, 60
clarification process, *see* final settling
cleaning operations, 252

Clostridium spp., 54, 136, 197
coagulation processes
to eliminate viruses, 175
to increase oxygen supply, 258–259
phosphates, 90, 151, 182, 183, 233, 234, 235–236
simultaneous coagulation, 90, 91, 151, 183
special processes, 208
weighting thread-like organisms, 260
coal industry effluent, 85, 91, 166
cocci bacteria, 57, 58
COD (chemical oxygen demand), 28, 128, 251
coenzymes, 99, 100
Coleps hirtus, 83
coliform bacteria, 54, 197
colonisation of sludge flocs, 59, 78–87
counting colonies, 54
colouring agents as pollutants, 166
colour of sludge and biomass quality, 249–250
Colpidium campylum, 79
co-metabolism, 137
compact plants, *see* small plants
competitive inhibition, 127–128
compressed air systems, 22, 190–191, 195
illustration, 285
sludge deposits, 260
tank configuration, 23
computers, 261
concrete corrosion by sulphate, 212
continuous-feed test systems, 39
corrosion of tanks, 212
costs (*see also* power consumption)
of agricultural utilisation, 262
future prospects, 264
of increasing oxygen, 258–259
of trickling filters, 263
counting organisms, 46, 47, 53–54, 55
Coxsackie viruses, 174
Crabtree effect, 127
cryptosaprobic state, 50, 77
culturing, 55, 169, 264
cyanide, 127, 155, 165, 208
Cyclidium, spp., 83, 85
cysteine-casein nutrients, 54

dairy industry effluent, 89, 90, 208
decomposition rate, 103
decoupling agents, 127, 191
deep shaft tanks, 214
deflocculent growth (*see also* solitary bacteria), 192, 209–210, 254
deflocculation due to temperature, 132
dehydrogenase activity, measurement, 43
Dendromonas, turbulence inhibits, 83
denitrification, 16, 25, 29, 30, 141, 145–148, 230–233
advised when low alkalinity, 182
cause of flotation, 255
denitrification rate, 147, 148
determination by saprobic state, 79–80
in final settling tank, 223
interferes with BOD determination, 35

Index

monitoring and control, 29, 30
occurring when oxygen deficiency, 138
dinitrobacillus as denitrifiers, 146
deoxygination methods, 187
deoxyribonucleic acid, *see* DNA
Desulfovibrio spp. (bacteria), 136, 146
detergents, 98, 163–164, 192
detoxification (*see also* toxins), 145–155
 in anaerobic plants, 211, 212
 cyanide, 165, 208
 formaldehyde, 137, 208
 in primary settling tank, 253
 special processes, 33, 207–210, 227–238
 waterborne substances, 152–168
diarrhoeal diseases, 55, 174, 176
diauxy phenomenon, 16, 142, 205
dichloromethane, 159
Diffusion, Ficks Law of, 185
diffusion through cell wall, 96
digester liquor, 253–254, 260
dilution methods, 25, 35, 54, 204
2,4 dinitrophenol, 127
Diplogaster (nematode), 72, 74
Diplophrys archeri, 283
diseases, *see* pathogens; infectious diseases
disinfectants as toxins, 163, 167–168
disinfection processes, 236–237, 246
dissolved organic carbon (DOC), 29
distillery effluent, 89
Distomatinae, 63, 76
diurnal BOD_5 fluctuations, 11, 12, 192, 224, 250–251
DNA method of biomass measurement, 42–43
DOC (dissolved organic carbon), 29
Dortmund tanks, 178
drinking water, 229, 238
dyes as pollutants, 166
dysentery, 176, 177

echoviruses, 174
ecology laws, 76
ectoenzymes, 99
effluent (*see also* agricultural; industrial; municipal; septic)
 bacteria numbers in raw sewage, 113
 chemical compounds in, 133
 definition of sewage, 11
 final effluent quality, 264
 fluctuations in composition, *see* diurnal
 health risk from sewage (*see also* pathogens; toxins), 52
 organisms in raw sewage, 76
electrochemical methods, 27, 30
electron transport inhibition, 127
elimination rate, 103
Encentrum (rotifer) and temperature, 84
encephalitis, 174
Enchelymorpha vermicularis (ciliate), 76
endogenous respiration (OV_{end}), 37–38, 124
 measurement of oxygen input, 188
 oxygen demand, 34, 205
 surplus sludge stabilisation, 245

Endomyces lactis (fungi), 60, 61, 84, 89
energy production, 123, 124
 anaerobic fermentation, 139–140, 210
Entamoeba hystolytica, 177
enteritis, 176
Enterobacter (bacteria), metabolism, 135
Enterobius spp. (threadworm), 72, 74, 75, 179
Enterococci (bacteria), 54
enteroviruses, 55, 174
environmental pollution from aeration systems, 195–199
enzymes (*see also* substrate), 16, 99
 control systems, 101–102, 125
 conversion rates, 106–108
 denitrification, 147
 effect of toxins, 127–128
 extracellular, 99, 137
 hydrolytic, 97, 102, 134
 kinetics, 104–112
 phosphate cycle, 100, 149
 respiratory, 123, 125
 temperature effects, 131, 132
Epistylis, 80
Equilibrium Law (ecology), 76, 84–85
equipment failure, 257–258
Erysipelothrix insidiosa, 176
Escherichia spp., 135
 E. coli, 54, 96, 97, 129, 146, 176
Euplotes spp., 80, 85
eutrophication, 229
evaporation residue, 26
exoenzymes, 99, 137
eye infections (viral), 174

facultative anaerobic organisms, 125, 126–127
 bacteria, 123, 140, 210
faeces, organisms in, 55, 174, 176
fats and oils, animal (*see also* lipids), 32–33, 137
Fedapon washing process, 164
fermentation, anaerobic, 16, 139–140, 210
Fick's Law of Diffusion, 185
filtration
 filters, 21, 262, 263, 289, 338
 filtrable solids, 15, 25–26, 29, 251–252
 floc-filtration, 238
 influence on BOD, 35
final settling process (*see also* final settling tanks; hindered settling), 78, 215–223
 elimination of viruses, 175
 low temperatures, 132
 settling time, 218
 to eliminate worm eggs, 55
final settling tanks (*see also* final settling process; primary settling), 14, 218–221, 222
 biomass growth, 119
 dimensional design, 24
 organisms in outlet, 91–92
 sludge carry over from, 254, 255
fish industry effluent, 85, 165
fixation, 45
fixatives as toxins, 159
Flagellata (protozoa), 63, 64, 65

flagellates, 79, 81, 172
  found during start-up, 75, 87
  found in intake channel, 76
Flavobacterium, specific metabolism, 135
Flexibacter spp., 62
flocs
  colonisation of, 78–87, 98–99, 116
  crucial importance of, 119
  eliminate pathogens, 175, 177
  floc-filtration, 238
  floc geometry, 44, 77–78
  illustrations, 281, 282, 283
  lag-phase, 119
  metabolism, 113–118, 195
  pinhead flocs, 254
flotation, 55, 255
foaming, 157, 192, 256, 259
food chains, 80, 84–85, 86, 170–171, 172, 175
food industry effluent, 89
foot-and-mouth virus, 174, 175
formaldehyde elimination, 208
formalin disinfectants, 168
Francisella tularensis, 176
Franz pattern centrifugal aerators, 192
freezing problems, 213, 249, 263
fungi, 55, 60, 61, 62
  floc formation, 98
  growth in acidic medium, 84
  metabolic roles, 169–170
  in municipal plants, 76
  stabilise foam, 259
  thread-forming, 60, 62, 87, 170
fungicides, 163
Fusarium spp. (fungi), 60, 61
future prospects of sewage treatment, 262–264

gas-chromatography, 33
gases, volatile, 183–184, 192, 196
gas gangrene, 54
gastroenteritis, 55, 174, 176
generation times of organisms, 83
Giardia lamblia, 177
Glaucoma spp., 68, 76, 79, 83, 87
glucose and the Crabtree effect, 127
gram-negative and gram-positive bacteria, 94, 95, 96, 97
growth
  biomass, 43, 118–121
  fungi in acidic medium, 84
Gymnamoebae, 63, 66

H_2O_2, *see* hydrogen peroxide
H_2S, *see* hydrogen sulphide
Habrotrocha spp. and temperature, 84
Haliscomenobacter spp., 62, 90, 282
halogen compounds, 137, 158–159
Hartmaniella (rhizopods), 76, 79
health risks (*see also* pathogens; toxins), 52, 258
heat balance calculations, 133
heat resistance of organisms, 84
heavy metals, 152, 153, 154, 155, 156, 257
  adsorption, 208

chemical phosphate coagulation, 151
Heliozoa, 66, 67
Helkesimastix faecicola, 79, 81, 87
Helminth ova (worm eggs), 55, 72, 74, 75, 178–179
hepatitis, 174
herbicides, 33, 85
heterocyclic compounds, 134
hindered settling, 25, 110
history of biological treatment, 20–23
Holotricha, 67, 68, 69
hookworm, 75
horizontal rotors, 191, 287
human stomach worm, 75
hydrocarbons, *see* mineral oils
hydrogen and carbon metabolism, 33–34, 134, 137–140
hydrogen peroxide (H_2O_2), 236–237, 258, 260–261
hydrogen sulphide (H_2S), 27, 28
  formation, 138, 151–152, 211
  health risk if plant shut down, 258
  Kraus process, 208
hydrolases, 97, 102, 134
hydrophobic path (cell transport), 97
hygiene, personal, 173, 250
Hymenolepsis nana (tapeworm), 179
hypersaprobic state, 50, 91

ice, problems of, 213, 249, 263
immobilisation techniques, 23, 264
industrial effluent (*see also* alcohol industry;
  chemical factory; coal; fish; metal-working;
  petroleum; textile; pulp), 13, 224–226
  aerations sytems, 191
  anaerobic sludge, 211, 212, 264
  bulking sludge, 89, 90, 91, 226
  chemical oxygen demand, 28
  dimensions of sewage plant, 240
  diurnal BOD_5 fluctuations, 11, 224
  enzymes inhibited by toxins, 127–128
  filtrable solids, 251–252
  foaming, 259
  Magdeburg-P process, 23, 225
  odour reduction, 196–197
  organisms in, 77, 84, 85, 87, 169
  residence times, 19
  running-in phase, 248, 249
  salinity, 165
  sludge biomass growth, 120, 121
  special detoxification methods, 208
  temperature of, 84, 130, 131, 226
infectious diseases (*see also* pathogens), 173–179
  in aerosols, 197–199
influenza viruses, 175
Infusoria, 67
inhibitors, reversible and irreversible, 127–128
insects, 72, 74, 83, 92
intake channel, organisms in, 76–77
intracellular enzymes, 99
iodates, detoxification of, 209
ion exchange, 154, 183

ionising radiation techniques, 236, 237
iron deficiency and respiration, 123
iron salts, 85, 151
iron sulphide precipitation, 79, 182
irradiation of pathogens, 236, 237
irreversible inhibitors, 127, 128
irritants, 153
isomerases, 102
isosaprobic state, 50

ketones, 134
Kraus process, 207–208, 253
Krebs (citric acid) cycle, 121, 122, 123, 124, 127, 134
　decoupling agents, 127, 191

lag-phase, 119
lead as toxin, 153
Leptomitus lacteus (fungi), 84
Leptospira interrogans, 176
leucothrix spp. (bacteria), 62, 135
ligases, 102
Lionotus lamella, 68, 80
lipids (*see also* fats), 16, 32, 33
lipophilic substances, 32
liquid composition of sewage and sludge, 27–33
Listeria spp., 176
luxury uptake of phosphorus, 149, 201, 263
lyases, 102
Lyngbya (blue algae), 76

Mg ions and enzyme activity, 100
macrofauna identification, 46
Magdeburg-P process, 23, 225
maggot worms, 74
magnesium ions and enzyme activity, 100
manometric method of BOD analysis, 35
mass spectroscopy, 153
maw-worms, 74
mechanical failures, 257–258
meningitis, 174
mercury as toxin, 153, 156
mesophilic bacteria and temperature, 130
mesosaprobic state, 85
　α-mesosaprobic, 48, 50, 56, 77, 80, 87
　β-mesosaprobic, 48, 50, 56, 79
　visual appearance, 92
metabolic processes, 16
　effect of toxins, 85, 100, 127–128, 143, 144, 157
　fundamentals, 102–133
　materials conversion, 133–173
　specific functions of bacterial families, 135–136
metal ions, 100, 182–183
metallic activators, 99
metal salts, 90, 98–99, 100, 155
metals, heavy, *see* heavy metals
metal-working industry effluent, 91
metasaprobic state, 50
metazoa, 70–74
　food chain, 80, 172
　metabolic role, 170–173
methane (CH$_4$), 138, 140, 183, 211

methanemoglobinaemia, 229
Methanobacterium, 136
methanogenic bacteria, 137, 140, 257
methylene chloride, 159
methyl parathion, 159
microaerophilic bacteria, 123
Micrococcus spp. (bacteria), 58, 136, 146
microflora and fauna identification, 46
microprocessors, 261
Microscilla, 62
microscopic analysis, 44, 45, 53, 256
Microthrix parvicella, 62, 90, 91
mineralisation, 50
mineral oils, 155, 161–163, 184, 257
　polycyclic aromatic hydrocarbons (*see also* coal; petroleum), 166–167
mixed reactor types, 23, 80, 203, 204
mixotrophic bacteria, 168
Mohlmann index (sludge volume index SVI), 15, 26–27, 89, 250
Monas spp., 79, 81, 83, 87, 171
mucin formation, 157
Mucor spinosus (fungi), 60
multistage systems (*see also* cascade systems), 263
municipal effluent, 11–13, 19, 28
　bulking sludge, 90
　carcinogens, 166
　dimensional design, 239
　hygienic quality criteria, 241
　odour reduction, 196
　organisms found, 76, 84, 120, 121, 173, 177
　plant running-in phase, 248
　primary settling tanks, 253
　salinity, 165
　sorption processes, 183
　stormwater flows, 252
　temperature, 84, 130, 131
mutagens, 153, 236
mycelium-forming bacteria, 59
Mycobacterium spp., 55, 176, 177
Mycophyta, *see* fungi
myocarditus, 174

NO$_3$ (nitrate), 29, 30, 34–35
Navicula cryptocephala, 92
nematodes, 72, 74, 172
nickel as toxin, 153, 156
nitrate (NO$_3$), 29, 30, 34–35
nitrate-reducing bacteria, *see* denitrification
nitrification, 16, 29, 136, 141, 142–145, 194
　in carrousel systems, 201
　effects of high temperature, 84
　toxic effects, 128
nitrite, 29, 34, 121, 128
Nitrobacter, 136, 143
nitrogen gas, 183, 187
nitrogen compounds, harmful effects of, 229
nitrogen metabolism (*see also* ammonification; denitrification; nitrification)
　nitrogen fixation, 135, 226
Nitrosolobus (bacteria), 143

Index

Nitrosomonas, 83, 136, 143, 145
Nitrosospira (bacteria), 143
Nitzchia palae, 92
Nocardia, 23, 58, 62, 136
noise reduction, 192, 195–196
non-competitive inhibition, 128
Nostocoidea limicola, 62
number determination of organisms, 46, 47, 53–54, 55
nutrients (*see also* BOD$_5$ sludge loading)
  deficiency in industrial effluent, 23, 225
  removal, 228–236, 240, 263

O$_2$, *see* oxygen
ocean-going vessels, activated sludge plants on, 85
odour a problem of anaerobic plants, 211
Oicomonas mutabilis, 76
oil-refinery effluent, volatile gases, 184
oils, *see* fats and oils; mineral oils
Oligochaeta (worms), 72, 74
oligosaprobic state, 49, 50, 56
Oospora fragans (fungi), 84
Opercularia, 83
optic fibres and sludge level, 27
organic acids, 137, 183, 208
organic carbon, 28–29, 184
organic chlorine, 159
organic compounds
  cause bulking sludge, 90
  extended treatment of contaminants, 227
  non-biodegradable, 225, 263
  organic halogen compounds, 158–159
  organic peracids as disinfectants, 236–237
  organic polymers, 99
  principal groups, 134
  organic solvents, extraction with, 33
organisms, taxonomic identification, 56–76
organochlorine compounds, 33, 158–159
organohalogen compounds, 158–159
Oscillatoria spp., 92
osmogens (volatile gases), 183–184, 192, 196
outlet weir rating, 218
oxidation ditches, 205, 213–214, 287
  stabilisation of surplus sludge, 245
  temperature, 132, 213
oxidoreductase, 102
oxygen (*see also* biochemical oxygen demand; oxygen deficiency; oxygen input; oxygen method; respiration)
  electrochemical measurements, 27
  prevents odour, 196
  as volatile gas, 183
oxygen deficiency (*see also* anaerobic; anoxic), 77, 78, 79, 138–139, 258–259
  anaerobic count as unreliable indicator, 54
  correction by coagulation, 258–259
  high temperature as cause, 84, 187
  Pasteur effect, 125–127
oxygen input (*see also* aeration systems, 15
  controversy (aeration tanks), 78–79
  measurement, 187–189

  overriding importance of, 185
  pure oxygen, 184, 194–195
  regulation and control, 187, 192–194, 258–259
oxygen method of BOD analysis, 35
Oxystricha (ciliate), 84
ozone as disinfectant, 236–237

paper and pulp industry effluent, 89, 90, 91, 96, 282
paracetic acid, 236–237
Paramecium caudatum, 68, 85, 171
parasitic ova, *see* Helminth ova
parvoviruses, 174
passive transport through cell wall, 96
Pasteur effect, 125–127
pathogens (*see also* Helminth ova), 153
  in aerosol particles, 197–199
  bacteria, 52, 54–55, 57, 175–177
  disinfection, 236–237
  protozoa, 177
  viruses, 55, 173–175
Pelodictyon luteolum (bacteria), 117
Penicillium spp., 60
n-pentane, 184
peptide, definition, 16
perchlorates, 208
Peritricha, 70, 71
peroxide in disinfectants, 167
personal hygiene, 173, 250
pesticides, 152, 159, 160
petroleum, 85, 161, 162, 166, 184, 208
pH (*see also* acidic conditions; alkalinity), 30–32, 210
  effects floc formation, 98, 99
  limits tolerated, 30, 157–158
  lowered by nitrification, 143
  measurement, 30–32
  organisms found when low, 78, 84, 170
  regulation by biomass metabolism, 99, 179–182
phenolic disinfectants, 168
phenols as toxins, 155, 157
Phormidium (blue algae), 92
phosphates, 148–151
  coagulation of, 90, 151, 182, 183, 234, 235–236
  deficiency and respiration, 123
  enzymatic transfer, 100, 149
  metal–phosphate salts and bulking sludge, 90
phosphorus
  effect on bulking sludge, 90
  deficiency, 78
  liberation in final settling tank, 223
  phosphorus–nitrogen hypothesis, 90
  removal (*see also* coagulation), 23, 169, 233–236
psychrophilic bacteria, 129
pinhead flocs, 254
pinworm, *see* threadworm
pitworm (hookworm), 75
plant operation (*see also* activated sludge plants)
  automation, 261
  normal phase, 249–254
  operating costs, *see* costs

Index

operating instructions, 249
operational disturbances, 213, 254–261
running-in phase, 85, 87, 88, 163, 247–249
plant seed destruction, 246
pleomorphy, 59
plug-flow reactors
  bulking sludge, 89
  control methods, 200, 201, 202
  enzyme-catalysis, 111
  living communities, 77, 80, 83
  sorption processes, 183
Podiphyra spp., 84, 85
polarography, 27
polioviruses, 174, 175
polishing lagoons, 238
pollution, *see* toxins
polyauxy, 16, 107, 205
polycyclic aromatic hydrocarbons, 166–167
polysaprobic state, 48, 50, 56, 77, 85
  effect of oxygen deficiency, 79
  high temperatures, 84
  protozoa, 170
  start-up phase, 87
pork tapeworm, 75, 178
power consumption (*see also* costs)
  sludge stabilisation, 245, 246
  reduction of, 231
  submerged jet aerators, 192
power cuts, 257
precipitation reactions, 79, 182–183
predators, 84–85, 86, 170–171, 175
pressure fluctuations, effect on organisms, 84
primary settling tanks, 252–254
proteins, 16, 42, 134, 141
protozoa, 50, 63–70
  eliminate pathogens, 175–176
  feed on bacteria, 80
  metabolic roles, 170–173
  numbers in intake channel, 76
  salinity adaptation, 85
  survive oxygen deficiency, 79
Pseudomonas, spp., 23, 58, 83, 113, 146
  effects of oxygen deficiency, 79
  pathogens, 175, 176
  Pseudomona aeruginosa, 54, 96, 146
  P. proteus, 175
  P. stutzeri, 146
  specific metabolism, 135
pulp and paper industry, *see* pulp
pumps, failure of sludge recycle, 257–258
putrefaction, 43, 223, 259–260

radioactive substances, 50, 168
radioactive treatment of pathogens, 236, 237
radiosaprobic state, 50
reactor types (*see also* activated sludge plants; aeration tanks; final settling tanks; mixed reactors; multistage; plugflow; stirred tank), 200–214
  corrosion of tanks, 212
  Dortmund tanks, 178
  effect on BOD$_5$, 131–133, 201

recycled sludge, 200, 201, 221–222
  flowrate, 14, 218
  inlet point, 207
  oxygen deficiency, 78
  pump failure, 257–258
  reaeration, 14, 27, 91, 205–206
redox potential, (rH or Eh), 43–44, 138
regeneration, *see* recycled sludge
reoviruses, 55, 174
respiration (*see also* endogenous respiration), 35–38
  fundamentals, 121–127
  maximum respiration rate (OV$_{max}$), 37, 38, 115
  respiratory enzymes, 123, 125
  respirometric studies, 113–118, 283
respiratory illness (viral), 174
reversible inhibitors, 127–128
Rhabditis (nematode), 74
rhizopods, 56, 63, 66, 67
  food chain, 172
  found when oxygen deficiency, 79
  as function of BOD$_5$ loading, 81
  in municipal plants, 76
Rhodospirillum (bacteria), 135
Rhodotorula, 60
RNA (ribonucleic acid) method of biomass measurement, 42–43
rods (bacteria category), 58, 59
Rotatoria (metazoa), 70, 73
rotaviruses, 173, 174
rotifers, 70, 73, 79, 81, 84
running-in phase, 247–249
  detergents, 163
  living communities, 85, 87, 88

Saccharomyces (fungi), 60
salinity, *see* salt concentration
Salmonella spp., 55, 135, 175, 176, 177
  in aerosols, 197, 198
salt concentration, 85, 164–165
sampling, 45, 52
Sanitary Inspectorate, 257
saprophytes (fungi), 169–170
saprophytic bacteria counts, 54
saprobic system (*see also* mesosaprobic; polysaprobic), 44, 48–52
  in aeration tank, 77, 80
  denitrification systems, 79–80
  identification of organisms, 56
  saprobic index, 46, 49, 52, 80
  visual appearance, final settling, 92
Sarcina (bacteria), 58, 59, 79, 136
Schizomycetes, 57–60, 81
screens, 252
scum, 219, 254–255
seasonal fluctuations, 92, 192–193
sedimentation, *see* final settling process
seed destruction, 246
self-purification of natural waters, 11, 48
separator systems, 219–220
septic effluent, 91, 260
serial dilutions, 25, 54

Index

settling, *see* final settling process; primary settling tanks; solids
sewage, *see* effluent
sewage lagoons, 262–263, 289
sewer-slime, 76
Shigella spp. (bacteria), 176
ships, activated sludge plants on, 85
shock loadings and reactor type, 205
simultaneous coagulation, 90, 91, 151, 183
single tank systems, 207
sludge, *see* activated sludge; sludge age
sludge age, 15, 26–27, 83–84, 109
   denitrification, 147–148
   nitrification, 145
small or compact plants, 213, 214, 245, 286
smell reduction, 196–197
sodium azide, 127
soil infiltration method, 262
solids
   filtrable, 15, 25–26, 29, 251–252
   deposit putrefaction, 259
   non-settlable coarse particulate, 120
   settleable (SV), 29, 252–253
   sludge solids content (TS), 15, 25–26, 29, 251–252
solitary bacteria (*see alo* deflocculent growth), 59, 171
   effect on BOD_5 loading, 80, 82
   effect on bloc metabolism, 115
   growth, 209–210
soprtion processes, *see* absorption; adsorption
species
   Action Law, 76, 85
   identification and numbers, 46–48
Sphaerotilus spp., 62, 78, 135, 281
   S. natans, 79, 89, 90
Spirillae (bacteria), 59
Spirillum spp., 58, 79, 135, 146
Spirotricha, 67, 69
staining methods (bacteria), 48
standards (TGL and WAPRO), 280
Staphylococcus spp., 55, 58, 59, 136, 177
   S. aureus, 171
start-up phase, *see* running in
steel tanks, 214
stepped feed, 201–202, 203
stepwise aeration system, 22, 203
Stigeoclonium tenue, 92
stirred tank reactors, 202, 203–205
   bulking sludge, 89, 90
   control methods, 201
   enzyme catalysis, 111
   sorption processes, 183
stomach worm, human, 75
storage of samples, 45
stormwater flows, 252
Streptococcus (bacteria), 59, 61, 136, 175
Streptomycete spp., 165
Streptothrix (bacteria), 135
Strongylides, 75, 178
Stylonchia, 84
submerged jet aerators, 192, 260, 286

substrate (*see also* enzymes)
   deficiency, 124
   definition, 16
   industrial effluent, 225
   inhibition, 128
   respiration, 37
   specificity of enzyme, 101
   substrate–enzyme ratio, 109, 110
Suctoria, 70, 72
sugars, 134
sulphate
   anaerobic reduction, 209
   problems of, 211, 212
sulphide, 79, 91, 151, 182
sulphite addition for deoxygenation, 187
sulphur compounds (*see also* hydrogen sulphide, sulphate; sulphide; sulphite), 48, 151–152
surfactants (detergents), 98, 163–164, 192
surplus sludge, *see* waste sludge
swine fever, 176
swine pest virus, 174, 175
symbols used in sewage treatment, 13–17

Taenia spp., 75, 178
tanks, *see* reactor types
tapeworms, 75, 178–179
taxonomic identification of organisms, 56–76
TCC, *see* Krebs cycle
temperature
   as cause of pleomorphy, 59
   as cause of O_2 deficiency, 84, 187
   effect on living communities, 84, 130
   effect on metabolic processes, 84, 125, 128–133, 143, 144
   effect on odour, 196
   of industrial effluent, 84, 130, 131, 226
   thermal treatment of pathogens, 236
   for treatment of surplus sludge, 84, 245, 246
teratogenic substances, 153, 236
terms and symbols used in sewage treatment, 13–17
Testaceae, 66, 67
tetrachloroethylene, 159
Tetahymena spp., 79, 87, 175
textile industry, 91, 282
TGL standards, 280
Thecameobae, 66, 67
thermal treatment of pathogens, 236
thermophilic bacteria, 130
thin-layer chromatography, 33
Thiobacillus, 23, 136
thiocyanate removal, 165, 208
Thiothrix, 61, 62, 79, 135, 151
   cause bulking sludge, 90, 91
   T. nivea, 282, 283
thread-forming organisms, 80
   bacteria, 56, 58, 59–60, 61, 62, 87, 207, 254
   cause bulking sludge, 87, 89, 90, 91
   fungi, 60, 62, 87, 170
   impairment, 195, 260, 261
   prevention, 207, 254
threadworms, 72, 74, 75, 179

Index

titre, dimensions of, 54
toal organochlorine determination, 33
toluene, 184, 196
Torula (fungi), 60
toxins (*see also* carcinogens; detoxification; decoupling; mutagens)
   accumulation by organisms, 158
   aeration as cause of environmental pollution, 195–199
   cause foaming, 256
   effect on floc formation, 98–99
   effect on metabolic processes, 85, 100, 127–128, 143, 144, 157
   investigations, 38–42, 52
   lethal and median concentrations (LC and LC_{50}), 38, 40
   identification, 256
   no-effect concentration (NEC), 38, 40
   non-biodegradable, 225, 263
   problems with activated sludge process, 44
   running-in phase, 248
   saprobic grade, 50
   threshold concentration (TC), 38, 40
   toxin-producing bacteria, 55
   waterborne substances, 152–168
transferases, 102
Trepomonas agilis, 79, 81, 84
tricarboxylic acid cycle, *see* Krebs cycle
Trichina, 74, 75
trichloroethylene, 159
trichloromethane, 159
Trichosporan spp. (fungi), 60, 84
Trichuris spp., 71, 75, 178
trickling filters, 21, 262, 263, 289
2,3,5-triphenyltetrazolium chloride (TTC), 43
tuberculosis, 55, 176
tubular separator, 219, 220
turbidity
   correlation with protozoa, 170
   of final effluent, 255–257
   measurement, 30, 31
   in pure oxygen systems, 195
   sing of operational failure, 254
   sign of O_2 deficiency (aeration tank), 258
   sign of toxins, 128
turbulence
   causes odour, 196
   effect on enzyme activity, 101
   effect on organisms in aeration tank, 83–84
   effect on respiration rate, 115
   to prevent settling, 189
turnover rate, 103
two-stage systems, 263

typhoid, 176
typhus, 55

ultrasaprobic state, 50
ultrasonic radiation treatment of pathogens, 236
ultra-violet absorption, 33
ultra-violet treatment of pathogens, 236
uncompetitive inhibition, 128
upflow velocity, 218
urea and floc formation, 98
urine, pathogens in, 174, 176
UV, *see* ultra-violet

vaccinia viruses, 175
Vahlkampfia (fungi), 64, 76, 79, 81
viruses, 173–175
   in aerosols, 199
   consumed by protozoa, 171
   investigations, 55
   taxonomic identification, 57, 58
volatile gases (osmogens), 183–184, 192, 196
vomiting, viruses caused by, 174
Vorticella spp., 56, 80, 81, 83, 84, 171
   V. convallaria, 71
   V. microstroma, 76, 79, 83, 84, 87

WAPRO standards, 24, 244, 280
waste activated sludge, 15, 20
   growth, 118–121
   treatment, 244–246
   two-stage systems, 206–207
   withdrawal, 250
water
   classification of flowing (quality), 51
   drinking water, 229, 238
   quality protection during start-up, 248
   self-purification of natural, 11, 48
waterborne toxins, 152–168
Water Quality Inspectorate, 239, 257
weighting of thread-forming organisms, 260
whipworm (Trichina), 75
WHO (World Health Organization), 21
Winkler method, 27
wood pulp effluent (*see also* pulp), 282
World Health Organization (WHO), 21
worm eggs, *see* Helminth ova

Yersinia spp. (bacteria), 176

Zigerli process, 22
zinc as a toxin, 153, 156
Zoogloea (bacteria), 81, 135
zoonosis, 175, 176